C0-AZX-619

ADVANCED CALCULUS FOR USERS

Advanced Calculus for Users

ALAIN ROBERT
University of Neuchâtel
Switzerland

Presses Académiques Neuchâtel Ltd liability Co

1989

NORTH-HOLLAND – Amsterdam · New York · Oxford · Tokyo

ISBN: 2-88333-000-X (Switzerland)
ISBN: 0-444-87324-4 (Outside Switzerland)

Sole distributors for Switzerland:

PAN, Presses Académiques Neuchâtel Ltd liability Co
Case postale 1420
CH-2001 Neuchâtel
Switzerland

Sole distributors for the U.S.A. and Canada:

ELSEVIER SCIENCE PUBLISHING COMPANY, INC.
655 Avenue of the Americas
New York, N.Y. 10010
U.S.A.

For all other countries:

ELSEVIER SCIENCE PUBLISHERS B.V.
P.O. Box 103
1000 AC Amsterdam
The Netherlands

PRINTED IN THE NETHERLANDS

to Ann

About the Author

A. Robert was born in 1941 and studied in Neuchâtel where he acquired his Ph.D. He visited Paris (1967/68), Princeton University (1968/70), Princeton IAS (1970/71), Rio de Janeiro PUC (1977), Berkeley (1983/84) and Queen's University (Kingston, Ont.) on several occasions. He is thus familiar with the university education system of several countries.

His research activities range from algebra to analysis and he is the author of several books and monographs:

— Elliptic Curves, Lecture Note, Springer-Verlag (2nd ed. 1986);
— Representation Theory, Cambridge University Press (1983);
— Nonstandard Analysis, John Wiley (1988) — (first appeared in French at PPR in 1985).

He is currently full professor at the University of Neuchâtel.

FOREWORD

Mathematics are made to be *understood*, *enjoyed* and *applied*...

This text grew out of a course that I gave several years consecutively. It contains more material than can be covered in a two semesters course and, in particular, a selection of topics was made each year. It is *applications* oriented and —I hope— will be consulted by users (physicists, electrotechnicians, microtechnicians, engineers, ...). For this last purpose, I have added a few appendices and lists of formulae, even if they are not proved in the text.

I believe that **excellent** textbooks are now available for the deductive point of view of analysis. Let me only mention

▷ W.Rudin "Principles of mathematical analysis"
and "Real and complex analysis" (on a higher level),

▷ H.Cartan "Cours de calcul différentiel",

▷ S.Lang "Analysis I"

(cf. Bibliography at the end of this volume).

My attempt is not to duplicate these, but rather delve into the wealth of applications (including historical ones).

Consequently, even if my formalization (or axiomatization) is not pushed to its maximum, I hope that users will be able to grasp the *meaning* of the mathematical concepts developed. Like proofs, applications can enlighten the comprehension of a mathematical result. An example will illustrate this point. Stokes' theorem can be understood through its proof (via partitions of unity on manifolds, simplices, showing the way to homology...). But in this optic, the Archimedes principle is lost and most modern treatises avoid the *surface element* $d\vec{\sigma}$ and its meaning (thus certainly losing an important part of its applications).

My attempt here has been to recover these classical applications and to present them in an updated fashion.

* * *

The central idea in this calculus course is that of

LINEARIZATION.

It occurs in the notion of derivative (tangent linear map) and differential linear forms (fields of linear forms). In these first two parts, *finite dimensional* vector spaces play the central role (although $\Omega^k(\mathbb{R}^n)$ is already *infinite dimensional*). The third part constitutes an introduction to *functional analysis* and thus, many *infinite dimensional function spaces* are introduced and examined. Convergence (and in particular uniform convergence) for sequences and series of functions is studied more or less systematically : the importance of these concepts in analysis cannot be overemphasized (definition of functions by means of series or parametric integrals, to mention only these). Finally, the fourth part on Fourier series studies the linear operators which associate to a periodic function f its Fourier sequence $(c_k(f))_{k\in\mathbb{Z}}$, and to a sequence $(c_k)_{k\in\mathbb{Z}}$ the series $\sum_{k\in\mathbb{Z}} c_k e^{ikt}$.

Since students were supposed to have already followed a first calculus course and a linear algebra course, I have taken for granted the Cauchy-Schwarz inequality in $\mathbb{R}^n$ (and this, from the first chapter onwards). However this inequality, in the more general context of scalar products, is proved later on (chapter 17). In a similar vein, I have already used the possibility of differentiating a parametric integral in the first part, although the Leibniz rule is only given and proved in chapter 16. I believe that these transgressions to a strict deductive order will not cause any difficulty. My purpose was to make a reasonably short book with many applications of the theory. This forced me to raise the level of sophistication within each part. For example, PART 3 on functional spaces starts quite elementarily with the notion of uniform convergence for continuous functions and ends with a few notions and results for the Lebesgue spaces L^p.

* * *

Let me now explain a few *C O N V E N T I O N S*.

For simplicity, I have adopted a single numbering for all sections. Thus, 8.5 refers to Chap.8, sec.5 (potentials : definitions). Important results are given a special name, and if a section contains several theorems, this name will help in finding which particular result is referred to (my experience shows that three —or four!— figure cross references are awkward and difficult to remember whereas names are more suggestive). Figures are numbered separately.

The symbol ■ denotes the end of a proof, or the end of a statement whose proof has already been given or whose proof will not be given. The symbol □ indicates the end of a statement whose proof follows.

The term *canonical* is used for algorithmic constructions (independent of arbitrary choices). For example, *the canonical basis* of $\mathbb{R}^n$ is $\mathbf{e}_1 = {}^t(1,0,\dots,0)$, $\mathbf{e}_2 = {}^t(0,1,0,\dots,0)$, ... But there is no canonical basis in the two-dimensional vector space of solutions of the differential equation $y'' + y = 0$. Similarly, the term *intrinsic* refers to a construction or definition which is made independently from choices of bases.

In the statement of a theorem, the numeration i), ii),... is reserved for *equivalent* properties (this is repeated in each case) and thus *different* assertions would be numbered differently e.g. as a), b),... or 1), 2),...

* * *

Finally, it is a pleasure to acknowledge the help that I got during the writing of this book.

Let me thank especially D. Straubhaar and M. Lanz who helped me with first versions, D. Jeandupeux who carefully read the proofs of the final version and suggested a few improvements, and last but not least, my wife Ann who checked my English and corrected many misprints (but I take full responsibility for remaining mistakes...).

Neuchâtel, August 1988 A. Robert

PREREQUISITES AND NOTATIONS

Although it would be difficult to make an exhaustive list of prerequisites for this book (I have already mentioned that the level of sophistication increases somehow inside each part), it may be helpful to list some terminology used throughout the book.

Generalities

injective = one to one into (i.e. $f(x) = f(y) \Longrightarrow x = y$).

${}^tA$ denotes the transpose of a matrix, and in particular, if $\mathbf{x} \in \mathbb{R}^n$ denotes a column vector, ${}^t\mathbf{x} = (x_1, \dots, x_n)$ is a row vector with n components and $\mathbf{x} = {}^t(x_1, \dots, x_n)$. In particular ${}^{tt}A = {}^t({}^tA) = A$.

$f|_X$: restriction of a map $f : E \longrightarrow F$ to a subset $X \subset E$.

id_E : identity function $E \longrightarrow E$.

If f is a map, $x \longmapsto f(x)$ is the correspondence for elements.

The basic numerical sets are denoted by

$\mathbb{N} = \{0,1,2,3,\dots\}$: natural numbers,

$\mathbb{Z} = \{\dots,-1,0,1,2,\dots\}$ ring of rational integers,

$\mathbb{Q}$, $\mathbb{R}$, $\mathbb{C}$: fields of rational, resp. real and complex numbers.

$I \times J$ denotes a Cartesian product (e.g. a rectangle in $\mathbb{R}^2 = \mathbb{R} \times \mathbb{R}$)

sinx, cosx, tan x, cot x = 1/tan x, Arctan $x \in]-\pi/2, \pi/2[$ denote the usual trigonometric functions whereas Shx, Chx, Thx denote the hyperbolic functions (e.g. $\mathrm{Ch}^2x - \mathrm{Sh}^2x = 1$).

Linear algebra

$\mathbb{R}^n$: Euclidean space of n-tuples of real numbers (column vectors)

$\mathbb{C}^n$: vector space consisting of n-tuples of complex numbers

All vector spaces E to be considered have field of scalars $\mathbb{R}$ or $\mathbb{C}$ (and the context should always make it clear which!)

Form = scalar valued homogeneous function $E \longrightarrow \mathbb{R}$ (or $\mathbb{C}$) (e.g. linear form, quadratic form,...)

E' : dual of E, space of linear forms on E,
E" = (E')' : dual of E', bi-dual of E
$\varepsilon : E \longrightarrow E''$ Dirac evaluation map, $\varepsilon_a(\varphi) = \varphi(a)$

Kronecker symbol $\delta_{ij} = \delta^i_j = \delta^j_i$ (= 0 if i≠j and = 1 if i=j)

$\mathcal{L}\langle a : a \in A\rangle$ linear span of A (in a vector space E)

$\mathcal{L}(E,F) = \mathrm{Hom}(E,F)$: space of linear maps $E \longrightarrow F$,

$\mathbb{M}_n(\mathbb{R})$, $M_n(\mathbb{C})$: ring of $n \times n$ matrices with real (resp. $\mathbb{C}$) entries

$\mathbb{G}\ell(E) \subset \mathcal{L}(E) = \mathcal{L}(E,E)$: group of invertible linear transformations of E in itself

$\mathbb{G}\ell_n(\mathbb{R}) \subset \mathbb{M}_n(\mathbb{R})$: group of n by n invertible matrices with real entries (similarly for $\mathbb{G}\ell_n(\mathbb{C}) \subset \mathbb{M}_n(\mathbb{C})$ for complex entries)

Matrix representation of an operator in a basis : $A = (a^i_j)$

The jth column of A is made up with the components of the jth basis vector : $A(\mathbf{e}_j) = \sum_i a^i_j \mathbf{e}_i$,

$Tr(A) = \sum_i a^i_i$: trace of A ,

det(A) : determinant of A .

Some familiarity with orthonormal bases, eigenvectors and characteristic vectors (Jordan reduced form) is assumed.

Analysis

Derivative at 0 of a function $f : [0,1] \longrightarrow \mathbb{R}$ means

right derivative (similarly at 1 : *left* derivative)

$o(x^k)$ represents a function f such that $|f(x)/x^k| \longrightarrow 0$ for $x \longrightarrow a$ (and a is given explicitely in each context).

$O(x^k)$ represents a function f such that $|f(x)/x^k|$ remains bounded for $x \longrightarrow a$ (and a is given explicitely in each context).

$f(x) \sim g(x)$ means $f(x)/g(x) \longrightarrow 1$, (for $x \longrightarrow a$ as before)

Topology

We assume that the reader is familiar with the intuitive notions of open and of closed subsets of $\mathbb{R}^n$. Neighborhoods of points and limits of sequences are also used here without comment. A compact set in $\mathbb{R}^n$ is simply a closed and bounded subset : on a compact set, a continuous function always attains a maximum (hence is bounded).

This book has been typed with *CHI-WRITER,* produced and distributed by

Horstman Software Design Corp.
P.O.Box 5039
SAN JOSE, CA 95150 U.S.A.

TABLE OF CONTENTS

Chapter 5 : SCALAR FIELDS : THE GRADIENT

Chapter 6 : JACOBIANS

Chapter 7 : SECOND ORDER DERIVATIVES

Chapter 8 : VECTOR FIELDS IN THE USUAL PHYSICAL SPACE $\mathbb{R}^3$

Chapter 9 : CURVILINEAR COORDINATES

Chapter 10 : PHYSICAL APPLICATIONS

PART II : INTEGRATION OF DIFFERENTIAL FORMS

PART III : FUNCTION SPACES

PART IV: FOURIER SERIES

Chapter 20 : TRIGONOMETRIC SERIES AND PERIODIC FUNCTIONS

Chapter 21 : APPLICATIONS

Chapter 22 : FINER CONVERGENCE RESULTS

Chapter 23 : CONVERGENCE IN QUADRATIC MEAN

Chapter 24 : HISTORICAL APPLICATIONS

APPENDIX A : CONVERGENCE, DEFINITIONS AND RESULTS

APPENDIX B : S A S PROGRAMS FOR A FEW FIGURES

E X E R C I S E S

I N D E X

STRUCTURE DIAGRAM FOR FIRST PART

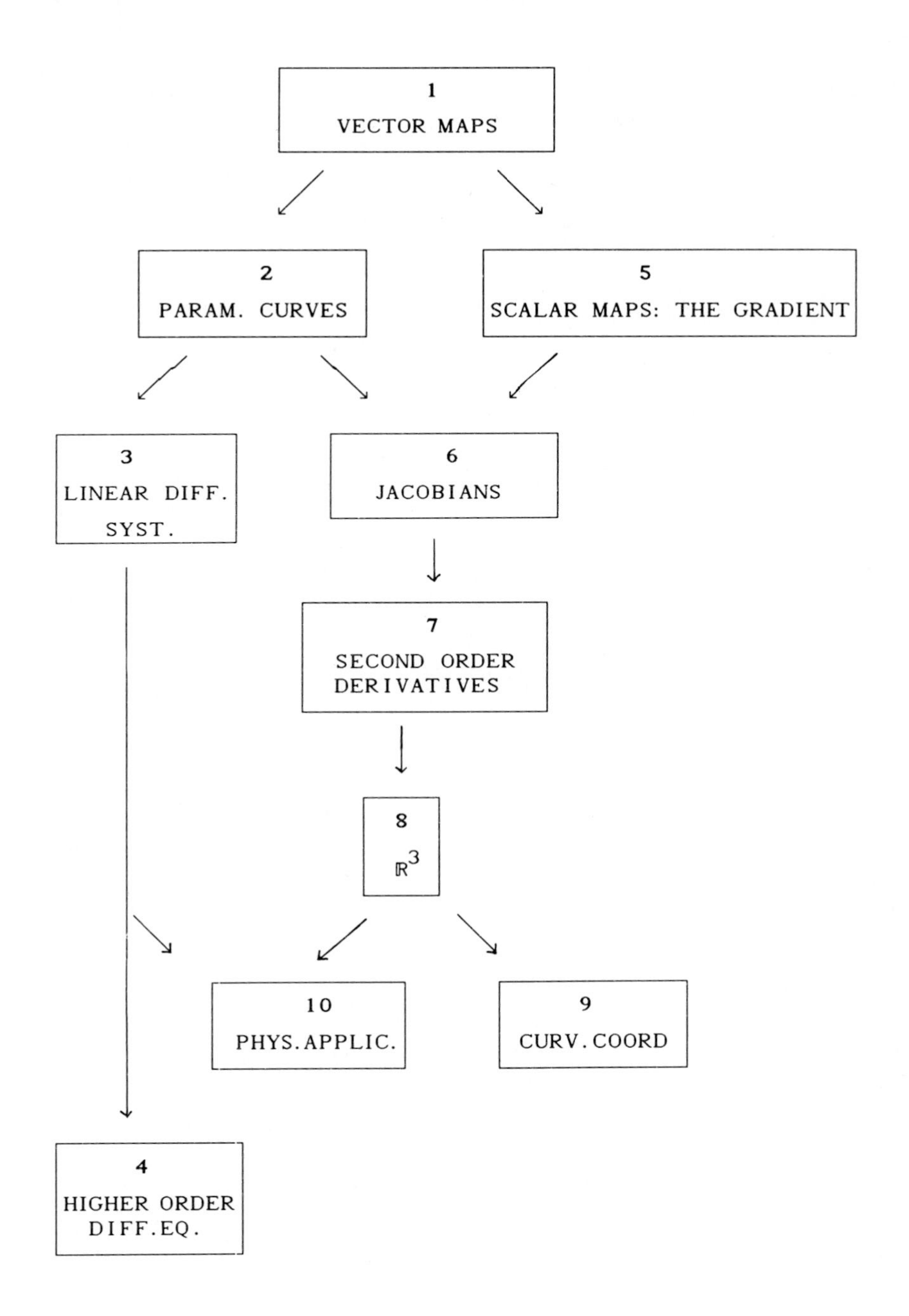

Table - 7

PART ONE

DIFFERENTIABILITY

CHAPTER 1
VECTOR MAPPINGS

1.1 CONVENTIONS AND NOTATIONS

In this chapter we shall study mappings from a subset $U \subset \mathbb{R}^n$ to some other $\mathbb{R}^m$. The vector spaces $\mathbb{R}^n$ and $\mathbb{R}^m$ will be considered as real vector spaces whose elements are *column vectors*

$$\mathbf{x} = \begin{pmatrix} x_1 \\ \vdots \\ x_n \end{pmatrix} = {}^t(x_1, \ldots, x_n) \in \mathbb{R}^n \quad (x_i \in \mathbb{R}).$$

(For typographical reasons, we shall often use the row notation preceded by the upper "t" meaning "transpose" instead of the column notation.) Special values of the exponent n (or m) will lead to interesting applications. For example, n = 3 leads to the usual physical space $\mathbb{R}^3$ whose elements will more conveniently be denoted by

$$\vec{r} = \mathbf{r} = \begin{pmatrix} x \\ y \\ z \end{pmatrix} = {}^t(x,y,z) \in \mathbb{R}^3.$$

Occasionally, we shall even identify a vector $\mathbf{r}$ to its extremity P once a fixed origin O has been chosen. Thus the components of $\mathbf{r} = \overrightarrow{OP}$ are taken as coordinates of the point P.

We shall also have to use row vectors $\mathbf{a} = (a_1, \ldots, a_n)$ and denote by $\mathbb{R}_n$ their vector space (observe the position of the index n in this vector space). Thus this space $\mathbb{R}_n$ is also a real vector space of dimension n, but its elements have a different representation from those of $\mathbb{R}^n$. We shall identify row vectors $\mathbf{a} \in \mathbb{R}_n$ to linear forms on $\mathbb{R}^n$:

$$\mathbf{a} : \mathbf{x} = \begin{pmatrix} x_1 \\ \vdots \\ x_n \end{pmatrix} \longmapsto (a_1, \ldots, a_n)\begin{pmatrix} x_1 \\ \vdots \\ x_n \end{pmatrix} = a_1x_1 + \ldots + a_nx_n.$$

In other words, we identify the vector space $\mathbb{R}_n$ of row vectors to the *dual* of the vector space $\mathbb{R}^n$ of column vectors.

The vector spaces $\mathbb{R}^n$ will also be endowed with their usual scalar product

$$\mathbf{x}\cdot\mathbf{y} = x_1y_1 + \ldots + x_ny_n \quad (\mathbf{x} \text{ and } \mathbf{y} \in \mathbb{R}^n).$$

This scalar product gives the *Euclidean structure* of the space $\mathbb{R}^n$. Let us also recall that the length of a vector $\mathbf{x} \in \mathbb{R}^n$ is given by its *norm*

$$\|\mathbf{x}\| = \sqrt{(\mathbf{x}\cdot\mathbf{x})} = (x_1^2 + \ldots + x_n^2)^{1/2} \geq 0.$$

Although we shall re-establish the *Cauchy-Schwarz inequality* in a more general context (cf.17.5), we shall already have to use it for $\mathbb{R}^n$ in this first chapter. This basic inequality states that for all vectors $\mathbf{x}$ and $\mathbf{y} \in \mathbb{R}^n$

$$|\mathbf{x}\cdot\mathbf{y}| \leq \|\mathbf{x}\|\|\mathbf{y}\|.$$

In words: the absolute value of the scalar product is majorized by the product of the lengths of the two vectors. Thus, the real number $\mathbf{x}\cdot\mathbf{y}/\|\mathbf{x}\|\|\mathbf{y}\|$ belongs to the interval $[-1,+1]$ and could be interpreted as the cosine of the angle between the two vectors

$$-1 \leq \cos\alpha = \frac{\mathbf{x}\cdot\mathbf{y}}{\|\mathbf{x}\|\,\|\mathbf{y}\|} \leq +1 , \quad \text{where} \quad 0 \leq \alpha \leq \pi.$$

This defines the usual angle in $\mathbb{R}^3$ and can be used to define the notion of angle in $\mathbb{R}^n$.

The vector mappings to be considered will always be defined in a (non empty) subset $U \subset \mathbb{R}^n$ which could be assumed to be an open subset of $\mathbb{R}^n$. Typically, U could be

- ▷ an open ball, or the interior of a parallelepiped,
- ▷ the complement of a point or of an affine subspace, (e.g. the complement of a coordinate axis)
- ▷ a region defined by restricting the variation of some coordinate(s) (e.g. $-1 < x_2 < 1$).

Such vector mappings will be denoted by capital letters as in

$$F : U \longrightarrow \mathbb{R}^m \quad (U \subset \mathbb{R}^n), \text{ or } F : \mathbb{R}^n \supset U \longrightarrow \mathbb{R}^m.$$

However, to simplify the theoretical considerations, we shall often assume that the open set in which the function F is defined is the whole space $\mathbb{R}^n$. In most cases, we shall not be interested in the topological properties of open sets $U \subset \mathbb{R}^n$, but rather in the local properties of the functions. Thus we shall simply consider functions $F : \mathbb{R}^n \longrightarrow \mathbb{R}^m$.

1.2 VISUALIZING VECTOR MAPPINGS

Let us recall the classical terminology used for mappings and the interpretation that can be given for them in some cases. Thus, we take a few particular cases of vector maps $F : \mathbb{R}^n \longrightarrow \mathbb{R}^m$.

Case $n = m = 1$. This is the usual case and it is customary to draw the *graph* of the function $F = f : \mathbb{R} \longrightarrow \mathbb{R}$. This graph is the set

$$\mathrm{Gr}(f) = \{(x,y) : y = f(x)\} \subset \mathbb{R}^2.$$

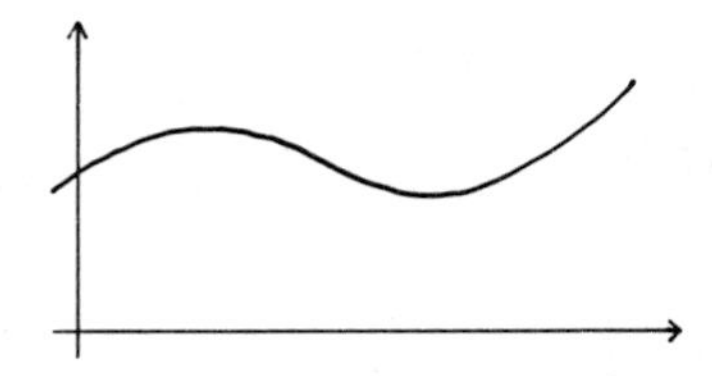

Fig.1.1

It is a *curve* (at least when f is sufficiently regular). We assume that the reader is familiar with this representation.

Case $n = 2$, $m = 1$. We still denote by $F = f : \mathbb{R}^2 \longrightarrow \mathbb{R}$ such a function. In this case, it is still possible to get an idea of the map by sketching its graph. This is the set

$$Gr(f) = \{(x,y,z) : z = f(x,y)\} \subset \mathbb{R}^3.$$

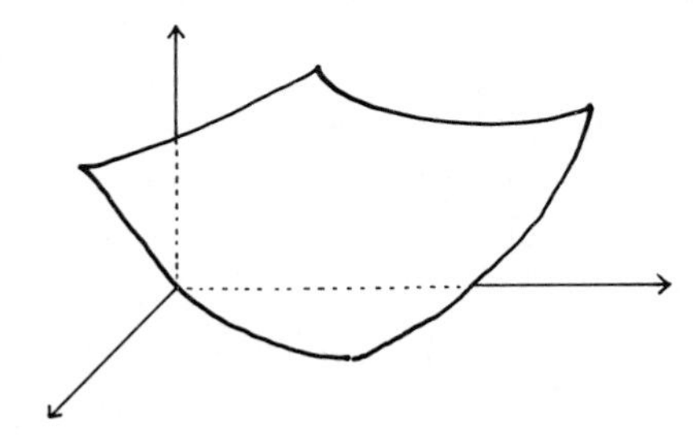

Fig.1.2

It is a *surface* (at least when ...). An image of (a portion of) this surface is usually very helpful in the study of the function f. For example, continuity of f at the origin can best be apprehended if a picture of the surface is available...

Case n arbitrary, $m = 1$. Such functions $f : \mathbb{R}^n \longrightarrow \mathbb{R}$ are called *scalar functions* simply because they take scalar values. It is no more possible to make a picture of the graph of f (this would be a subset of $\mathbb{R}^n \times \mathbb{R} = \mathbb{R}^{n+1}$). But —at least when $n \leq 3$— it may be informative to get an idea of the *level surfaces* of f (if $n = 3$) or the *level curves* of f (if $n = 2$). Observe that this way of describing scalar mappings $f : \mathbb{R}^2 \longrightarrow \mathbb{R}$ is systematically used in weather forecasting (cf. Fig.1.3). The atmospheric pressure on a certain geographical area is indeed a scalar function, and meteo-

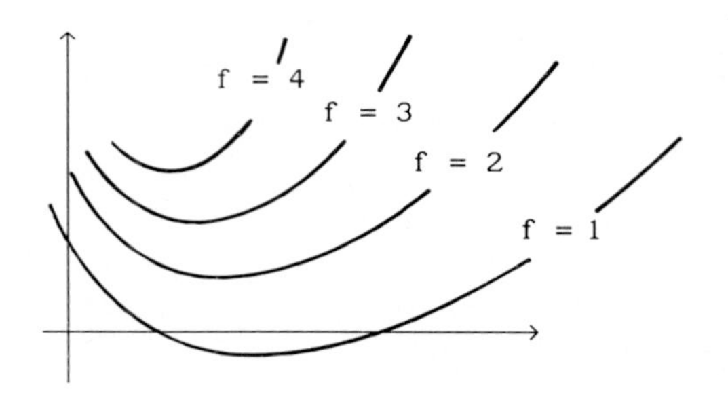

Fig.1.3

rological institutes establish a map containing a few level curves with constant pressure (isobaric curves).

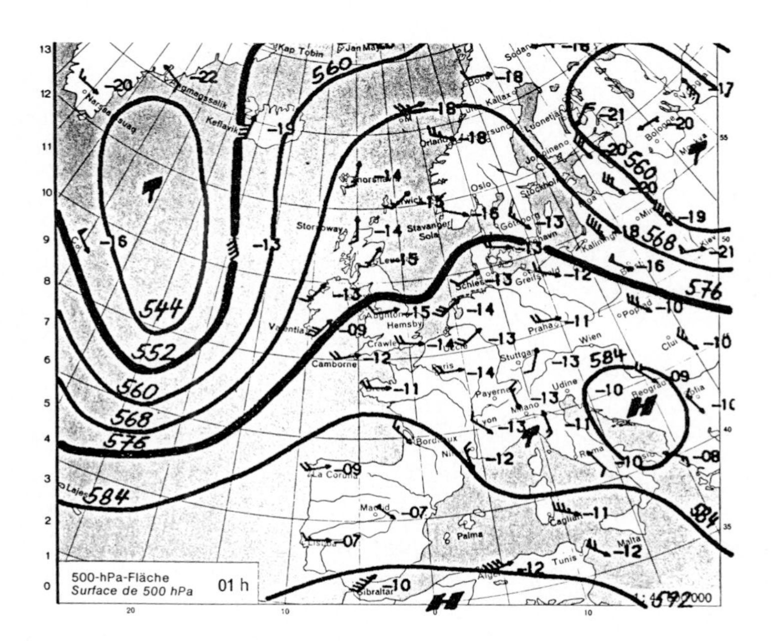

Example of isobaric curves

(Meteorological Bulletin 8.8.88: Swiss Meteo.Inst. CH-8044 Zürich)

Fig.1.4

Similar situations abound with repartition of temperatures represented by isothermal curves (or surfaces) etc.

Case $n = 3$, $m = 1$. Consider the following function

$$F(x,y,\lambda) = x + y^3 - 3\lambda y$$

as scalar function $F : \mathbb{R}^3 \longrightarrow \mathbb{R}$. To visualize it, we draw its level surfaces. Let us only consider here the special $F = 0$ level surface. It is defined by the equation

$$x = 3\lambda y - y^3.$$

It is easy to draw a few of these cubics for particular values of the parameter λ: these curves represent the intersection of the surface with planes parallel to the (x,y)-coordinate plane. Here we see an example of a *cusp catastrophe*: the surface possesses vertical tangent planes above the curve

$$x = 2\lambda^{3/2} \quad \text{(cuspidal curve)}.$$

These points project on the vertical (λ,y)-plane on the parabola $\lambda = y^2$. This example also shows an example of *bifurcation* : at the origin, the negative λ-axis is connected to the parabola $y^2 = \lambda$ in the shape of a *pitchfork*!

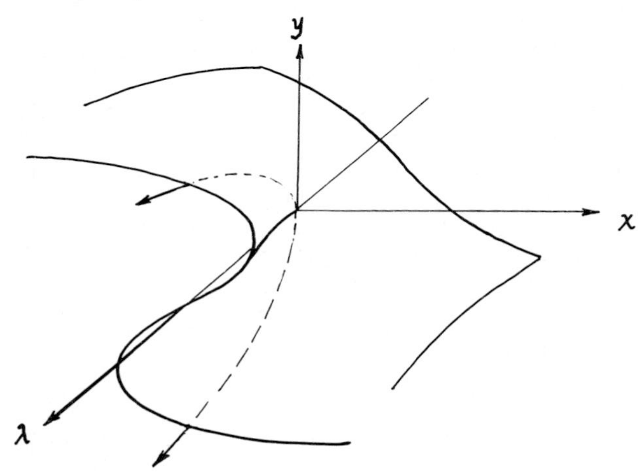

Fig.1.5

Case $n = 1$, $m = 2$ or 3. A mapping $F : \mathbb{R} \longrightarrow \mathbb{R}^2$ (or $\mathbb{R}^3$) is usually viewed as *parameterized plane* (resp. *space*) *curve*. It can be given by a couple (resp. triple) of scalar maps $f_i : \mathbb{R} \longrightarrow \mathbb{R}$:

$$F = \begin{pmatrix} f_1 \\ f_2 \end{pmatrix} : t \longmapsto F(t) = \begin{pmatrix} f_1(t) \\ f_2(t) \end{pmatrix}.$$

A few values for $t \in \mathbb{R}$ will lead to a few points on the representative parameterized curve described by F. We have adopted $t \in \mathbb{R}$ as the name for the (scalar) variable simply because this situation occurs naturally in the description of the movement of a point in the course of time. A point mass moving in the usual physical space $\mathbb{R}^3$, under the effect of certain forces, will describe such a parameterized curve.

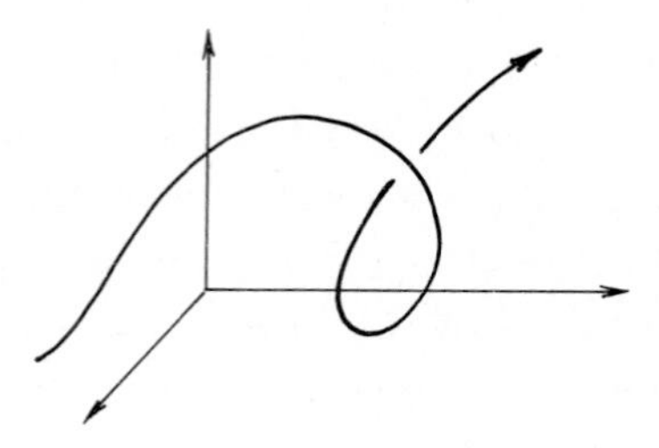

Fig.1.6

Case n = 1, m arbitrary. It is interesting to note that the description of a couple of point masses in $\mathbb{R}^3$ is conveniently represented by a map

$$F = \begin{pmatrix} F_1 \\ F_2 \end{pmatrix} : t \longmapsto F(t) = \begin{pmatrix} F_1(t) \\ F_2(t) \end{pmatrix}$$

where each $F_i : \mathbb{R} \longrightarrow \mathbb{R}^3$ represents the movement of a single particle. Thus the global movement of the couple is represented by a vector map

$$F : \mathbb{R} \longrightarrow \mathbb{R}^3 \times \mathbb{R}^3 \cong \mathbb{R}^6.$$

More generally, the description of a system of N particles evolving in the usual physical space is described by a map

$$F : \mathbb{R} \longrightarrow \mathbb{R}^{3N} \quad \text{(3N is the degree of freedom).}$$

Other situations still lead to generalized parameterized curves. For example, a 3 × 3 matrix $A(t) \in M_3(\mathbb{R})$ varying with time is described by its nine function coefficients and we are led to consider a mapping $\mathbb{R} \longrightarrow \mathbb{R}^9$ in this case. We would still call this a *generalized parameterized curve*, with values in the space $M_3(\mathbb{R}) \cong \mathbb{R}^9$ (we shall study more systematically such a situation in a later chapter).

1.3 DIFFERENTIABILITY AT A POINT

We come to the first basic **definition.** Let $F : \mathbb{R}^n \longrightarrow \mathbb{R}^m$ be a vector function. We shall say that F is **differentiable** at a point $\mathbf{a} \in \mathbb{R}^n$ if there exists a linear map $L = L_{\mathbf{a}} : \mathbb{R}^n \longrightarrow \mathbb{R}^m$ giving rise to a **limited expansion of the first order** at **a**, namely if

$$F(\mathbf{a} + \mathbf{h}) = F(\mathbf{a}) + L(\mathbf{h}) + \|\mathbf{h}\| \cdot \Phi(\mathbf{h})$$

$$\text{with } \Phi(\mathbf{h}) \longrightarrow 0 \text{ when } \mathbf{h} \longrightarrow 0 \text{ (i.e. } \|\mathbf{h}\| \longrightarrow 0).$$

Obviously, it would be enough to assume that the vector function F is defined in a neighborhood of the point **a**, and then to consider only increments **h** such that **a** + **h** is still in the domain of definition of F. As we already mentioned, such obvious generalizations will be ignored in our treatment!

In a first order limited expansion of the above type, the terms $F(\mathbf{a}) + L(\mathbf{h})$ represent an *affine linear map* giving a first approximation of F for $\mathbf{x} = \mathbf{a} + \mathbf{h}$ close to **a**. The term $\|\mathbf{h}\|\cdot\Phi(\mathbf{h})$ is defined as the difference between $F(\mathbf{a} + \mathbf{h})$ and $F(\mathbf{a}) + L(\mathbf{h})$ and is required to tend to 0 *faster* than $\|\mathbf{h}\|$. This difference, also called the *remainder* of the limited expansion, divided by $\|\mathbf{h}\|$ should tend to 0 when **h** goes to 0. In a more intuitive way, the difference between F and its linear approximation should go to zero *when expressed in percent* of the length of the increment. It is quite common to represent this requirement by the notation of Landau

$$F(\mathbf{a} + \mathbf{h}) - F(\mathbf{a}) - L(\mathbf{h}) = \|\mathbf{h}\|\cdot\Phi(\mathbf{h}) = o(\|\mathbf{h}\|).$$

For example, with this notation, we can write

$$(1 + x)^{-1} = 1 - x + o(x),$$
$$\log(1 + x) = x + o(x),$$
$$e^{x} = 1 + x + o(x),$$
$$\sin x = x + o(x), \quad \text{etc.}$$

The reader will have recognized the most classical first order limited expansions. This list is by no means complete and we should add

$$(1 + x)^{a} = 1 + ax + o(x) \qquad (a \in \mathbb{R} \text{ or } \mathbb{C}),$$
$$\text{Sh } x = x + o(x).$$

To be more precise, one should mention that the notation $o(x)$ is relative to the condition $x \longrightarrow 0$.

Let us show a **uniqueness** result. If there is a linear map L with

$$F(\mathbf{a} + \mathbf{h}) = F(\mathbf{a}) + L(\mathbf{h}) + \|\mathbf{h}\|\cdot\Phi(\mathbf{h}) = F(\mathbf{a}) + L(\mathbf{h}) + o(\|\mathbf{h}\|)$$

then L is uniquely determined by this condition. Thus let us assume that we have two linear maps L and M giving rise to limited expansions

$$F(\mathbf{a} + \mathbf{h}) = F(\mathbf{a}) + L(\mathbf{h}) + \|\mathbf{h}\|\cdot\Phi(\mathbf{h}),$$
$$F(\mathbf{a} + \mathbf{h}) = F(\mathbf{a}) + M(\mathbf{h}) + \|\mathbf{h}\|\cdot\Psi(\mathbf{h}).$$

By subtraction we get

$$L(\mathbf{h}) - M(\mathbf{h}) = \|\mathbf{h}\| \, [\Psi(\mathbf{h}) - \Phi(\mathbf{h})].$$

We would like to prove L = M, i.e. $L(\mathbf{v}) = M(\mathbf{v})$ for all $\mathbf{v} \in \mathbb{R}^n$. For this purpose, we let $\mathbf{h} = t\mathbf{v}$ where $t > 0$. We get

$$L(t\mathbf{v}) - M(t\mathbf{v}) = t\|\mathbf{v}\| \, [\Psi(t\mathbf{v}) - \Phi(t\mathbf{v})]$$

and by linearity of L and M

$$t\,[L(\mathbf{v}) - M(\mathbf{v})] = t\|\mathbf{v}\| \, [\Psi(t\mathbf{v}) - \Phi(t\mathbf{v})].$$

Simplifying by $t > 0$, we find

$$\|\mathbf{v}\| \, [\Psi(t\mathbf{v}) - \Phi(t\mathbf{v})] = L(\mathbf{v}) - M(\mathbf{v}) \text{ independent of } t.$$

We evaluate this constant by letting $t \longrightarrow 0$ on the left hand side. By assumption, $\Psi(t\mathbf{v})$ and $\Phi(t\mathbf{v}) \longrightarrow 0$. Thus, the only possibility for the right hand side is $L(\mathbf{v}) - M(\mathbf{v}) = 0$. Since $\mathbf{v}$ is arbitrary, this proves that L = M as expected. ∎

Definition. We say that a vector map $F : \mathbb{R}^n \longrightarrow \mathbb{R}^m$ is differentiable at the point $\mathbf{a} \in \mathbb{R}^n$ when there exists a limited expansion of the first order of F

$$F(\mathbf{a} + \mathbf{h}) = F(\mathbf{a}) + L(\mathbf{h}) + o(\|\mathbf{h}\|) \quad (\mathbf{h} \longrightarrow 0)$$

valid in a neighborhood of **a**. The linear homogeneous part in this limited expansion is uniquely determined (by F and $\mathbf{a} \in \mathbb{R}^n$) and is called the **differential of F at a** and denoted by $dF(\mathbf{a})$. This linear mapping is also called **derivative of F at a** and denoted by $F'(\mathbf{a})$. It might be better to reserve the terminology of derivative for the representative matrix of this linear map *with respect to the canonical bases of* $\mathbb{R}^n$ *and* $\mathbb{R}^m$. In the case n = m = 1, this is coherent with a common practice : the differential is a linear form whereas the derivative is a number giving a percentage increment and is thus the coefficient in the (1 by 1) representative matrix of the linear form.

Since the vector spaces that we are dealing with are endowed with canonical bases, we shall often identify the two notions of differential and derivative. But in some cases, it might be important to distinguish them, and the terminology of the definition has been carefully selected. Even in the case n = m = 1 it is sometimes useful to make the preceding distinction. The differential of a function $f : \mathbb{R} \longrightarrow \mathbb{R}$ at $a \in \mathbb{R}$ is the linear form $df(a) : \mathbb{R} \longrightarrow \mathbb{R}$ given by

$$h \longmapsto f'(a)h \text{ , (or equivalently } dx \longmapsto f'(a)dx).$$

It is represented by a 1×1 matrix with a single entry $f'(a)$. This number is the increment quotient, or the slope, of f at a and is the usual derivative of f at a.

1.4 TANGENT LINEAR MAPS

The importance of the notion of differentiability justifies a geometrical interpretation of its meaning. As an introduction, let us quickly recall the notion of tangent plane curves. Assume that two differentiable functions $f,g : \mathbb{R} \longrightarrow \mathbb{R}$ take the same value at $x = a : f(a) = g(a)$. Their graphs will represent two smooth curves intersecting above $x = a$. These curves will be tangent when the two derivatives $f'(a) = g'(a)$ are equal. Namely

$$\frac{f(a+h) - g(a+h)}{h} \longrightarrow 0 \quad \text{when} \quad h \longrightarrow 0.$$

With the notation of Landau, this could be written

$$f(a+h) = g(a+h) + o(h) \quad \text{for } h \longrightarrow 0 \ .$$

The *percentage difference* between f and g goes to zero when $x \longrightarrow a$.

In a quite general context, we can define the notion of **tangent maps** $F,\ G : \mathbb{R}^n \longrightarrow \mathbb{R}^m$ (at a point $\mathbf{a} \in \mathbb{R}^n$) simply by requiring $F(\mathbf{a} + \mathbf{h}) = G(\mathbf{a} + \mathbf{h}) + o(\|\mathbf{h}\|)$ for $\|\mathbf{h}\| \longrightarrow 0$.

More explicitly, this means

$$\frac{F(\mathbf{a} + \mathbf{h}) - G(\mathbf{a} + \mathbf{h})}{\|\mathbf{h}\|} \longrightarrow 0 \quad \text{for } \|\mathbf{h}\| \longrightarrow 0.$$

The *equivalence relation* $F \underset{\mathbf{a}}{\simeq} G$ meaning that F is tangent to G at **a** can be introduced in the set of *all* maps $\mathbb{R}^n \longrightarrow \mathbb{R}^m$. Equivalence classes can be formed. The uniqueness result proved in the last section asserts that in a given equivalence class, there is at most *one* affine linear map. When there is indeed one, all functions in this class are differentiable at the point **a** with the same differential (equal to the homogeneous linear part of this affine map). More concisely, we shall say that a map F defined in a neighborhood of $\mathbf{a} \in \mathbb{R}^n$ is differentiable at the point **a** when it admits a **tangent linear map**. The differential itself $dF(\mathbf{a})$ will also be called "tangent linear map of F at **a**". This is an abuse of terminology since the differential is only the homogeneous part of the tangent affine map. We hope that the reader will never be confused by this abuse (we would not like to be held responsible for the fact that the term *linear* is used in two different contexts: it covers the *affine* context where constants can be added, but it is also used in its stricter meaning in linear algebra where *homogeneity* is assumed!).

To illustrate the situation, let us take an example. Let us consider the function $F : \mathbb{R}^n \longrightarrow \mathbb{R}^n$ $(n = m)$ defined as follows.

When supplied with an n-tuple of real entries, F furnishes the same n-tuple rearranged in decreasing order. Thus

$$F(x_1, \dots, x_n) = {}^t(y_1, \dots, y_n) \quad \text{with } y_1 = \text{Max } (x_i) \geq y_2, \dots$$

Fig.1.7

Is this function differentiable at a given sequence $\mathbf{a} = (a_i)$? (The reader should convince himself first that it is continuous, not linear!) Let us take a particular sequence $\mathbf{a} = (a_i)$ with n distinct entries $a_i \neq a_j$ for $i \neq j$. I claim that F is differentiable at such a point. In this case, there is indeed a permutation σ of the indices such that $F(a_1,\dots,a_n) = {}^t(a_{\sigma(1)},\dots,a_{\sigma(n)})$. Let L denote the linear map defined by

$$L : \mathbf{h} = {}^t(h_1,\dots,h_n) \longmapsto {}^t(h_{\sigma(1)},\dots,h_{\sigma(n)})$$

for all $\mathbf{h} \in \mathbb{R}^n$. Obviously

$$F(x_1,\dots,x_n) = {}^t(x_{\sigma(1)},\dots,x_{\sigma(n)})$$

whenever $\mathbf{x}$ is close to $\mathbf{a}$ (the permutation which reorders $\mathbf{a}$ in decreasing order will also reorder $\mathbf{x}$ in decreasing order if $\mathbf{x}$ is in the vicinity of $\mathbf{a}$). More precisely, if $r = \underset{i \neq j}{\text{Min}} |a_i - a_j| > 0$, we shall have

$$F(\mathbf{a} + \mathbf{h}) = L(\mathbf{a} + \mathbf{h}) = L(\mathbf{a}) + L(\mathbf{h}) = F(\mathbf{a}) + L(\mathbf{h})$$

whenever $\|\mathbf{h}\| < r/2$, and this certainly proves

$$F(\mathbf{a} + \mathbf{h}) = F(\mathbf{a}) + L(\mathbf{h}) + o(\|\mathbf{h}\|).$$

But the function F would not be differentiable at a point $\mathbf{a}$ having different components with the same value.

1.5 EXAMPLES OF SURFACES

Let us examine a few cases of surfaces in the usual physical space $\mathbb{R}^3$. More particularly, let us examine a few scalar fields $f : \mathbb{R}^2 \longrightarrow \mathbb{R}$ and their representing graphs in $\mathbb{R}^3$. In this case, the notion of differentiability at a point is well described by the existence of a tangent plane of the corresponding point on the graph.

Case $f(x,y) = xy/(x^2 + y^2)$ for $(x,y) \neq (0,0)$ and $f(0,0) = 0$.

The choice of special value at the origin is suggested by the fact that f vanishes identically on the axes $y = 0$ and $x = 0$. If we let

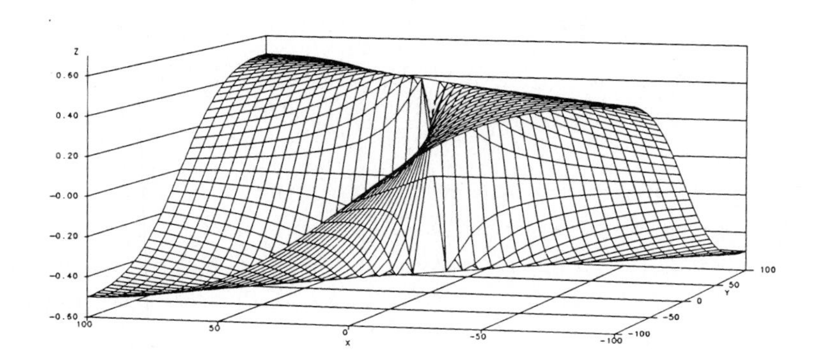

Fig.1.8

the point P : (x;y) vary on the line $y = mx$, we see that

$$z = f(x,mx) = m/(1 + m^2) \text{ is constant (if } x \neq 0)$$

and f is not even continuous at the origin. F does not possess a limited expansion of the 0th order at that point and differentiability is out of the question.

Case $f(x,y) = x^2y/(x^4 + y^2)$ for $(x,y) \neq (0,0)$ and $f(0,0) = 0$.

Again, the choice of value at the origin has been suggested by the fact that f vanishes identically on the two axes. But now, if we look at points of the graph above a line $y = mx$

$$z = f(x,mx) = mx/(x^2 + m^2) \longrightarrow 0 \quad \text{when } x \longrightarrow 0$$

This proves that f is continuous *when restricted to any radius going through the origin.* BUT... look at Fig. 1.9 ! The picture indeed suggests that f has a level curve in the shape of a parabola. Indeed, if we study the values of f when the point P:(x;y) moves on $y = x^2$ we find

$$z = f(x,y) = x^4/2x^4 = 1/2 \quad \text{constant (if } x \neq 0).$$

This function f is not continuous at the origin either! A fortiori it is not differentiable at this point. But observe that *when res-*

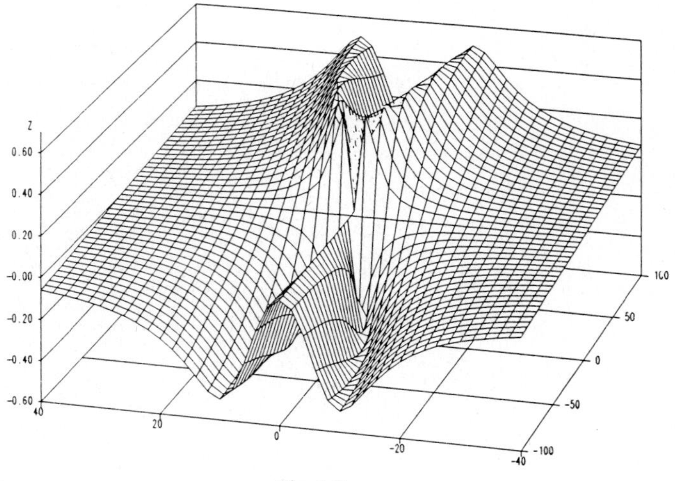

Fig.1.9

tricted to lines going through the origin, f is derivable. The first computation shows that $f(x,mx) = mx/(m^2 + x^2)$ even if $x = 0$, and in particular this function admits the limited expansion

$$f(x,mx) = m^{-1}x(1 + x^2/m^2)^{-1} = m^{-1}x(1 - x^2/m^2 + \dots) = x/m + o(x)$$

This example reveals part of the subtlety of the notion of differentiability. Only requiring differentiability along lines does not imply existence of a tangent plane.

Case $f(x,y) = x^2y/(x^2 + y^2)$ for $(x,y) \neq (0,0)$ and $f(0,0) = 0$.

It is convenient to introduce polar coordinates since $\mathbf{r} = {}^t(x,y)$ has norm $r = \|\mathbf{r}\| = (x^2 + y^2)^{1/2}$. Thus we introduce

$$x = r\cos\vartheta \ , \ y = r\sin\vartheta$$

in the definition of f :

$$f(x,y) = f(r\cos\vartheta, r\sin\vartheta) = r\sin\vartheta \cdot \cos^2\vartheta \quad \text{(even if } r = 0)$$

On this expression, we see that f is continuous at the origin and more precisely

$$|f(\mathbf{r})| \leq \|\mathbf{r}\| = r \ .$$

However, the expression $f(x,y) = r\sin\vartheta \cdot \cos^2\vartheta$ is *not* a limited expansion of the first order at the origin (it *is* a limited expansion of the 0th order!). Consider the following Fig. 1.10. It is useful to observe that this function is homogeneous :

$$f(\lambda x, \lambda y) = \lambda f(x,y) \qquad (\lambda \in \mathbb{R}).$$

Consequently, the graph of f is made up of lines going through the origin. When $r = 1$, namely $x^2 + y^2 = 1$, say $x = \cos\vartheta$ and $y = \sin\vartheta$

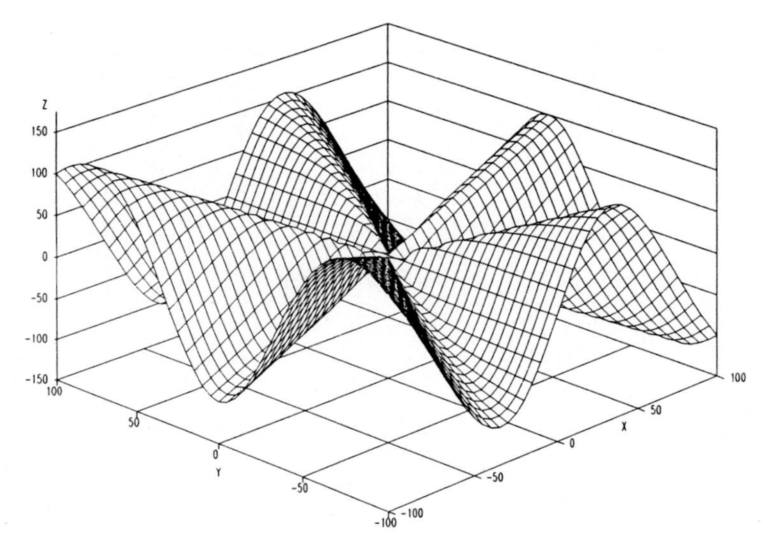

Fig.1.10

we find

$$f(x,y) = \sin\vartheta \cdot \cos^2\vartheta .$$

It is easy to plot a few values of f over the unit circle in $\mathbb{R}^2$ and get a good view of the surface: it is a conical surface generated by a line constantly passing through the origin (vertex of the cone) and leaning on the preceding curve. As a consequence, it is easy to show that f is differentiable along *any differentiable curve going through the origin.* More precisely, if

$$t \longmapsto \gamma(t) = {}^t(x(t),y(t))$$

is a parameterized curve with existing and non vanishing initial speed (we shall come back systematically to this notion in a later chapter) $\mathbf{v} = \gamma'(0)$ exists and is not $\mathbf{0}$, then f is differentiable along γ. By this we mean that the functions $t \longmapsto f(\gamma(t))$ are differentiable at $t = 0$. This is explained intuitively by the fact that the points $(\gamma(t),f(\gamma(t)))$ make up a curve on the surface which is tangent to the generatrix of the cone lying above the line of direction $\mathbf{v}$.

Case $f(x,y) = \mathrm{Re}(x + iy)^{2k+1}/(x^2 + y^2)^k$ (still with $f(0,0) = 0$), k denoting a positive integer. Still working in polar coordinates, we find

$$f(x,y) = \mathrm{Re}(r\cos\vartheta + ir\sin\vartheta)^{2k+1}/r^{2k} =$$

$$= r\ \mathrm{Re}(e^{i\vartheta})^{2k+1} = r\ \cos(2k+1)\vartheta.$$

This function is still homogeneous of degree one and represents a conical surface having the origin as vertex. Such functions are not differentiable for $k \geq 1$ and proved useful in antiquity

(How uncomfortable it must be!)

Fig.1.11

Case $f(x,y) = |xy|^{3/4}$ (resp. $|xy|^{1/2}$). The first function is differentiable at the origin since

$$|xy|^{3/4} \leq (r^2)^{3/4} = r^{3/2} = r \cdot r^{1/2} = o(r).$$

This limited expansion even shows that the differential (and the derivative) of f at the origin vanishes. The second function is not differentiable at the origin (cf. Fig.1.12)

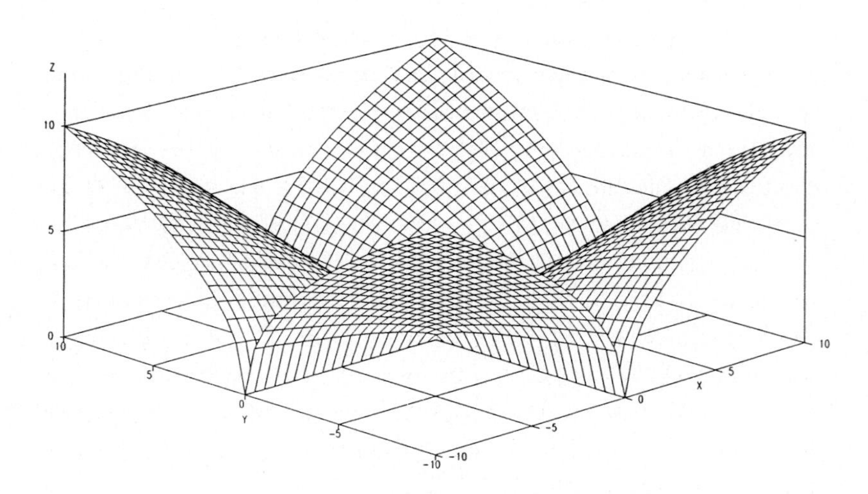

Fig.1.12

1.6 MORE EXAMPLES

By now, it should be obvious that a good definition of differentiability can only be given using limited expansions. Differential quotients cannot even be formed with vector increments, and partial derivatives (or more generally, directional derivatives) are insufficient to imply differentiability. Moreover, it is often simpler to write a limited expansion than to compute partial or directional derivatives. We give two examples.

Let us consider first the map defined by taking the square of a matrix X. Thus $F : M_n(\mathbb{R}) \longrightarrow M_n(\mathbb{R})$ is defined by $X \longmapsto X^2$. We are looking for the differential of F at a matrix A (if it exists) and consider thus the expansion

$$F(A + H) = (A + H)^2 = A^2 + AH + HA + H^2,$$

from which it is obvious that

$$F(A + H) = F(A) + AH + HA + o(\|H\|)$$

for example with the norm of a matrix defined by the identification of $M_n(\mathbb{R})$ with $\mathbb{R}^{n^2}$: $\|H\|^2 = \sum(h^i_j)^2$. The linear part of this limited expansion of the first order is the mapping

$$F'(A) : M_n(\mathbb{R}) \longrightarrow M_n(\mathbb{R}) , \; H \longmapsto AH + HA$$

and F is differentiable at the matrix A. Similar considerations apply for the powers $F_k : X \longrightarrow X^k : M_n(\mathbb{R}) \longrightarrow M_n(\mathbb{R})$ and lead to

$$F_k'(A) : M_n(\mathbb{R}) \longrightarrow M_n(\mathbb{R}) , \; H \longmapsto H \cdot A^{k-1} + A \cdot H \cdot A^{k-2} + \ldots + A^{k-1} \cdot H.$$

In the second **example**, we consider the following situation. A firm produces a product P (say chocolate) using a basic ingredient A (say cocoa or sugar). A unit production (say one ton) of P requires a certain quantity a of A costing α dollars per ton. Thus the cost of ingredient A for the production of one ton of P is $a\alpha$. For some reason, the firm has to change of supplier. Both quality and cost of the product A vary. With the new supplier, a unit production of P requires a quantity x of A at new unit price ξ. The new cost for ingredient A (still allowing the production of one ton of P) is now $x\xi$. Can one compute a percentage of increase in cost production due separately to increase of price (α replaced by ξ) and to variation of quality (a replaced by x) of A? The analysis of the situation is simple enough.

Let us write

$$x = a + h, \ \xi = \alpha + \eta$$

and

$$x \cdot \xi = (a + h)(\alpha + \eta) = a\alpha + \alpha h + a\eta + h\eta.$$

The linear term representing the cost increase is

$$(h,\eta) \longmapsto \alpha h + a\eta$$

and the respective percentage increases $h = x - a$, $\eta = \xi - \alpha$. BUT observe the presence of the quadratic term $h\eta$! A single change in price of A (no variation of quality, no difference in quantity needed) produces a linear variation in cost of A. Similarly, a single change in quantity of A also produces a linear variation in cost of A. However, simultaneous variations of a and α produce *non linear* variations in cost. It is best to give a picture of the surface representing the graph of $f(x,y) = xy$. This surface contains two lines going through each point (but is not a plane).

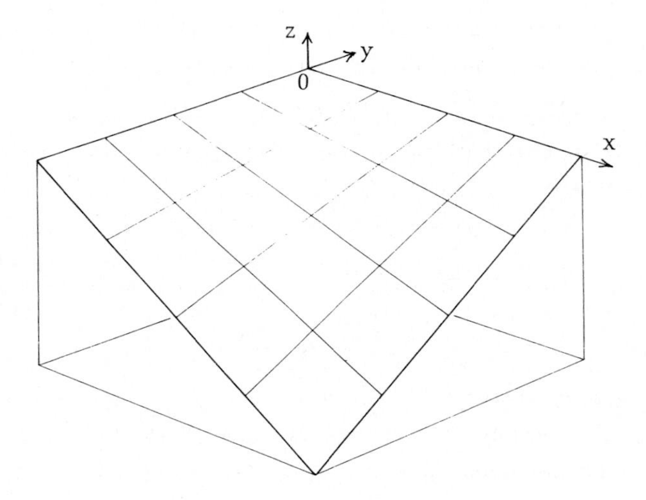

Fig.1.13

Nevertheless, a linear approximation can be acceptable for small variations, and this is the whole idea behind derivatives and limited expansions of the first order!

CHAPTER 2
PARAMETERIZED CURVES

2.1 TANGENT VECTOR : VELOCITY

Continuous mappings from $\mathbb{R}$ to $\mathbb{R}^n$ are viewed as *parameterized curves.* In particular, it is then convenient to imagine that the scalar variable is *time* and to denote it by t.

Let $t \longmapsto F(t)$ be such a map. Differentiability of F at some point a requires the existence of a limited expansion

$$F(a+h) = F(a) + L_a(h) + \varphi(h) \qquad (\varphi(h) = o(|h|) \text{ for } h \to 0).$$

The tangent linear map $L_a : \mathbb{R} \longrightarrow \mathbb{R}^n$ can simply be identified to a vector $\mathbf{v} = L_a(1)$, so that $L_a(h) = h\mathbf{v}$ and

$$F(a+h) = F(a) + \mathbf{v}h + o(h).$$

In this case, the variable t is a scalar number and differential quotients can be formed. The limited expansion gives

$$\frac{F(a+h) - F(a)}{h} - \mathbf{v} = h^{-1}o(h) \longrightarrow 0 \quad \text{when} \quad h \longrightarrow 0$$

and if F is differentiable at the point t = a with derivative $\mathbf{v}$ $(= L_a = F'(a))$

$$\mathbf{v} = \lim_{h \to 0} \frac{F(a+h) - F(a)}{h}.$$

Definition. When $F : \mathbb{R} \longrightarrow \mathbb{R}^n$ is differentiable at a point a, the derivative

$$\mathbf{v} = L_a = F'(a) = \lim_{h \to 0} \frac{F(a+h) - F(a)}{h}$$

is the **velocity vector** of F at a. ∎

We shall mainly be interested in the case where $F \in \mathcal{C}^1$, namely, F is differentiable at every point and $t \longmapsto F'(t)$ is a continuous mapping $\mathbb{R} \longrightarrow \mathbb{R}^n$. Here, F and F' have the same mathematical nature : both are parameterized curves.

Practically, it is obvious that when F is given in components $F = {}^t(f_1, f_2, \ldots, f_n)$, the derivative is computed in components too

$$\mathbf{v} = F'(a) = {}^t(f'_1(a), f'_2(a), \ldots, f'_n(a)).$$

It is important to realize that the image of a continuously differentiable curve can have *corners*. For **example**, let us consider the following plane curve $\mathbb{R} \longrightarrow \mathbb{R}^2$ given in components by

$$f_1(t) = \begin{cases} 0 & \text{if } t \le 0 \\ t^2 & \text{if } t > 0 \end{cases}, \quad f_2(t) = \begin{cases} t^2 & \text{if } t \le 0 \\ 0 & \text{if } t > 0 \end{cases}.$$

Then

$$f'_1(t) = \begin{cases} 0 \text{ if } t \le 0 \\ 2t \text{ if } t > 0 \end{cases}, \quad f'_2(t) = \begin{cases} 2t \text{ if } t \le 0 \\ 0 \text{ if } t > 0 \end{cases}.$$

In these formulas, we see that $F' = {}^t(f_1, f_2)$ is continuous. But the image of F is not *smooth* (cf.Fig.).

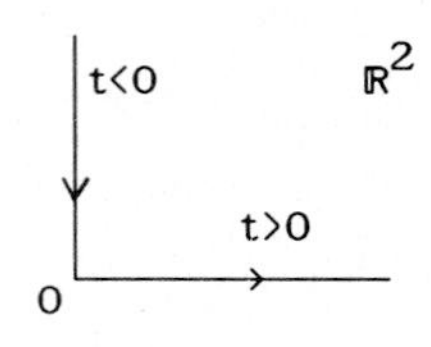

Fig.2.1

However, the only points where a $\mathcal{C}^1$-parameterized curve F has a non smooth image are the points where the velocity vector vanishes $\mathbf{v} = F'(t) = 0$. Indeed, if $v = F'(t) \neq 0$, the image of F is a curve admitting $\mathbf{v}$ as tangent vector.

Although we do not want to insist on pathology, let us indicate that a *continuous* curve $F : \mathbb{R} \longrightarrow \mathbb{R}^2$ can fill the plane. Such examples are known under the name of **Peano curves**. But if a parameterized curve $F : \mathbb{R} \longrightarrow \mathbb{R}^n$ is $\mathcal{C}^1$, its image $F(\mathbb{R})$ is *negligible* in $\mathbb{R}^n$ (provided $n > 1$). This is a consequence of **Sard theorem** for which we refer to specialized books in differential geometry. In particular, a $\mathcal{C}^1$-parameterized curve is never a Peano curve.

Here is the basic result for $\mathcal{C}^1$-parameterized curves.

Theorem. Let $F : \mathbb{R} \longrightarrow \mathbb{R}^n$ be a $\mathcal{C}^1$-parameterized curve. Then

$$\|F(b) - F(a)\| \le (b - a) \sup_{a \le t \le b} \|F'(t)\|$$

for all couples $a \le b$. □

Proof. This theorem is easily proved using the similar result (finite increments theorem) in the scalar case. For each vector $\mathbf{v} \in \mathbb{R}^n$, consider the scalar function $f = f_{\mathbf{v}}$ obtained by scalar product with $\mathbf{v}$: $f(t) = F(t)\cdot\mathbf{v}$. Eventually, we shall take for $\mathbf{v}$ the vector $\mathbf{v} = F(b) - F(a)$, but we can argue quite generally at first. For this scalar function f, there will be a point $\tau = \tau_{\mathbf{v}}$ in the interval $[a,b]$ with

$$f(b) - f(a) = (b - a)f'(\tau) = (b - a)F'(\tau)\cdot\mathbf{v}.$$

The Cauchy-Schwarz inequality gives

$$|f(b) - f(a)| = |(b - a)F'(\tau)\cdot\mathbf{v}| \le (b - a) \|F'(\tau)\| \|\mathbf{v}\|.$$

In particular,

$$|(F(b) - F(a))\cdot \mathbf{v}| \leq (b - a) \sup_{a\leq t\leq b} \|F'(\tau)\| \;\|\mathbf{v}\|.$$

If $F(b) = F(a)$, there is nothing to prove. Otherwise we can take $\mathbf{v} = F(b) - F(a)$ and we get

$$\|F(b) - F(a)\|^2 \leq (b - a) \sup_{a\leq t\leq b} \|F'(\tau)\| \;\|F(b) - F(a)\|.$$

Since $\|F(b) - F(a)\| \neq 0$, we can simplify by this factor and get the announced result. ■

If we interpret a differential quotient $\dfrac{F(b) - F(a)}{b - a}$ as *average velocity* in the interval $[a,b]$ and its norm as *average speed* in the same interval, the statement of the above theorem is the following :

the average speed in an interval $[a,b]$ of a parameterized

curve cannot exceed the maximum of the speeds $\|\mathbf{v}\|$ over the same interval.

Observe that in the preceding proof, if we take for $\mathbf{v}$ a vector $\mathbf{e}_i$ of the canonical basis of $\mathbb{R}^n$, the corresponding scalar function f_i is the ith-component of F and we have an intermediate value $\tau_i \in [a,b]$ with

$$f_i(b) - f_i(a) = (b - a)f'_i(\tau_i).$$

But usually *no intermediate value of* τ *in* $[a,b]$ *will work simultaneously for all components* f_i. This means that —in general— no intermediate value of τ in $[a,b]$ will work for the vector valued function F.

Example. Let $F : \mathbb{R} \longrightarrow \mathbb{R}^3$ be a parameterized **helix** :

$$x = \cos t\ ,\ y = \sin t\ ,\ z = t.$$

The velocity vector has components

$$v_1 = -\sin t,\ v_2 = \cos t,\ v_3 = 1.$$

In particular, this vector is never vertical but $F(2\pi) - F(0) = \begin{pmatrix} 0 \\ 0 \\ 2\pi \end{pmatrix}$ is vertical. In this example, one can show that

$$F(b) - F(a) = (b - a)F'(\tau)$$

implies $a = b$ so that there is no intermediate point τ giving the precise value for $F(b) - F(a)$.

2.2 RECTIFIABLE CURVES

Estimates of length of curves

$$F : I = [a,b] \longrightarrow \mathbb{R}^n \quad \text{(F continuous)}$$

are given by finite sums of length of *secants*

$$\sum_{1\le i\le m} \|F(t_{i+1}) - F(t_i)\|$$

corresponding to a finite subdivision $a = t_1 < t_2 < \ldots < t_{m+1} = b$.

Definition. A (continuous) curve $F : I = [a,b] \longrightarrow \mathbb{R}^n$ is called **rectifiable** when the supremum over all finite subdivisions

$$\text{Sup} \sum_{1\le i\le m} \|F(t_{i+1}) - F(t_i)\|$$

is finite. The value of this supremum is by definition the **length** of the curve.

Theorem. A $\mathcal{C}^1$-parameterized curve F is rectifiable and its length is given by $\int_I \|F'(t)\|dt$. □

Proof. Let $F : [a,b] \longrightarrow \mathbb{R}^n$ be $\mathcal{C}^1$. The derivative F' of F is continuous on [a,b] (recall that by convention F'(a) and F'(b) denote the right, resp. left, derivatives at these end points). For each finite subdivision as above, we compute the sequences $P_i = F(t_i)$,

$$\mathbf{u}_i = \overrightarrow{P_iP_{i+1}} = F(t_{i+1}) - F(t_i),$$

$$\mathbf{v}_i = F'(t_i)(t_{i+1} - t_i) = F'(t_i)\Delta t_i \ .$$

By definition of the integral, the sums $\sum \|\mathbf{v}_i\|$ tend to $\int_a^b \|F'(t)\|dt$ when the subdivisions get finer, namely when $\text{Sup}_i |t_{i+1}-t_i| \longrightarrow 0$.

To estimate $\|\mathbf{u}_i - \mathbf{v}_i\|$, we shall apply the main theorem of 2.1 to the $\mathcal{C}^1$-function $t \longmapsto \Phi(t) = F(t) - F(t_i) - F'(t_i)(t - t_i)$ on the interval $[t_i,t_{i+1}]$. The derivative of this function is

$$\Phi'(t) = F'(t) - F'(t_i),$$

and $\Phi(t_i) = 0$ so that $\|\mathbf{u}_i - \mathbf{v}_i\| = \|\Phi(t_{i+1})\| = \|\Phi(t_{i+1}) - \Phi(t_i)\|$.

Since F' is continuous on the compact interval [a,b], it is *uniformly continuous* on this interval and $\|\Phi'\|$ will be small as soon as the subdivision is fine enough. Precisely, if $\varepsilon > 0$ is given, there exists a positive δ such that

$$\|F'(s) - F'(t)\| \le \varepsilon \quad \text{as soon as } |s - t| \le \delta.$$

In particular, if we assume that the subdivision is fine enough, say $\text{Sup}_i |t_{i+1}-t_i| \le \delta$, then

$$\|\mathbf{u}_i - \mathbf{v}_i\| \le \sup_{t_i \le t \le t_{i+1}} \|\Phi'(t)\| \,(t_{i+1} - t_i) \le \varepsilon\, \Delta t_i \,.$$

Finally, we estimate the difference

$$\left|\sum \|\mathbf{u}_i\| - \sum \|\mathbf{v}_i\|\right| = \left|\sum (\|\mathbf{u}_i\| - \|\mathbf{v}_i\|)\right| \le \sum\left|\|\mathbf{u}_i\| - \|\mathbf{v}_i\|\right| \le$$

$$\le \sum \|\mathbf{u}_i - \mathbf{v}_i\| \le \sum \varepsilon\, \Delta t_i = \varepsilon(b - a).$$

This proves that the sums $\sum \|\mathbf{u}_i\|$ are arbitrarily close to corresponding sums $\sum \|\mathbf{v}_i\|$ tending to $\int_a^b \|F'(t)\|dt$. ■

Corollary. Let $F \in \mathcal{C}^1$ and denote by $s = s(t)$ the length of the curve $F|_{[a,t]}$. Then $t \longmapsto s(t)$ is also $\mathcal{C}^1$ with $s'(t) = \|F'(t)\|$. ■

In particular, the parameter t coïncides with the **special parameter** s given by the length of the curve exactly when the *speed* of the parametrization is equal to 1 : $\|F'(t)\| \equiv 1$.

In the course of the proof of the above theorem, we have used the inequality $|\;\|\mathbf{u}\| - \|\mathbf{v}\|\;| \le \|\mathbf{u} - \mathbf{v}\|$. Let us quickly recall its proof. The triangle inequality gives

$$\|\mathbf{u}\| = \|\mathbf{u} - \mathbf{v} + \mathbf{v}\| \le \|\mathbf{u} - \mathbf{v}\| + \|\mathbf{v}\|,$$

hence $\|\mathbf{u}\| - \|\mathbf{v}\| \le \|\mathbf{u} - \mathbf{v}\|$. Interchanging the roles of **u** and **v**, we get $\|\mathbf{v}\| - \|\mathbf{u}\| \le \|\mathbf{u} - \mathbf{v}\|$. These two inequalities can be combined using the absolute value as stated.

Application. When $A \ne B$ are two fixed points in $\mathbb{R}^n$, the length of all $\mathcal{C}^1$-parameterized curves linking A to B is $\ge \|A - B\|$. In particular, straight lines minimize the length between two points. □

In spite of its intuitive content, the preceding assertion deserves a **proof** ! Let us consider an arbitrary $\mathcal{C}^1$-parametrization F of a curve $\mathcal{C}$ linking A and B, say with $F(a) = A$, $F(b) = B$. Take moreover $\mathbf{v} = \overrightarrow{AB} = F(b) - F(a)$ and compute

$$\int_a^b F'(t)\cdot\mathbf{v}\; dt = \Big[F(t)\cdot\mathbf{v}\Big]_a^b = [F(b) - F(a)]\cdot\mathbf{v} = \|\mathbf{v}\|^2.$$

On the other hand, the Cauchy-Schwarz inequality gives

$$\left|\int_a^b F'(t)\cdot\mathbf{v}\; dt\right| \le \int_a^b |F'(t)\cdot\mathbf{v}|\; dt \le \|\mathbf{v}\| \int_a^b \|F'(t)\|\; dt.$$

Together, the two results obtained prove

$$\|\mathbf{v}\|^2 \le \|\mathbf{v}\| \text{ length}(\mathcal{C}) \quad \text{hence} \quad \|\mathbf{v}\| \le \text{length}(\mathcal{C})$$

after simplification by $\|\mathbf{v}\| \ne 0$. ■

2.3 CURVES IN MATRIX GROUPS

The theory of n-dimensional curves that we have developed so far is quite general. It includes curves $\mathbb{R} \longrightarrow \mathbb{M}_d(\mathbb{R})$ with values in matrix groups, simply since $\mathbb{M}_d(\mathbb{R})$ can be identified to $\mathbb{R}^n$ with $n = d^2$. However, some interesting features can be given in this context.

If F and G are two $\mathcal{C}^1$-parameterized curves $\mathbb{R} \longrightarrow \mathbb{M}_n(\mathbb{R})$, we can consider the "product" $F \cdot G : t \longmapsto F(t) \circ G(t)$ as a parameterized curve $\mathbb{R} \longrightarrow \mathbb{M}_n(\mathbb{R})$. This is a $\mathcal{C}^1$-curve with

$$(F \cdot G)' = F' \cdot G + F \cdot G'.$$

This result is obvious since the derivative is computed coefficient by coefficient, each of them being a sum of products

$$\Big(\sum_k a^i_k\ b^k_j\Big)' = \sum_k (a^i_k)'\ b^k_j + \sum_k a^i_k\ (b^k_j)'.$$

Let us denote by $\mathbb{G}\ell_n(\mathbb{R})$ the *group* of n by n real invertible matrices

$$\mathbb{G}\ell_n(\mathbb{R}) = \{\ M \in \mathbb{M}_n(\mathbb{R}) : \det M \neq 0\ \}.$$

For example, the identity matrix I of size $n \times n$ is in $\mathbb{G}\ell_n(\mathbb{R})$, and for each $M \in \mathbb{G}\ell_n(\mathbb{R})$, M^{-1} exists and is also in $\mathbb{G}\ell_n(\mathbb{R})$. A $\mathcal{C}^1$-parameterized curve in $\mathbb{G}\ell_n(\mathbb{R})$ is simply a $\mathcal{C}^1$-curve $F : \mathbb{R} \longrightarrow \mathbb{M}_n(\mathbb{R})$ with $F(t) \in \mathbb{G}\ell_n(\mathbb{R})$ (namely F(t) invertible) for all $t \in \mathbb{R}$. The known formulas for the computation of the coefficients of $F(t)^{-1}$ show that $F^{-1} : t \longmapsto F(t)^{-1}$ is also $\mathcal{C}^1$. Take $G = F^{-1}$ and differentiate the identity $F \cdot G = I$ according to the product rule just explained. We get

$$0 = F' \cdot G + F \cdot G' \quad \text{hence} \quad F \cdot G' = -F' \cdot G \quad \text{and} \quad G' = -F^{-1} \cdot F' \cdot G.$$

Recalling that $G = F^{-1}$, we have proved

$$(F^{-1})' = -F^{-1} \cdot F' \cdot F^{-1}.$$

This formula is to be compared with the usual rule for taking the derivative of a scalar function f

$$(1/f)' = -f'/f^2$$

of which it constitutes the correct generalization in our context (the matrix product is not *commutative*).

Lemma. Let $F : \mathbb{R} \longrightarrow \mathbb{G}\ell_n(\mathbb{R})$ be $\mathcal{C}^1$. Then the derivative of det F(t) is given by the following formula

$$(d/dt) \det F(t) = \mathrm{Tr}[F'(t) \circ F(t)^{-1}] \cdot \det F(t). \qquad \square$$

Proof. Write $\Delta(t) = \det F(t)$ and

$$F(t+h) = F(t) + hM_{t,h} \text{ with } M_{t,h} \longrightarrow F'(t) \text{ (when } h \to 0).$$

Then

$$\Delta(t+h) = \det[F(t) + hM_{t,h}] = \Delta(t)\cdot\det[I + hM_{t,h}F(t)^{-1}].$$

Let us show that quite generally (A is any square matrix)

$$\det(I + hA) = 1 + h \operatorname{Tr}A + h^2(\ldots).$$

Indeed, the terms with a low power of h involved in the computation of the determinant

$$\begin{vmatrix} 1+ha^1_1 & ha^1_2 & \ldots & ha^1_n \\ ha^2_1 & 1+ha^2_2 & \ldots & ha^2_n \\ \ldots & & \ldots & \\ ha^n_1 & ha^n_2 & \ldots & 1+ha^n_n \end{vmatrix}$$

arise from products containing many terms in the principal diagonal. Starting with the product of the terms in the main diagonal we get

$$\det(I + hA) = (1 + ha^1_1)(1 + ha^2_2)\ldots(1 + ha^n_n) + h^2(\ldots)$$

(since non written terms contain at least *two* non diagonal terms, hence are divisible by h^2), and hence

$$\det(I + hA) = 1 + h(a^1_1 + a^2_2 + \ldots + a^n_n) + h^2(\ldots)$$

$$= 1 + h \operatorname{Tr}A + h^2(\ldots).$$

Coming back to the expression for $\Delta(t+h)$, we see that

$$\Delta(t+h) = \Delta(t)\left[1 + h \operatorname{Tr} M_{t,h}F(t)^{-1} + h^2(\ldots)\right].$$

In particular, we get the limited expansion

$$\Delta(t+h) = \Delta(t) + h\,\Delta(t) \operatorname{Tr} M_{t,h}F(t)^{-1} + o(h)$$

on which the value of the derivative at the point t is immediately read :

$$\Delta'(t) = \Delta(t) \operatorname{Tr} M_{t,h}F(t)^{-1}$$ ∎

Another more special —but very useful— result concerns curves in the orthogonal group $\mathbb{O}_n(\mathbb{R}) \subset \mathbb{G}\ell_n(\mathbb{R})$. Recall that an orthogonal matrix $M \in \mathbb{G}\ell_n(\mathbb{R})$ is an invertible matrix for which

$$M^{-1} = {}^tM \text{ (transpose of M).}$$

If M is an orthogonal matrix, so is its inverse. Moreover, if M and N are two orthogonal matrices, so is their product since

$$(MN)^{-1} = N^{-1}M^{-1} = {}^tN\,{}^tM = {}^t(MN).$$

Taking the determinant of the identity ${}^tM\cdot M = I$ we get

$$1 = \det({}^tM\cdot M) = \det({}^tM)\cdot\det M = (\det M)^2$$

for any orthogonal matrix (since M and its transpose have the same determinant). This proves that orthogonal matrices always have determinant ±1.

Theorem. Let $F : \mathbb{R} \longrightarrow \mathbb{G}\ell_n(\mathbb{R})$ be a $\mathcal{C}^1$-parameterized curve.

a) If $\det F(t) = 1$ for all $t \in \mathbb{R}$, then

$$F'(t)\circ F(t)^{-1} \text{ has trace } 0 \text{ for all } t \in \mathbb{R}$$

and in particular, $F'(\tau)$ has trace 0 if $F(\tau) = I$,

b) If $F(t)$ is an orthogonal matrix for all $t \in \mathbb{R}$ (namely F is a $\mathcal{C}^1$-parameterized curve in the orthogonal group $\mathbb{O}_n(\mathbb{R})$), then

$${}^tF(t)'\circ F(t) = -\,{}^tF(t)\circ F(t)' \quad \text{for all } t \in \mathbb{R},$$

$$F(t)'\circ {}^tF(t) = -\,F(t)\circ {}^tF(t)' \quad \text{for all } t \in \mathbb{R},$$

and in particular $F'(\tau)$ is antisymmetric for each value of the parameter τ where $F(\tau) = I$. □

Proof. Part a) results immediately from the lemma with $\Delta(t) \equiv 1$ hence $\Delta'(t) \equiv 0$. To obtain part b), it is enough to differentiate the identity ${}^tF(t)\circ F(t) = I$, obtaining

$${}^tF(t)'\circ F(t) + {}^tF(t)\circ F(t)' = 0.$$ ■

If we define $A = F'\circ F^{-1}$, then A is a continuous curve in $\mathbb{G}\ell_n$ and F satisfies the *differential equation* $F' = A(t)\circ F$. Let us summarize the statements of the theorem in this form.

Corollary. Let $F: \mathbb{R} \longrightarrow \mathbb{G}\ell_n(\mathbb{R})$ be a $\mathcal{C}^1$-parameterized curve. Define $A = F'\circ F^{-1}$ so that F is a solution of the differential equation

$$F' = A(t)\circ F.$$

a) If $\det F(t) = 1$ for all t, then $\operatorname{Tr} A(t) = 0$ for all t,

b) If $F(t) \in \mathbb{O}_n(\mathbb{R})$ for all t, then $A(t)$ is antisymmetric. □

Proof. Part a) has been seen in the theorem. Let us check part b)

$${}^tA = {}^t(F'\circ F^{-1}) = {}^tF^{-1}\circ {}^tF' = F\circ {}^tF' = -F'\circ {}^tF = -F'\circ F^{-1} = -A.$$ ■

Let us observe that a solid in movement determines a curve

$$\mathbb{R} \longrightarrow \mathbb{O}_n(\mathbb{R}).$$

Indeed, if we choose an orthonormal basis attached to the solid (e.g. attached to its *center of gravity*), the motion of the solid will determine a variation of the basis with respect to a fixed basis. Writing the components of this variable basis either in lines or columns, we get a curve of the announced type.

Differential equations of the type $F' = A(t)\circ F$ will be encountered and studied more systematically in the next chapter.

2.4 SERRET-FRENET FORMULAS FOR CURVES IN $\mathbb{R}^3$

In this section, we are going to consider **regular curves**

$$\gamma : \mathbb{R} \longrightarrow \mathbb{R}^3$$

namely $\mathcal{C}^1$-parameterized curves in $\mathbb{R}^3$ *with* $\gamma'(t) \neq 0$ for all t.

For such a curve, the velocity vector $\mathbf{v} = \gamma'(t)$ furnishes a tangent vector at each point of $\mathcal{C} = \mathrm{Im}(\gamma)$ and we shall use the unit tangent vector $\mathbf{t} = \mathbf{v}/\|\mathbf{v}\|$ at a point $P = \gamma(t) \in \mathcal{C}$. It will be convenient to assume that the parameter t coïncides with the special parameter s given by arc length (computed from a fixed reference point). Thus, we shall write $\mathbf{t} = \mathbf{t}(s)$.

Let us assume now that γ is $\mathcal{C}^2$, hence $s \longmapsto \mathbf{t}(s)$ is $\mathcal{C}^1$. Differentiating the identity $\mathbf{t}\cdot\mathbf{t} = \|\mathbf{t}\|^2 = 1$ with respect to the special parameter s, we get

$$\mathbf{t}'\cdot\mathbf{t} + \mathbf{t}\cdot\mathbf{t}' = 0 \quad \text{i.e. } 2\mathbf{t}\cdot\mathbf{t}' = 0 \quad \text{and} \quad \mathbf{t}' \perp \mathbf{t}.$$

If not zero, this *acceleration* vector $\mathbf{t}'$ furnishes a *normal* vector to the curve.

Definition. The **curvature** at a regular point P of a $\mathcal{C}^2$-curve $\mathcal{C}$ is the number $\kappa = \kappa(P)$ defined as follows. Choose a parametrization γ of $\mathcal{C}$ by a special parameter s (there are two such parametrizations with $s = 0$ at P) and compute the unit tangent vector map $s \longmapsto \mathbf{t}(s)$ and its derivative $\mathbf{t}'$. Then $\kappa = \|\mathbf{t}'\|$ at the point P. When the curvature $\kappa = \kappa(P)$ does not vanish, we define the **normal** of $\mathcal{C}$ at P to be the unit vector $\mathbf{n} = \mathbf{t}'/\kappa$. The **binormal** of $\mathcal{C}$ at P is the unit vector $\mathbf{b} = \mathbf{t} \wedge \mathbf{n}$. ∎

In this way, to each regular point P of a $\mathcal{C}^2$-curve having nonzero curvature, we associate an orthonormal basis $(\mathbf{t}, \mathbf{n}, \mathbf{b})$. This basis is called the **Serret-Frenet frame** of $\mathcal{C}$ at the point P. When the point P moves on the curve $\mathcal{C}$, this basis *varies*. When the curve $\mathcal{C}$ is *three times continuously differentiable*, it is possible to control the variation of these bases. For example, we have

$$\mathbf{t}' = \kappa\mathbf{n} \tag{1}$$

by definition of the curvature κ. Also, differentiating the identity $\mathbf{n}\cdot\mathbf{n} = \|\mathbf{n}\|^2 = 1$, we get $2\mathbf{n}\cdot\mathbf{n}' = \mathbf{n}'\cdot\mathbf{n} + \mathbf{n}\cdot\mathbf{n}' = 0$, hence $\mathbf{n}'$ is orthogonal to $\mathbf{n}$ and can be expressed as a linear combination of the two unit vectors $\mathbf{t}$ and $\mathbf{b}$, say $\mathbf{n}' = \alpha\mathbf{t} + \beta\mathbf{b}$. The coefficients can be determined by scalar product, e.g.

$$\alpha = \mathbf{n}'\cdot\mathbf{t} = -\mathbf{n}\cdot\mathbf{t}' \quad \text{(since } \mathbf{n}\cdot\mathbf{t} = 0 \text{ is constant)}$$
$$= -\mathbf{n}\cdot\kappa\mathbf{n} = -\kappa.$$

The coefficient β is by definition the **torsion** of the curve at the point P and denoted by τ

$$\tau = \mathbf{n}'\cdot\mathbf{b} = -\mathbf{n}\cdot\mathbf{b}' \quad \text{(since } \mathbf{n}\cdot\mathbf{b} = 0 \text{ is constant)}$$

and

(2) $$\mathbf{n}' = -\kappa\mathbf{t} + \tau\mathbf{b}.$$

Similarly, we can determine the variation of the binormal vector **b**. Differentiating the definition $\mathbf{b} = \mathbf{t} \wedge \mathbf{n}$ we obtain

$$\mathbf{b}' = \mathbf{t}' \wedge \mathbf{n} + \mathbf{t} \wedge \mathbf{n}' = \kappa\mathbf{n} \wedge \mathbf{n} + \mathbf{t} \wedge (-\kappa\mathbf{t} + \tau\mathbf{b}) = \tau\, \mathbf{t} \wedge \mathbf{b}.$$

Since the Frenet frame $(\mathbf{t}, \mathbf{n}, \mathbf{b})$ is direct (i.e. $\mathbf{b} = \mathbf{t} \wedge \mathbf{n}$), we also have by cyclic permutation

$$\mathbf{t} = \mathbf{n} \wedge \mathbf{b} \quad \text{and} \quad \mathbf{n} = \mathbf{b} \wedge \mathbf{t}.$$

Thus we have found

(3) $$\mathbf{b}' = -\tau\mathbf{n}.$$

We have thus proved the following result.

Theorem(Serret-Frenet formulas). Let $\mathcal{C}$ be a $\mathcal{C}^3$-curve parameterized by the special parameter s (arc length) and $P \in \mathcal{C}$ a regular point with nonzero curvature κ. Then the Serret-Frenet frame $(\mathbf{t}, \mathbf{n}, \mathbf{b})$ is a direct orthonormal basis $\mathbf{b} = \mathbf{t} \wedge \mathbf{n}$ with variation given by the following three formulas

(1) $$\mathbf{t}' = \kappa\mathbf{n}\ ,$$

(2) $$\mathbf{n}' = -\kappa\mathbf{t} + \tau\mathbf{b}\ ,$$

(3) $$\mathbf{b}' = -\tau\mathbf{n}\ .$$ ■

Remark. With the above assumptions and notations, we could have introduced the components t_i of **t** (in the canonical basis), resp. n_i of **n** and b_i of **b**. Then the Serret-Frenet formulas could be written for the variation of the matrix

$$C = C(s) = \begin{pmatrix} t_1 & t_2 & t_3 \\ n_1 & n_2 & n_3 \\ b_1 & b_2 & b_3 \end{pmatrix} \in \mathbb{O}_3(\mathbb{R})\ .$$

The relation $\mathbf{t}' = \kappa\mathbf{n}$ is equivalent to $t_i' = \kappa n_i$ (i=1,2 or 3) and implies a following matrix relation

$$\begin{pmatrix} t_1' & t_2' & t_3' \\ . & . & . \\ . & . & . \end{pmatrix} = \begin{pmatrix} 0 & \kappa & 0 \\ . & . & . \\ . & . & . \end{pmatrix} \begin{pmatrix} . & . & . \\ n_1 & n_2 & n_3 \\ . & . & . \end{pmatrix}.$$

Altogether, the Serret-Frenet relations give the matrix relation

$$\frac{dC}{ds} = \begin{pmatrix} 0 & \kappa & 0 \\ -\kappa & 0 & \tau \\ 0 & -\tau & 0 \end{pmatrix} \cdot C$$

of the form

$$C' = A \cdot C \quad \text{where} \quad A = A(s) \text{ is antisymmetric.}$$

This is another illustration of the general fact explained in the preceding section. Let us examine more closely the canonical form of a curve near a regular point (where its curvature is not zero). Translating and rotating the curve if necessary, there is no restriction in generality if we only consider the case where the point P is the origin and the Serret-Frenet frame at P is the canonical basis of $\mathbb{R}^3$.

Theorem (Canonical form of a curve near a regular point). Let the origin be a regular point on a $\mathcal{C}^3$-curve, with nonzero curvature κ at this point. Then a local parametrization of this curve —using the arc length counted from the origin— has the following form

$$\begin{cases} x(s) = s - (\kappa^2/6)s^3 + o(s^3), \\ y(s) = (\kappa/2)s^2 + (\kappa'/6)s^3 + o(s^3), \\ z(s) = (\kappa\tau/6)s^3 + o(s^3). \end{cases}$$ □

Proof. Let $s \longmapsto F(s) = \mathbf{r}(s)$ denote a $\mathcal{C}^3$-parametrization by arc length as supposed in the statement. We have a limited expansion of the third order (valid component by component, hence as in the scalar case)

$$\mathbf{r}(s) = \mathbf{r}(0) + \mathbf{r}'(0)s + \mathbf{r}''(0)s^2/2! + \mathbf{r}^{(3)}(0)s^3/3! + o(s^3)$$

with

$$\mathbf{r}(0) = 0,\ \mathbf{r}'(0) = \mathbf{t}(0) = \mathbf{e}_1,\ \mathbf{r}''(0) = \mathbf{t}'(0) = \kappa\mathbf{n}(0) = \kappa\mathbf{e}_2 .$$

Moreover,

$$\mathbf{r}^{(3)}(0) = \kappa'\mathbf{n}(0) + \kappa\mathbf{n}'(0) =$$

$$= \kappa'\mathbf{n} + \kappa(-\kappa\mathbf{t} + \tau\mathbf{b}) = -\kappa^2\mathbf{e}_1 + \kappa'\mathbf{e}_2 + \kappa\tau\mathbf{e}_3 .$$

The above limited expansion is thus given by

$$\mathbf{r}(s) = \mathbf{e}_1 s + \kappa\mathbf{e}_2 s^2/2! + (-\kappa^2\mathbf{e}_1 + \kappa'\mathbf{e}_2 + \kappa\tau\mathbf{e}_3)s^3/3! + o(s^3) .$$

In components, this reads

$$x(s) = s - \kappa^2 s^3/3! + o(s^3) .$$

This is the first announced formula and the other ones are extracted by similar means from the vector formula. ■

From this theorem it follows that conversely $s = x + O(x^3)$ (inversion theorem in the one variable case) and also

$$y = (\kappa/2)x^2 + O(x^3),$$
$$z = (\kappa\tau/6)x^3 + o(x^3).$$

The following figure gives the projections of the curve near the origin. Neglecting all remainders, we get a canonical curve having a *contact of order 3* with the given curve, and depending only on the two parameters $\kappa > 0$ (curvature) and τ (torsion).

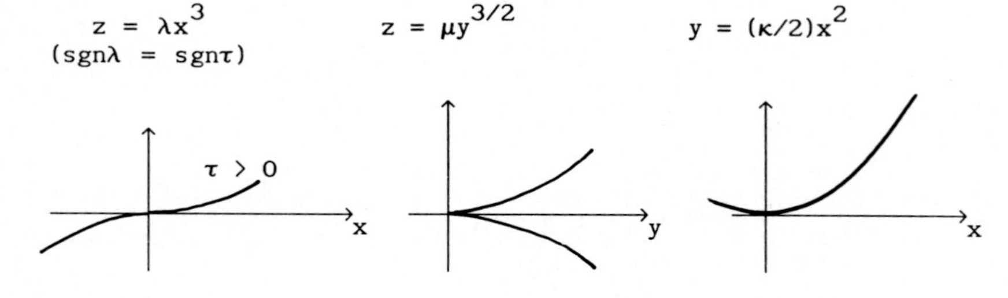

Fig.2.2

Observe that the parabola with equation $y = (\kappa/2)x^2$ has a *curvature circle* (at the origin) with radius $\rho = 1/\kappa$.

2.5 GENERALIZATION OF THE SERRET-FRENET FORMULAS IN $\mathbb{R}^n$

Let $F : \mathbb{R} \longrightarrow \mathbb{R}^n$ be a *smooth* parameterized curve (we shall need derivatives of F up to order n included).

Assumption. We are going to study *generic* points F(t), namely points $P = F(t)$ where the n vectors $F'(t)$, $F''(t)$, ..., $F^{(n)}(t)$ are *linearly independent*.

At such a generic point, we can construct an orthonormal basis $(\varepsilon_i)_{1\le i\le n}$ inductively as follows :

$\varepsilon_1 = F'(t)/\|F'(t)\|$ is a unit tangent vector,

$\varepsilon_2 = \alpha_2 F''(t) + \beta_2 F'(t)$ normed and $\perp$ to ε_1, ...

$\varepsilon_i = \alpha_i F^{(i)}(t) + \beta_i F^{(i-1)}(t) + \ldots + \lambda_i F'(t)$

normed, $\perp \varepsilon_j$ for all $j < i$, etc.

To make this construction completely canonical, we shall require $\alpha_i > 0$, for $i < n$, and $\det(\varepsilon_i) > 0$ (this imposes $\det(\varepsilon_i) = +1$ and the corresponding α_n will be positive precisely when the basis $(F^{(i)}(t))_{1\le i\le n}$ has the same *orientation* as the canonical basis).

The preceding construction is the well-known **Gram-Schmidt** orthonormalization procedure and we shall have to review it in more detail in a later chapter.

As observed in sec.2.3, the matrix M formed with the components of the ε_i (written in line) defines a curve in $\mathbb{O}_n(\mathbb{R})$ and satisfies a differential equation

$$M' = \Omega(t) \circ M \quad \text{where} \quad \Omega(t) \text{ is antisymmetric (for all t).}$$

Observing that since ε_i is a linear combination of

$$F'(t),\ F''(t),\ \ldots\ ,\ F^{(i)}(t)$$

we infer that ε_i' will be a linear combination of

$$F'(t),\ F''(t),\ \ldots\ ,\ F^{(i)}(t) \text{ and } F^{(i+1)}(t)$$

(don't forget to differentiate the coefficients in this variable linear combination!). Consequently,

$$\varepsilon_i' \text{ is a linear combination of } \varepsilon_1, \ldots, \varepsilon_{i+1}$$

and the coefficients a_{ij} of Ω vanish for $j \geq i+2$.

Define $\kappa_i = \omega_{i,i+1}$ for $1 \leq i \leq n-1$. Then the antisymmetric matrix $\Omega = \Omega(t)$ has the simple form

$$\Omega = \begin{pmatrix} 0 & \kappa_1 & 0 & \ldots & 0 \\ -\kappa_1 & 0 & \kappa_2 & & \vdots \\ 0 & -\kappa_2 & 0 & \ddots & \\ \vdots & & \ddots & & \kappa_{n-1} \\ 0 & \ldots & & & 0 \end{pmatrix}.$$

It is easy to check that $\kappa_i > 0$ for $1 \leq i \leq n-2$ but the sign of the last κ_{n-1} depends on the orientation of the frame

$$F'(t),\ \ldots\ ,\ F^{(n)}(t).$$

The positive numbers κ_i $(1 \leq i \leq n-2)$ play the role of *generalized curvatures* and $\kappa_{n-1} = \tau$ is a *generalized torsion* at the generic point P of the given curve in $\mathbb{R}^n$.

2.6 REVIEW OF FORMULAS FOR THE CROSS PRODUCT IN $\mathbb{R}^3$

0. Definition

$$\mathbf{a} \wedge \mathbf{b} = \begin{vmatrix} \mathbf{e}_1 & a_1 & b_1 \\ \mathbf{e}_2 & a_2 & b_2 \\ \mathbf{e}_3 & a_3 & b_3 \end{vmatrix} = \begin{pmatrix} a_2b_3 - a_3b_2 \\ \dots \\ \dots \end{pmatrix} \qquad \text{(cyclic permutations)}$$

1. Antisymmetry : $\quad \mathbf{b} \wedge \mathbf{a} = -\mathbf{a} \wedge \mathbf{b} \quad (= 0 \text{ iff } \mathbf{a} \mathbin{/\!/} \mathbf{b})$

2. Matrix formulation :

$$M_{\mathbf{a}} = \text{Matrix of } \mathbf{a} \wedge \dots = \begin{pmatrix} 0 & -a_3 & a_2 \\ a_3 & 0 & -a_1 \\ -a_2 & a_1 & 0 \end{pmatrix} \quad \text{in the canonical basis}$$

$$[M_{\mathbf{a}}, M_{\mathbf{b}}] := M_{\mathbf{a}}\, M_{\mathbf{b}} - M_{\mathbf{b}}\, M_{\mathbf{a}} = M_{\mathbf{a}\wedge\mathbf{b}}$$

$$\mathbf{a} \wedge (\mathbf{b} \wedge \mathbf{c}) + \mathbf{b} \wedge (\mathbf{c} \wedge \mathbf{a}) + \mathbf{c} \wedge (\mathbf{a} \wedge \mathbf{b}) = 0 \qquad \text{(Jacobi)}$$

3. Mixed product : $\mathbf{a}\cdot(\mathbf{b} \wedge \mathbf{c}) = \begin{vmatrix} a_1 & b_1 & c_1 \\ a_2 & b_2 & c_2 \\ a_3 & b_3 & c_3 \end{vmatrix} := (\mathbf{a},\mathbf{b},\mathbf{c})$

$(\mathbf{a},\mathbf{b},\mathbf{c}) = (\mathbf{b},\mathbf{c},\mathbf{a}) = (\mathbf{c},\mathbf{a},\mathbf{b})$ (cyclic permutations)

$(\mathbf{a},\mathbf{b},\mathbf{c}) = \pm$ volume of parallelepiped generated by **a**, **b** and **c**

$= 0$ when **a**, **b** and **c** are linearly dependent

4. Gibbs formula : $\mathbf{a} \wedge (\mathbf{b} \wedge \mathbf{c}) = \mathbf{b}(\mathbf{a}\cdot\mathbf{c}) - \mathbf{c}(\mathbf{a}\cdot\mathbf{b})$

$$(\mathbf{a} \wedge \mathbf{b})\cdot(\mathbf{c} \wedge \mathbf{d}) = \begin{vmatrix} \mathbf{a}\cdot\mathbf{c} & \mathbf{b}\cdot\mathbf{c} \\ \mathbf{a}\cdot\mathbf{d} & \mathbf{b}\cdot\mathbf{d} \end{vmatrix} =$$

$$= (\mathbf{a} \wedge \mathbf{b},\mathbf{c},\mathbf{d}) = (\mathbf{d},\mathbf{a} \wedge \mathbf{b},\mathbf{c})$$

$$\|\mathbf{a} \wedge \mathbf{b}\|^2 = \|\mathbf{a}\|^2\|\mathbf{b}\|^2 - |\mathbf{a}\cdot\mathbf{b}|^2$$

$$(\Longrightarrow\ |\mathbf{a}\cdot\mathbf{b}|^2 \le \|\mathbf{a}\|^2\|\mathbf{b}\|^2 \ \text{Cauchy-Schwarz})$$

$$\|\mathbf{a} \wedge \mathbf{b}\| = \|\mathbf{a}\|\,\|\mathbf{b}\| \sin\vartheta \quad (\text{since } |\mathbf{a}\cdot\mathbf{b}| = \|\mathbf{a}\|\,\|\mathbf{b}\| \cos\vartheta)$$

5. Double cross product

$$(\mathbf{a} \wedge \mathbf{b})\wedge(\mathbf{c} \wedge \mathbf{d}) = \begin{cases} \mathbf{c}(\mathbf{a},\mathbf{b},\mathbf{d}) - \mathbf{d}(\mathbf{a},\mathbf{b},\mathbf{c}) \\ \text{or} \\ -\mathbf{a}(\mathbf{b},\mathbf{c},\mathbf{d}) + \mathbf{b}(\mathbf{c},\mathbf{d},\mathbf{a}) \end{cases}$$

$$\mathbf{a}(\mathbf{b},\mathbf{c},\mathbf{d}) - \mathbf{b}(\mathbf{c},\mathbf{d},\mathbf{a}) + \mathbf{c}(\mathbf{d},\mathbf{a},\mathbf{b}) - \mathbf{d}(\mathbf{a},\mathbf{b},\mathbf{c}) = 0$$

(linear dependence relation between four vectors of $\mathbb{R}^3$)

CHAPTER 3
LINEAR DIFFERENTIAL SYSTEMS

3.1 INTRODUCTION : THE NOTION OF DIFFERENTIAL EQUATION

The most general first order differential equation for an unknown scalar function $y = f(x)$ has the form

$$\Phi(x,y,y') = 0.$$

It gives an *implicit relation* between a function and its derivative. A solution consists in a $\mathcal{C}^1$-function $y = f(x)$ such that

$$\Phi(x,f(x),f'(x)) = 0$$

identically (in $\mathbb{R}$ or some sub-interval $I \subset \mathbb{R}$).

One should stress that the function and its derivative have to be computed at the *same* point. If not, the relation is a *functional equation* of a more general type. For example, the relation $f'(x) = f(x+1)$ is *not* a differential equation.

The precise formulation of the relation $\Phi(x,y,y') = 0$ is given by a (continuous) function $\Phi : \mathbb{R}^3 \longrightarrow \mathbb{R}$ (more generally, Φ could be defined only in some open subset of $\mathbb{R}^3$), say

$${}^t(x,y,z) \longmapsto \Phi(x,y,z),$$

and we are looking for a $\mathcal{C}^1$-curve $y = f(x)$ such that all points

$$x,\ y = f(x),\ z = f'(x)$$

are in the subset $\Phi = 0$ of $\mathbb{R}^3$.

The first step towards finding such functions consists in solving the implicit relation $\Phi(x,y,y') = 0$ for y', namely in finding an equivalent relation of the simpler form

$$y' = f(x,y).$$

This means that we would like to find a continuous function

$$f : \mathbb{R}^2 \longrightarrow \mathbb{R}$$

such that

$$\Phi(x,y,z) = 0 \quad \text{is equivalent to} \quad z = f(x,y)$$

(at least perhaps for the triples x, y, z in some open subset of $\mathbb{R}^3$). The theoretical possibility of finding such an f is given in some cases by the implicit function theorem (which will be explained in sec.6.5).

A very useful generalization of differential equation concerns vector functions (in particular, we are going to show how a general differential equation of order n —for an unknown scalar function— can be brought down to a *first order vector* differential equation in Chap.4).

Consider indeed a parameterized curve $x \longmapsto Y(x)$ in some $\mathbb{R}^n$. It is quite natural to look at differential equations

$$Y' = F(x,Y)$$

satisfied by Y. Such an equation will be meaningful if F is a continuous mapping

$$F : \mathbb{R} \times \mathbb{R}^n \longrightarrow \mathbb{R}^n, \ (x,Y) \longmapsto F(x,Y).$$

Let us adopt the notations used in the preceding chapter in this context too. That is, let us denote by $t \in \mathbb{R}$ (instead of x) the scalar variable, and by $X = X(t)$ (instead of $Y = Y(x)$) the parameterized curve. The differential equation in question is thus

$$X' = F(t,X).$$

As we know from our experience, such an equation has usually (infinitely) many solutions. It is customary to require one more condition, e.g. to prescribe an initial data in the form

$$X(0) = X_o \quad \text{a priori given in } \mathbb{R}^n.$$

The conjunction of the two requirements

$$X' = F(t,X) \quad \text{and} \quad X(0) = X_o$$

is a *well posed problem* : it will often admit a unique solution in the form of a parameterized curve $t \longmapsto X(t)$ defined in some neighborhood of the origin in $\mathbb{R}$.

We shall only be interested in the linear case, namely the case where $X \longmapsto F(t,X)$ *is affine linear.* In this case, we can write

$$F(t,X) = A(t)\cdot X + \mathbf{b}(t)$$

where A is a matrix valued parameterized curve and **b** a vector valued parameterized curve. Obviously $\mathbf{b}(t) = F(t,0)$ and in the general case of a $\mathcal{C}^1$-function F (in X), we might try to use a limited expansion of F in X

$$F(t,X) = F(t,0) + L_t\cdot X + o_t(\|X\|)$$

$$L_t : \mathbb{R}^n \longrightarrow \mathbb{R}^n, \quad \text{derivative of } X \longmapsto F(t,X) \text{ at } X = 0 .$$

A first approach to the general problem might be given by a consideration of the *linear approximation* with $A(t) = L_t$.

We are going to take for granted the following fundamental result from the theory of differential equations.

Theorem. Let $A : \mathbb{R} \longrightarrow \mathbb{M}_n(\mathbb{R})$ and $\mathbf{b} : \mathbb{R} \longrightarrow \mathbb{R}^n$ be given continuous functions. Then, for each given initial data $X_o \in \mathbb{R}^n$, there is one and only one $\mathcal{C}^1$-parameterized curve $X : \mathbb{R} \longrightarrow \mathbb{R}^n$ such that

$$X' = F(t,X) \quad \text{and} \quad X(0) = X_o. \qquad \blacksquare$$

We have to insist on the fact that the parameterized curves which are solutions of these linear differential equations are defined and $\mathcal{C}^1$ (hence *continuous!*) for all values of $t \in \mathbb{R}$.

Let us give two **examples** exhibiting the crucial role of linearity to obtain the conclusion of the theorem.

Example 1. Let us consider the (scalar) differential equation

$$x' = dx/dt = x^2.$$

This differential equation corresponds to the continuous function

$$f(t,x) = x^2.$$

It is easy to get an idea of the *slopes* in the (t,x)-plane for this scalar field in $\mathbb{R}^2$. For each given point (t_o,x_o), x_o^2 represents the slope of a possible solution $x = x(t)$ going through this point (cf. Fig.3.1). On the other hand, apart from the obvious solution $y \equiv 0$, we find

$$dx/x^2 = dt, \quad d(-1/x) = dt, \quad 1/x = c - t$$

and finally the family of solutions $x = 1/(c - t)$ which represent *hyperbolas* with a vertical asymptote at $t = c$.

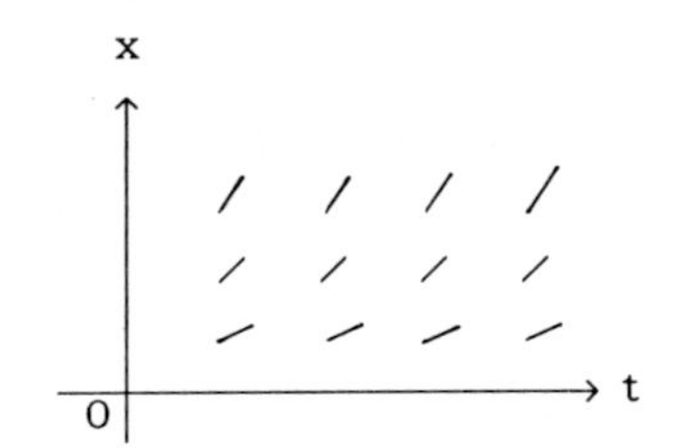

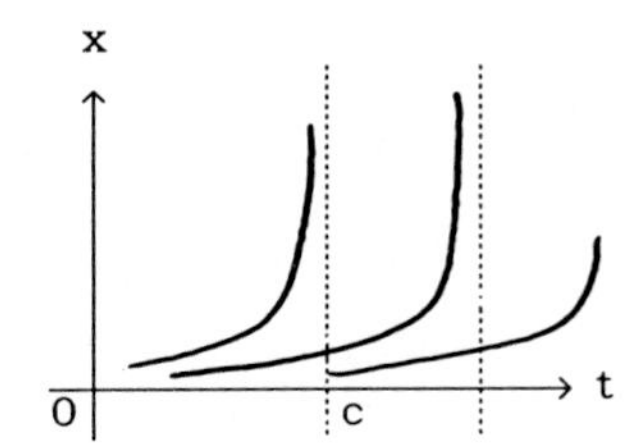

Fig.3.1

The point is that all nontrivial solutions have a discontinuity and are *not* defined over the whole real line. *Existence* of a $\mathcal{C}^1$-solution defined over the whole real line fails for all initial data $x(0) = x_o \neq 0$. *This phenomenon occurs because this differential equation is not linear* ($f(t,x) = x^2$ is *not linear* in the variable x !).

Example 2. Let us consider the differential equation

$$x' = dx/dt = f(t,x) = 2|x|^{1/2}.$$

Here again, the function f is continuous. As before, we see the trivial solution $x \equiv 0$ and if we look for solutions in the upper half plane $x > 0$, we can solve the equation in an elementary way

$$dx/2\sqrt{x} = dt, \quad \sqrt{x} = t - c \quad \text{and} \quad x = (t - c)^2.$$

This is a family of parabolas touching the t-axis and if we only consider the solutions for which $x \geq 0$ and $x' = 2\sqrt{|x|} \geq 0$, we must select only one half of each parabola (cf. Fig.).

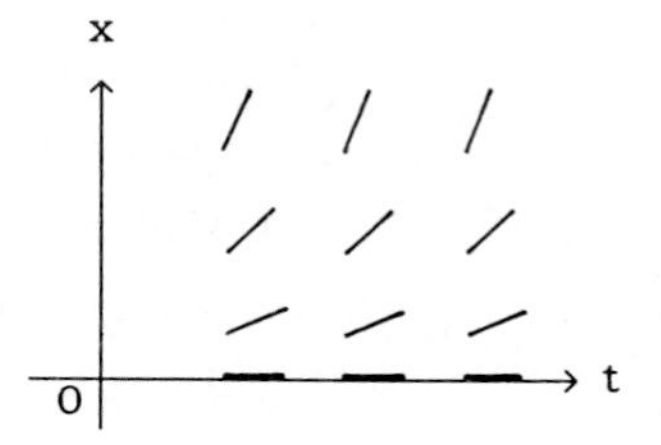

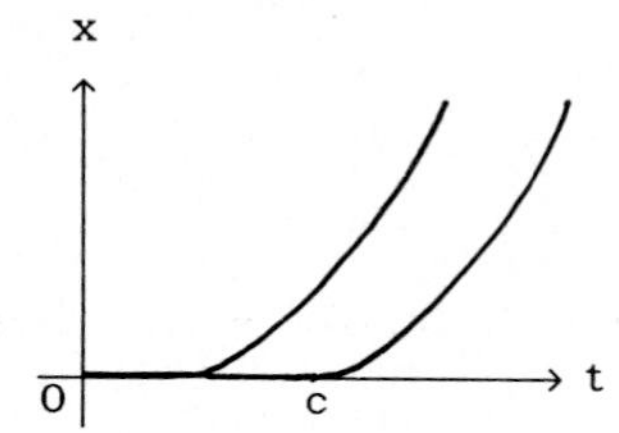

Fig.3.2

With the initial condition $x(0) = 0$, we find infinitely many solutions, namely for any value of $c > 0$

$$x_c(t) = \begin{cases} 0 \text{ for } t \leq c \\ (t-c)^2 \text{ for } t > c \end{cases}$$

is a $\mathcal{C}^1$-function which is a solution of our problem

$$x' = 2\sqrt{|x|} \quad \text{and} \quad x(0) = 0.$$

In other words, *uniqueness* fails for this problem. This happens because the function $f(x,y) = 2\sqrt{|x|}$ is *not linear* in x !.

3.2 LINEAR HOMOGENEOUS EQUATIONS

Let $A : \mathbb{R} \longrightarrow \mathbb{M}_n(\mathbb{R})$, $t \longmapsto A(t)$ be a continuous matrix curve. We are going to study the **linear homogeneous** equation

$$X' = A(t) \cdot X .$$

For each initial data $X(0) = X_o \in \mathbb{R}^n$, *the* solution is given by a $\mathcal{C}^1$-parameterized curve X better denoted by $X(\ ,X_o)$ or $X(?,X_o)$:

$$\mathbb{R} \longrightarrow \mathbb{R}^n, \ t \longmapsto X(t,X_o) .$$

If we like to work in components, we can imagine that

$$A(t) = (a_{ij}(t)), \ X(t) = {}^t(x_i(t))$$

and the homogeneous differential equation is equivalent to the

coupled system of differential equations for the components

$$x_i' = dx_i/dt = \sum_j a_{ij}(t)\, x_j \qquad (i = 1, \dots, n).$$

All scalar functions $x_1, \dots, x_n$ of t are unknown, and e.g. the derivative of x_1 is linked to the (unknown) values of all other x_i! When this *coupling* is absent —this is the case for x_1 when all $a_{1j} \equiv 0$ for $j \geq 2$ — we obtain a usual linear homogeneous differential equation which can be solved elementarily.

Our purpose is to find —if possible— a general method of solution. However, the uncoupled case is interesting enough to deserve a closer look.

Proposition. Assume that the matrix curve $A : \mathbb{R} \longrightarrow \mathbb{M}_n(\mathbb{R})$ has a fixed eigenvector X_o, say $A(t)\cdot X_o = \lambda(t)X_o$ (all $t \in \mathbb{R}$). Then the solution of the homogeneous linear equation $X' = A(t)\cdot X$ for which $X(0) = X_o$ is $X(t) = e^{\Lambda(t)}X_o$ where $\Lambda(t) = \int_0^t \lambda(\tau)d\tau$ is the primitive of the function λ which vanishes at $t = 0$. □

Proof. Choose a constant invertible matrix S with $SX_o = \mathbf{e}_1$ first vector of the canonical basis of $\mathbb{R}^n$. Then

$$A(t)\cdot X_o = \lambda(t)X_o \implies SA(t)X_o = \lambda(t)SX_o \implies$$

$$\implies SAS^{-1}SX_o = \lambda SX_o \implies SAS^{-1}\,\mathbf{e}_1 = \lambda\mathbf{e}_1.$$

Let us denote by $B = B(t) = SA(t)S^{-1}$. The preceding equality proves that the first *column* of $B(t)$ has the entries $\lambda(t), 0, \dots, 0$. On the other hand, if we put $Y = SX$, the differential equation $X' = AX$ becomes

$$Y' = (SX)' = SX' = SAX = SAS^{-1}SX = SAS^{-1}Y = BY.$$

In components, we obtain

$$y_1' = \lambda(t)y_1 + \sum_{i>1} b_{1i}y_i \ ,$$

$$y_j' = \sum_{i>1} b_{ji}y_i \quad \text{if} \quad j \geq 2.$$

We guess a particular solution : we can take $y_i \equiv 0$ for $i \geq 2$ and there only remains the *uncoupled* equation $y_1' = \lambda(t)y_1$ for the first component. This last scalar differential equation has the general solution $y_1(t) = c \exp \Lambda(t)$ where Λ is a primitive of λ. In particular if $\Lambda(0) = 0$, $y_1(0) = c$. Taking $c = 1$, we have found the solution

$$Y(t) = \mathbf{e}_1 \exp \Lambda(t).$$

By definition, $X(t) = S^{-1}Y(t) = S^{-1}\mathbf{e}_1 \exp \Lambda(t) = X_o \exp \Lambda(t)$ is the required solution. ■

The preceding method gives one particular solution which is called **eigensolution** associated to the eigenvector X_o.

Quite generally, for a given matrix curve A, let us consider

$$V = \text{set of solutions of } \{X' = A(t)\cdot X\}.$$

Lemma. This set V is a *vector space*. □

Proof. We have to check that if X and X_1 are two solutions, $X + X_1$ and αX ($\alpha \in \mathbb{R}$) are solutions too. This is obvious since

$$(X + X_1)' = X' + X_1' = AX + AX_1 = A(X + X_1),$$
$$(\alpha X)' = \alpha X' = \alpha(AX) = A(\alpha X).$$ ■

It is easy to get an idea of the *size* of the vector space V. In fact, let $\varepsilon : V \longrightarrow \mathbb{R}^n$ denote the *evaluation operator* (also called *Dirac operator*) at time $t = 0$

$$X \longmapsto \varepsilon(X) = X(0).$$

This map is obviously *linear*. The basic theorem of 3.1 shows that ε is one to one (uniqueness of a solution with given initial condition) onto (existence of a solution for any initial data). Hence ε defines an *isomorphism* $V \longrightarrow \mathbb{R}^n$. The inverse of evaluation ε is the isomorphism $\mathbb{R}^n \longrightarrow V$ which associates to an initial condition X_o *the* solution $X = X(\ ,X_o)$ with value X_o at time $t = 0$

$$\Phi = \varepsilon^{-1} : \mathbb{R}^n \longrightarrow V\ ,\ X_o \longmapsto X(\ ,X_o).$$

The preceding considerations prove the following result.

Theorem. The map Φ: $\mathbb{R}^n \longrightarrow V : X_o \longmapsto X(\ ,X_o)$ is an isomorphism of vector spaces. ■

This theorem shows that the vector space V is *parameterized* by $\mathbb{R}^n$ in a natural way. As was observed in its proof, the inverse isomorphism $\Phi^{-1} = \varepsilon$ is the evaluation map $X \longmapsto X(0)$. Other isomorphisms between V and $\mathbb{R}^n$ abound : for any value τ of t, evaluation at τ produces an isomorphism

$$\varepsilon_\tau : V \longrightarrow \mathbb{R}^n\ ,\ X \longmapsto X(\tau).$$

The composite $C(\tau) = \varepsilon_\tau \circ \Phi = \varepsilon_\tau \circ \varepsilon_o^{-1} : \mathbb{R}^n \longrightarrow \mathbb{R}^n$ is an isomorphism having the following action

$$C(\tau) : X(0) \longmapsto X(\tau) \quad (\text{or } X_o \longmapsto X(\tau,X_o)).$$

It is the **evolution operator** corresponding to the differential equation $X' = A(t)\cdot X$. Knowing it for all τ is equivalent to knowing the complete solution of the problem. More generally, if σ and τ are two fixed values of time,

$$\varepsilon_\tau \circ \varepsilon_\sigma^{-1} : \mathbb{R}^n \longrightarrow \mathbb{R}^n\ ,\ X(\sigma) \longmapsto X(\tau)$$

is an evolution operator $C(\sigma,\tau)$ describing the transition between times σ and τ.

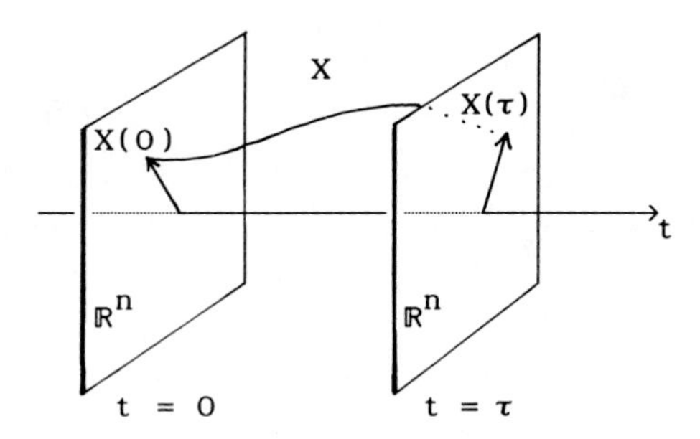

Fig.3.3

Corollary. If two solutions are colinear at t = 0, they remain colinear for all t. If three solutions are coplanar for t = 0, they remain coplanar at all time t. □

Proof. Any non trivial linear dependence relation for the initial conditions is transported by the isomorphism Φ on a similar non trivial linear dependence relation in V (for the corresponding solutions). The evaluation isomorphism ε_t will similarly prove the same linear dependence relation at time t. ■

The evolution operator is still called **Riemann mapping**, or **matrix fundamental solution** of $X' = A(t)\cdot X$ (for a reason to be explained soon). Knowledge of this mapping C(t) for all t gives the solution of the *problem*

$$X' = A(t)\cdot X \quad \text{and} \quad X(0) = X_o$$

in the form $X(t) = C(t)\cdot X_o$ or more precisely

$$X(t,X_o) = C(t)\cdot X_o \ .$$

The columns of the matrix C(t) are given by the images of the basic vectors, namely the solutions $\Xi_i = X(\ ,\mathbf{e}_i)$ going through the basic vectors $\mathbf{e}_i$ for t = 0. If $X_o = \sum a_i\mathbf{e}_i$, obviously

$$X(t,X_o) = C(t)\cdot X_o = \sum a_iC(t)\mathbf{e}_i = \sum a_i\Xi_i(t).$$

3.3 DETERMINATION OF THE EVOLUTION OPERATOR.

Let us still denote by C the evolution operator corresponding to the homogeneous linear differential equation $X' = A(t)\cdot X$. To be able to determine the matrix function C(t), let us compute its derivative. From

$$X(t) = C(t)\cdot X(0) = C(t)\cdot X_o \quad \text{and} \quad X'(t) = A(t)\cdot X(t)$$

we deduce

$$C'(t)\cdot X_o = (C(t)\cdot X_o)' = X'(t) = A(t)\cdot X(t) = A(t)\circ C(t)\cdot X_o$$

for all initial data $X_o \in \mathbb{R}^n$. Hence the identity $C'(t) = A(t)\circ C(t)$.

If we come back to the definition of the evolution operator C, we see immediately that $C(0) = I$ (identity map in $\mathbb{R}^n$). This proves that the *matrix parameterized curve* C is the solution of the following problem

(*) $$C' = A(t)\circ C \quad \text{and} \quad C(0) = I.$$

In this formulation, the initial condition is *fixed*. Another advantage comes from the fact that both C and A are matrices : they are similar mathematical objects, they belong to the *algebra* of matrices (in the *vector* differential equation $X' = A(t)\cdot X$, X is a vector and A is a matrix : they have a different mathematical nature). We shall use this to give a formula for C.

One idea to solve (*) is to proceed by iteration. More precisely, we can look for a sequence of matrix functions

$$C_o = I,\ C_1,\ C_2,\ \ldots,\ C_n,\ \ldots$$

satisfying

$$C'_n = A(t)\circ C_{n-1} \quad \text{and} \quad C_n(0) = I \quad (\text{for } n \geq 1).$$

Let us start :

$$C'_1 = A(t)\circ C_o = A(t) \quad \text{and } C_1(0) = I$$

implies

$$C_1 = \text{primitive of } A \text{ with value } I \text{ for } (t = 0)$$
$$= I + \int_0^t A(s)ds = I + B(t)\ .$$

Since derivation of a matrix with respect to the scalar parameter t simply means *coefficientwise* derivation, similarly, integration with respect to the scalar parameter s means *coefficientwise* integration.

Let us go on :

$$C'_2 = A(t)\circ C_1 = A(t)\circ(I + B(t)) = A(t) + A(t)\circ B(t) \quad \text{and} \quad C_2(0) = I.$$

Hence

$$C_2 = I + B(t) + \text{primitive of } A\circ B \text{ vanishing for } (t = 0).$$

Recalling that $A = B'$, we might be tempted to think that the primitive in question is simply $B^2/2$ ... but

$$(B^2)' = (B\circ B)' = B'\circ B + B\circ B' = A\circ B + B\circ A$$

and our expectation is only valid if A and B commute. This is why we are led to make an *additional assumption*.

Hypothesis. We shall assume that A(t) is permutable with its primitive vanishing for t = 0 : $A(t)\circ B(t) = B(t)\circ A(t)$.

Under this assumption, we shall have

$$C_2 = I + B(t) + B^2(t)/2 .$$

Continuing in the same vein, we guess that

$$C_n = I + B(t) + \dots + B^n(t)/n!,$$

a formula which is immediately proved by induction. Thus we hope that

$$C(t) = \sum_{n\geq 0} B^n(t)/n! = \exp B(t) = e^{B(t)}$$

is the evolution operator (when the hypothesis is satisfied).

Theorem. The evolution operator C is the solution of

$$C' = A(t)\circ C \quad \text{and} \quad C(0) = I.$$

Provided A(t) and $B(t) = \int_0^t A(s)ds$ commute, this evolution operator is given by the convergent series

$$C(t) = \sum_{n\geq 0} B^n(t)/n! = \exp B(t). \qquad \square$$

Proof. First step: we show that the exponential series converges for matrices. Let us fix real matrices $M = (m^i_j)$, $N = (n^i_j) \in \mathbb{M}_n(\mathbb{R})$ and define the positive real numbers

$$\mu = \mu(M) = \sum_{i,j} |m^i_j| \ , \ \nu = \mu(N) = \sum_{i,j} |n^i_j| .$$

Then the following majorations will hold for the sum of the absolute values of the coefficients c^i_j of $M\circ N$

$$\mu(MN) = \sum_{i,j} |c^i_j| = \sum_{i,j} |\sum_k m^i_k n^k_j| \leq \sum_{i,j,k} |m^i_k n^k_j| \leq$$

$$\leq \sum_{i,j,k,\ell} |m^i_k n^\ell_j| = \sum_{i,j} |m^i_k| \cdot \sum_{k,\ell} |n^\ell_j| = \mu\nu .$$

Taking M = N, we find

$$\mu(M^2) \leq \mu^2$$

and using induction (taking $N = M^{n-1}$)

$$\mu(M^n) \leq \mu^n.$$

In particular, each coefficient $c^i_j(M^n)$ of M^n will satisfy

$$|c^i_j(M^n)| \leq \mu^n.$$

The exponential series $\sum M^n/n!$ converges coefficientwise at least as fast as $\sum \mu^n/n!$. The series for e^μ is a majorant for the convergence of each coefficient in exp(M). This certainly proves the convergence of the exponential series of *any matrix*.

Second step. The matrix function $\exp B(t) = \sum B^n(t)/n!$ is well defined for all values of t. Since B(t) and its derivative B'(t) commute,

$$(B^n(t))' = B'B^{n-1} + BB'B^{n-2} + \dots + B^{n-1}B' \overset{!}{=} nB'B^{n-1} = A\cdot nB^{n-1}.$$

If we differentiate term by term (formally) the series for exp B, we find

$$\textstyle\sum_{n\geq 0} (B^n/n!)' = \sum_{n\geq 1} A\cdot nB^{n-1}/n! = A \sum_{n\geq 1} B^{n-1}/(n-1)! = A \exp B.$$

This derivation term by term leads to a series which converges at least as fast as the exponential series (for each coefficient). Hence the derivation term by term is *legitimate* (more will be said on uniform convergence in the chapter devoted to this topic!). ■

Remark. When the commutation relation between A and its primitive B vanishing at $(t = 0)$ is *not* satisfied, an expression for the evolution operator $C(t)$ is given by the following formula

$$C(t) = \textstyle\sum_{k\geq 0} \frac{t^k}{k!} \left(A(t) - \frac{d}{dt}\right)^k .$$

This expression is to be understood as follows. For each initial data X_o, the corresponding solution $X = X(\ ,X_o)$ is given by

$$\begin{aligned} X(t) &= C(t)\cdot X_o = X_o + t(A(t) - d/dt)X_o + \dots \\ &= X_o + tA(t)X_o + (t^2/2)(A(t) - d/dt)A(t)X_o + \dots \\ &= X_o + tA(t)X_o + (t^2/2)(A(t)^2 - A'(t))X_o + \dots \end{aligned}$$

It is easy to verify *formally* this formula :

$$\left(A(t) - \frac{d}{dt}\right)C(t)X_o = A \textstyle\sum_{k\geq 0} \frac{t^k}{k!} \left(A - \frac{d}{dt}\right)^k - \frac{d}{dt} \sum_{k\geq 0} \frac{t^k}{k!} \left(A - \frac{d}{dt}\right)^k =$$

$$= A \textstyle\sum_{k\geq 0} \frac{t^k}{k!} \left(A - \frac{d}{dt}\right)^k - \sum_{k\geq 1} \frac{t^{k-1}}{(k-1)!}\left(A - \frac{d}{dt}\right)^k - \sum_{k\geq 0} \frac{t^k}{k!} \frac{d}{dt}\left(A - \frac{d}{dt}\right)^k =$$

$$= \textstyle\sum_{k\geq 0} \frac{t^k}{k!} A\left(A - \frac{d}{dt}\right)^k - \sum_{k\geq 0} \frac{t^k}{k!} \left(A - \frac{d}{dt}\right)^{k+1} - \sum_{k\geq 0} \frac{t^k}{k!} \frac{d}{dt}\left(A - \frac{d}{dt}\right)^k = 0.$$

Terminological comment. The evolution operator $C(t)$ is also called **matrix fundamental solution of** $X' = A(t)X$. This comes from the fact that the discontinuous matrix function

$$E(t) = \begin{cases} 0 & \text{for } t < 0 \\ C(t) & \text{for } t \geq 0 \end{cases}$$

has a derivative

$$E'(t) = \left\{\begin{matrix} 0 & \text{for } t < 0 \\ C' = A(t)C & \text{for } t > 0 \end{matrix}\right\} = A(t)E(t) \text{ for } t \neq 0.$$

It can be shown that in a suitable sense

$E' - A(t)E = \delta I$ where δ is the Dirac *distribution*.

Consequently, E is the *fundamental solution* of the differential operator $d/dt - A(t)$ (operating on matrices).

3.4 NON HOMOGENEOUS EQUATION

Once the homogeneous equation $X' = A(t)X$ is solved, the non homogeneous one

$$X' = A(t)X + \mathbf{b}(t)$$

can be solved by the **method of variation of constants**. Here it is. The solution of the homogeneous equation can be written in the form $X = C(t)X_o$ where C is the evolution operator of the homogeneous equation and X_o is a constant vector (the initial condition for t = 0). Let us look for a solution of the non homogeneous equation of the form

$$X = C(t)X^{\sim}(t)$$

with a variable X~(t) instead of the constant X_o. The differential equation imposes the condition

$$C'X^{\sim} + CX^{\sim}{}' = X' = AX + \mathbf{b} = ACX^{\sim} + \mathbf{b}.$$

Since $C' = AC$, the preceding condition reduces to

$$CX^{\sim}{}' = \mathbf{b} \quad \text{namely} \quad X^{\sim}{}' = C^{-1}\mathbf{b}$$

(recall that C(t) is an *isomorphism* for every t, hence is invertible). We see that $X^{\sim}$ is a primitive of $C^{-1}\mathbf{b}$ and we shall select *the* one which takes the value X_o for t = 0. It is given by

$$X^{\sim}(t) = X_o + \int_0^t C^{-1}(s)\mathbf{b}(s)ds.$$

In this way, we obtain the solution

$$X = C(t)X^{\sim}(t) = C(t)X_o + C(t)\int_0^t C^{-1}(s)\mathbf{b}(s)ds.$$

The composition $C(t)\circ C(s)^{-1}$ represents the isomorphism transforming

initial data given at time s $\longmapsto$ value of solution at time t

(for the homogeneous equation). We shall denote by C(t,s) this isomorphism. By definition

$$C(t) = C(t,0).$$

We have obtained the following result.

Theorem. Let $X' = A(t)X + \mathbf{b}(t)$ be a vector linear equation with continuous coefficients. Assuming that the evolution operator C(t) of the associated homogeneous equation $X' = A(t)X$ is known, the solution of the non homogeneous equation taking the value X_o at time t = 0 is given by

$$X = C(t)X^{\sim}(t) = C(t)X_o + \int_0^t C(t,s)\mathbf{b}(s)ds. \qquad \blacksquare$$

Observe that

▷ $C(t)X_o$ is the *general* solution of the homogeneous equation,

▷ $\int_0^t C(t,s)\mathbf{b}(s)ds$ is a *particular solution* of the non homogeneous equation (the one which vanishes for t = 0).

In particular, this integral term has to be interpreted as a *perturbation* that has to be added to the solution of the homogeneous equation in order to get a solution of the non homogeneous one.

3.5 CONSTANT COEFFICIENTS CASE

A vector linear system $X' = AX + \mathbf{b}$ can always be solved by the preceding method when A is constant. Indeed, in this case, the primitive of A vanishing at time 0 is $B(t) = At$ and this matrix commutes with A so that the basic assumption of 3.3 is satisfied. The evolution operator of the associated homogeneous linear equation is simply

$$C(t) = \exp At.$$

It is easy to see (cf.exercise) that if two matrices M and N *commute*, then $\exp(M + N) = (\exp M)\circ(\exp N)$. Applying this observation to $M = At$ and $N = A\tau$, we get

$$C(t + \tau) = C(t)\circ C(\tau).$$

In particular,

$$C(-t) = C(t)^{-1} \quad \text{and} \quad C(t)C(s)^{-1} = C(t - s).$$

If X_o is an eigenvector of A, say $AX_o = \lambda X_o$, then X_o will also be an eigenvector of all powers A^n of A (with resp. eigenvalues λ^n) and of exp A (eigenvalue e^λ). This shows that

$$C(t)X_o = \exp At\cdot X_o = e^{\lambda t}X_o,$$

in perfect accordance with the notion of eigensolution developed in 3.2 .

More generally, for an eigenvalue λ of A take any vector X_o for which

$$(A - \lambda I)^k X_o = 0 \quad \text{for some integer } k \geq 1$$

(The case of eigenvectors corresponds to k = 1.) Then we can write

$$C(t)X_o = e^{At}X_o = e^{\lambda It + (A-\lambda I)t}X_o = e^{\lambda t}e^{(A-\lambda I)t}X_o =$$

$$= e^{\lambda t}\sum_{n\geq 0}(A-\lambda I)^n X_o t^n/n!.$$

Since $(A - \lambda I)^k X_o = 0$, all terms with $n \geq k$ in the sum vanish and we obtain a particularly simple result

$$C(t)X_o = e^{\lambda t} \sum_{0 \le n < k} (A-\lambda I)^n X_o t^n/n! = P_{k-1}(t)e^{\lambda t}$$

with a vector polynomial P of degree < k.

Definition. A **characteristic vector** for the eigenvalue λ of the matrix A is a non zero vector **v** satisfying an equation

$$(A - \lambda I)^k \mathbf{v} = 0 \quad \text{for some integer } k \ge 1. \qquad \blacksquare$$

Proposition. For each characteristic vector X_o relative to the eigenvalue λ of the constant matrix A, the solution $X(,X_o)$ of $X' = AX$ with initial value X_o at time 0 has the form

$$X(t,X_o) = P_{k-1}(t)e^{\lambda t}$$

where P_{k-1} is a vector polynomial of degree at most k-1. ■

Alternative method. It is also possible to derive the above result without a knowledge of the evolution operator. Here is this other way of treating the constant coefficients case. It is based on a simple sequence of observations.

Observation 1. If the parametrized curve $X : t \longmapsto X(t)$ satisfies the homogeneous equation $X' = AX$ where A is a constant matrix, then $X' = DX$ —where D is the differentiation operator— is still a solution of the same equation. □

Proof. The verification is immediate

$$(X')' = (AX)' = AX'. \qquad \blacksquare$$

Observation 2. If the parametrized curve $X : t \longmapsto X(t)$ satisfies the homogeneous equation $X' = AX$ where A is a constant matrix, then $p(D)X = p(A)X$ is still a solution of the same equation for any polynomial p. □

Proof. The preceding observation shows that $DX = X' = AX$ so that $D : V \longrightarrow V$ has the same action as the matrix A. ■

Observation 3. Let X_o be a characteristic vector of A, say

$$(A - \lambda I)^m X_o = 0.$$

Then the solution $X = X(,X_o)$ of $X' = AX$ with value X_o at time 0 also satisfies $(D - \lambda I)^m X = 0$. □

Proof. The second observation shows that $(D - \lambda I)^m X \in V$ is the solution $(A - \lambda I)^m X$. This solution takes the value $(A - \lambda I)^m X_o = 0$ for $t = 0$. Hence it must be *the trivial* solution. ■

Observation 4. Let F be any smooth parameterized curve. Then

$$(D - \lambda I)^m(e^{\lambda t}F) = e^{\lambda t}D^m F = e^{\lambda t}F^{(m)}. \qquad \square$$

Proof. We have

$$D(e^{\lambda t}F) = \lambda e^{\lambda t}F + e^{\lambda t}F',$$

hence $(D - \lambda I)(e^{\lambda t}F) = e^{\lambda t}F'$. An obvious induction leads to the announced result. ■

Concluding observation. With the assumptions and notations from observation 3, let us take $F = e^{-\lambda t}X$ (or equivalently $X = e^{\lambda t}F$). We then have

$$e^{\lambda t}F^{(m)} = (D - \lambda I)^m(e^{\lambda t}F) = (D - \lambda I)^m X = 0.$$

This proves that $F^{(m)} = 0$ (the exponential $e^{\lambda t}$ never vanishes !) and F must be a polynomial of degree at most m–1. Denoting this vector polynomial by P_{m-1}, we have obtained

$$X(t) = X(t,X_o) = e^{\lambda t}P_{m-1}(t).$$ ■

A word of **caution** is probably adequate here. This polynomial $P_{m-1}(t)$ is determined by its m vector coefficients. To furnish a solution, these *cannot be arbitrary*. If we take *any* vector polynomial P of degree $\leq$ m–1, the parameterized curve $Y = e^{\lambda t}P$ will obviously satisfy $(D - \lambda I)^m Y = 0$ *without being a solution*. In particular, $(D - \lambda I)^m Y$ will not be equal to $(A - \lambda I)^m Y$! If the polynomial P has to furnish a solution according to $X = e^{\lambda t}P$, its coefficients will have a particular form. They can either be determined by the method of indeterminate coefficients, expliciting the condition

$$(e^{\lambda t}P)' = A(e^{\lambda t}P) \ , \ldots$$

or by writing the first m terms of the evolution operator (as shown previously).

3.6 AN EXAMPLE

Let us consider the following coupled linear differential system

$$\begin{cases} x' + y' - x - 2y + z = 0 \\ y' + z' - 2x - y - z = 0 \\ x' + z' - x - y = 0 \end{cases}$$

Here, we are looking for triples of scalar functions $t \longmapsto x(t)$, $t \longmapsto y(t)$, $t \longmapsto z(t)$ having derivatives linked by the three above conditions. We shall group these three scalar functions in a unique vector function

$$t \longmapsto X(t) = {}^t(x(t),y(t),z(t)).$$

To arrive at the solved form $X' = F(t,X)$, we have to isolate the derivatives and solve the system for x', y', z'. An easy computation leads to

$$x' = y - z\,,$$
$$y' = x + y\,,$$
$$z' = x + z\,.$$

These three relations are summarized in $X' = AX$ with matrix

$$A = \begin{pmatrix} 0 & 1 & -1 \\ 1 & 1 & 0 \\ 1 & 0 & 1 \end{pmatrix}.$$

This is a constant coefficients matrix and we look for its eigenvalues. Its characteristic polynomial is

$$\det(A - \lambda I) = \begin{vmatrix} -\lambda & 1 & -1 \\ 1 & 1-\lambda & 0 \\ 1 & 0 & 1-\lambda \end{vmatrix} = -\lambda(1-\lambda)^2 + (1-\lambda) - (1-\lambda) = -\lambda(1-\lambda)^2.$$

Its roots, the eigenvalues of A, are the solutions of

$$\lambda(1-\lambda)^2 = 0.$$

We get $\lambda = 0$ and $\lambda = 1$ (twice). An eigenvector corresponding to the eigenvalue $\lambda = 0$ is

$$\mathbf{v}_0 = {}^t(1,-1,-1)$$

and the corresponding eigensolution is

$$X = \begin{pmatrix} 1 \\ -1 \\ -1 \end{pmatrix} \text{ : constant function since } \lambda = 0 \text{ implies } e^{\lambda t} \equiv 1.$$

For the eigenvalue $\lambda = 1$, we still find an eigenvector

$$\mathbf{v}_1 = {}^t(0,1,1).$$

The corresponding eigensolution is now

$$X(t) = \begin{pmatrix} 0 \\ 1 \\ 1 \end{pmatrix} e^t.$$

Unfortunately, there is no second linearly independent vector for the eigenvalue $\lambda = 1$. We have to look for a characteristic vector, i.e. a solution of $(A - I)^2\mathbf{v} = 0$. A small computation shows that a solution (linearly independent from $\mathbf{v}_1$) is given by

$$\mathbf{v}_2 = {}^t(1,1,0).$$

The corresponding solution is

$$e^t\left[I + (A - I)t\right]\begin{pmatrix} 1 \\ 1 \\ 0 \end{pmatrix} = e^t\begin{pmatrix} 1 \\ 1 \\ 0 \end{pmatrix} + e^t\begin{pmatrix} 0 \\ t \\ t \end{pmatrix} = \begin{pmatrix} 1 \\ 1+t \\ t \end{pmatrix} e^t.$$

The *general solution* is a linear combination of these three solutions

$$c_1\begin{pmatrix}1\\-1\\-1\end{pmatrix}+c_2\begin{pmatrix}0\\1\\1\end{pmatrix}e^t+c_3\begin{pmatrix}1\\1+t\\t\end{pmatrix}e^t.$$

A variant would consist in determining the eigensolutions and then using indeterminate coefficients to find a third linearly independent solution. Since we know in advance that there are solutions of the form $P_1(t)e^t$ with a linear polynomial P_1 (the multiplicity m of the eigenvalue $\lambda = 1$ is 2, hence the degree of the polynomial is at most $m-1 = 1$), we would look for one of the form

$$\begin{pmatrix}a+\alpha t\\b+\beta t\\c+\gamma t\end{pmatrix}e^t \quad \text{independent from} \quad \begin{pmatrix}0\\1\\1\end{pmatrix}e^t.$$

Without loss of generality, removing from it the c multiple of the second, we can even assume $c = 0$. Substituting

$$\begin{pmatrix}a+\alpha t\\b+\beta t\\\gamma t\end{pmatrix}e^t$$

in the differential equation, we find the system

$$\begin{cases}a+\alpha+\alpha t = b+(\beta-\gamma)t\\b+\beta+\beta t = a+b+(\alpha+\beta)t\\\gamma+\gamma t = a+(\alpha+\gamma)t\end{cases}$$

(for all t). Hence we must have

$$\begin{cases}a+\alpha = b\\\beta = a\\\gamma = a\end{cases},\quad\begin{cases}\alpha = \beta-\gamma\\\alpha = 0\\\alpha = 0\end{cases}.$$

Finally,

$$\alpha = 0,\ \beta = \gamma = a = b\ (\text{e.g.} = 1)$$

and we have obtained a third independent solution $\begin{pmatrix}1\\1+t\\t\end{pmatrix}e^t$ (the same as above).

Having solved completely the equation $X' = AX$, it is not difficult to give its associated evolution operator (without computing the exponential series). The general solution has been given in a form by which it is obvious that

$$X(0) = \begin{pmatrix}c_1+c_3\\c_2-c_1+c_3\\c_2-c_1\end{pmatrix} = \begin{pmatrix}a\\b\\c\end{pmatrix} \in \mathbb{R}^3.$$

The columns of the evolution operator are the basic solutions Ξ_i going through the vectors $\mathbf{e}_i$ of the canonical basis for $t = 0$. We let successively

$$\begin{pmatrix} a \\ b \\ c \end{pmatrix} = \mathbf{e}_1, \mathbf{e}_2 \text{ and } \mathbf{e}_3.$$

The first case leads to the system

$$\begin{cases} c_1 + c_3 = 1 \\ c_2 - c_1 + c_3 = 0 \\ c_2 - c_1 = 0 \end{cases} .$$

Its solution is

$$c_1 = c_2 = 1, \ c_3 = 0$$

whence

$$\Xi_1(t) = \begin{pmatrix} 1 \\ -1 + e^t \\ -1 + e^t \end{pmatrix}.$$

The other columns are computed in the same way. Finally, we find

$$C(t) = \begin{pmatrix} 1 & -1 + e^t & 1 - e^t \\ -1 + e^t & 1 + te^t & -1 + e^t - te^t \\ -1 + e^t & 1 - e^t + te^t & -1 + 2e^t - te^t \end{pmatrix} .$$

The example just treated was homogeneous. With a non homogeneous part $\mathbf{b}(t)$, the variation of constants method would have to be used. In such a case, the best way of practising it would be to look for a solution in the form

$$c_1(t)\begin{pmatrix} 1 \\ -1 \\ -1 \end{pmatrix} + c_2(t)\begin{pmatrix} 0 \\ 1 \\ 1 \end{pmatrix} e^t + c_3(t)\begin{pmatrix} 1 \\ 1+t \\ t \end{pmatrix} e^t.$$

Substituting in the system $X' = AX + \mathbf{b}$, we would obtain a linear system for the unknown *functions* c_i'. Solving this system, we would obtain these derivatives as known functions of t. Integrating, we would find the functions c_i themselves... *It is often better to practice a variation of constants on the general solution of the homogeneous equation than to use the evolution operator.*

3.7 COMPLEX SOLUTIONS

So far, we have (mainly) been interested in *real valued* solutions of $X' = A(t)X + \mathbf{b}(t)$. But since eigenvalues of A play a crucial role, we have to emphasize that the methods explained are valid for complex valued solutions $\mathbb{R} \longrightarrow \mathbb{C}^n$, $t \longmapsto X(t)$ as well. Even if the matrix A and the vector $\mathbf{b}$ are real, eigenvalues and eigenvectors are to be looked for in the complex domain. Thus we shall denote by $V_{\mathbb{C}}$ the complex vector space formed of all solutions $X : \mathbb{R} \longmapsto \mathbb{C}^n$ of the associated homogeneous system $X' = A(t)X$. The existence and uniqueness result of 3.1 still holds in this context and shows that the evaluation map $\varepsilon : X \longmapsto X(0)$ is still an isomorphism $V_{\mathbb{C}} \longrightarrow \mathbb{C}^n$ and hence $\dim_{\mathbb{C}}(V_{\mathbb{C}}) = n$.

Theorem. Let $A \in \mathbb{M}_n(\mathbb{C})$ be a constant $n \times n$ matrix with complex entries. Then the vector space $V_{\mathbb{C}}$ is the direct sum of vector subspaces F_λ where λ ranges over the eigenvalues of A. If $m = m_\lambda$ is the algebraic multiplicity of the root λ of the characteristic polynomial of A, $\dim(F_\lambda) = m$. More precisely, F_λ has a basis of solutions of the form

$$X_k(t) = \mathbf{P}_{k-1}(t) \cdot e^{\lambda t} \qquad (k = 1, \dots ,m)$$

where $\mathbf{P}_{k-1}$ is a vector polynomial of degree $\leq k-1$. □

Proof. This result follows immediately from our study of the constant coefficients case, taking into account the following result of linear algebra. For any matrix A operating in a finite dimensional complex vector space $V_{\mathbb{C}}$, there is a direct sum decomposition of $V_{\mathbb{C}}$ as sum of *characteristic subspaces* F_λ (λ ranging over the eigenvalues of A). More precisely, for each $k \leq m = m_\lambda$ (algebraic multiplicity of λ), the vector space of solutions of $(A - \lambda I)^k \mathbf{v} = 0$ has dimension at least k (and $\leq$ m). Taking successively for initial conditions, independent vectors $\mathbf{v}$ solutions of

$$(A - \lambda I)\mathbf{v} = 0, \quad (A - \lambda I)^2\mathbf{v} = 0, \dots$$
$$(A - \lambda I)^m\mathbf{v} = 0,$$

we obtain a basis of solutions as indicated. ■

When the matrix A(t) is *real for all* t, it is interesting and easy to produce a *real basis* of V from a basis of $V_{\mathbb{C}}$.

Proposition. Let A be a real matrix curve and $X_{(1)}, \dots , X_{(n)}$ be a basis of the space $V_{\mathbb{C}}$ of complex solutions of $X' = A(t)X$. Then, among the 2n real functions $\mathrm{Re}(X_{(k)})$, $\mathrm{Im}(X_{(k)})$, there is a subset $\{Y_1, \dots ,Y_n\}$ consisting of n linearly independent solutions of the same homogeneous equation. In particular, this subset (Y_i) is a basis of V. □

Proof. The first observation to make is the following. If A is real, and X is a complex solution of $X' = AX$, then both Re(X) and Im(X) are solutions of the same system. Write indeed $X = X_1 + iX_2$ with real functions X_ν's. Separating real and imaginary parts in the identity

$$(X_1 + iX_2)' = A(X_1 + iX_2),$$

we get $X'_\nu = AX_\nu$ ($\nu = 1, 2$) since A is real. This proves

$$\mathrm{Re}(\mathbf{X}_{(k)}),\ \mathrm{Im}(\mathbf{X}_{(k)}) \in V \text{ (space of real solutions of } X' = AX).$$

Let $m \leq n$ be the dimension of the subspace of V generated by the solutions $\mathrm{Re}(\mathbf{X}_{(k)})$, $\mathrm{Im}(\mathbf{X}_{(k)})$. Take a maximal set of independent solutions Y_ℓ ($1 \leq \ell \leq m$) among them. Then it will be possible to find expressions

$$\mathrm{Re}(\mathbf{X}_{(k)}) = \sum a_{k\ell} Y_\ell ,$$
$$\mathrm{Im}(\mathbf{X}_{(k)}) = \sum b_{k\ell} Y_\ell ,$$

whence

$$\mathbf{X}_{(k)} = \mathrm{Re}(\mathbf{X}_{(k)}) + i\mathrm{Im}(\mathbf{X}_{(k)}) = \sum (a_{k\ell} + ib_{k\ell}) Y_\ell .$$

The dimension of the space generated by the $\mathbf{X}_{(k)}$ is $\leq m$. We have thus shown that $m = n$. ■

Caution. The real parts (or the complex parts) of a set of complex linearly independent solutions are usually not independent in V. For example, the homogeneous equation

$$X' = \begin{pmatrix} 0 & 1 \\ -1 & 0 \end{pmatrix} X$$

admits the two $\mathbb{C}$-independent solutions

$$\begin{pmatrix} 1 \\ i \end{pmatrix} e^{it}, \begin{pmatrix} 1 \\ -i \end{pmatrix} e^{-it}$$

which have same real part

$$\begin{pmatrix} \cos t \\ -\sin t \end{pmatrix}.$$

(The imaginary parts are not independent either !)

CHAPTER 4
HIGHER ORDER DIFFERENTIAL SYSTEMS

4.1 REDUCTION OF THE ORDER

A general differential equation of order n is given by a relation between an unknown function $y = f(x)$ and its derivatives up to order n (all computed at a same point x)

$$\Phi(x,y,y', \dots ,y^{(n)}) = 0.$$

When possible, we try to solve it in terms of the highest derivative $y^{(n)}$

$$y^{(n)} = f(x,y, \dots ,y^{(n-1)}).$$

Let us introduce the unknown functions

$$y_1 = y \ , \ y_2 = y', \ \dots \ , \ y_n = y^{(n-1)}.$$

These functions are linked by the relations

$$y_1' = y_2 \ , \ y_2' = y_3 \ , \ \dots \ , \ y_{n-1}' = y_n$$

and

$$y_n' = y^{(n)} = f(x,y, \dots ,y^{(n-1)}) = f(x,y_1, \dots ,y_n).$$

It is suitable to introduce the vector function

$$Y = \begin{pmatrix} y_1 \\ \vdots \\ y_n \end{pmatrix}$$

which satisfies the first order differential equation

$$Y' = \begin{pmatrix} y_1' \\ \vdots \\ y_n' \end{pmatrix} = \begin{pmatrix} y_2 \\ \vdots \\ f(x,\dots,y_n) \end{pmatrix} = \begin{pmatrix} y_2 \\ \vdots \\ f(x,Y) \end{pmatrix} = F(x,Y).$$

Our case of interest is the *linear case*, namely, F is linear in Y. Such a linear function arises precisely when the scalar function $f(x,Y)$ is linear in Y. This corresponds to a scalar equation of order n of the form

$$y^{(n)} + a_{n-1}(x)\ y^{(n-1)} + \dots + a_o(x)\ y = b(x).$$

Let us observe that the *general solution* of an equation of order n depends on n integration constants. This is compatible with the fact that the general solution of a vector equation of the first order depends on one vector constant (containing the same number of scalar entities).

Mixed cases also occur. For example in physics, the Lagrange equations for a system having f degrees of freedom constitute a system of f scalar differential equations of the second order concerning the generalized coordinates $q_k = q_k(t)$ of the system. Classical mechanics brings them down to the *first order* Hamilton system concerning the 2f scalar functions

$$q_k = q_k(t) , p_k = p_k(t).$$

Let us illustrate the preceding situation by the example of the *coupled pendulum* (yes, the terminology of *coupled* differential systems may well stem from this physical origin...). Consider the situation illustrated in the following Fig.

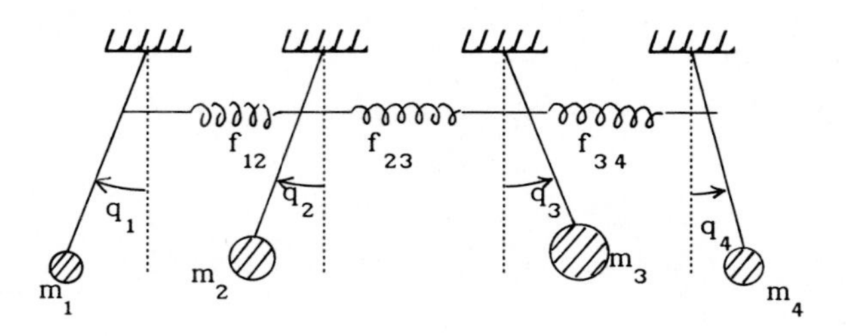

Fig.4.1

For given spring constants, masses, lengths,... the acceleration q_1'' of the first mass m_1 depends on the position of both m_1 and m_2 through the extension of the spring linking them. The four generalized coordinates q_k (which represent angular deviations from the vertical in this example) satisfy four second order scalar equations. The Hamilton theory furnishes eight first order equations which can be grouped in a single first order vector equation in $\mathbb{R}^8$ concerning the functions

$$p_1, \dots ,p_4, q_1, \dots ,q_4 .$$

Let us illustrate the method of reduction of the order for a linear system on the following particular **example**. Assume that three unknown functions of time t, say x, y and z are linked by the following coupled system of differential

$$x'' - x' + 2z - e^t = 0$$
$$y' - z' + 2x \sin t = 0$$
$$x' + y' + z' + 2t^2 = 0 .$$

To reduce its order, we define the new function $u = x'$ and solve the first order system

$$\begin{aligned} &x' - u = 0, \\ &u' - x' + 2z - e^t = 0, \\ &y' - z' + 2x \sin t = 0, \\ &x' + y' + z' + 2t^2 = 0 . \end{aligned}$$

The general method consists in solving this system for x', y', z' and u'. Here, $x' = u$, and substituting this value in the other equations

$$\begin{aligned} &u' = u - 2z + e^t, \\ &y' - z' = -2x \sin t, \\ &y' + z' = -u - 2t^2. \end{aligned}$$

Finally, we obtain the first order system in canonical form

$$\begin{aligned} &x' = u, \\ &y' = -\sin t\ x - u/2 - t^2, \\ &z' = -u/2 + \sin t\ x - t^2, \\ &u' = -\ 2z \ + u \ + e^t. \end{aligned}$$

In vector form, we have obtained the following first order linear equation

$$\begin{pmatrix} x \\ y \\ z \\ u \end{pmatrix}' = \begin{pmatrix} 0 & 0 & 0 & 1 \\ -\sin t & 0 & 0 & -1/2 \\ \sin t & 0 & 0 & -1/2 \\ 0 & 0 & -2 & 1 \end{pmatrix} \begin{pmatrix} x \\ y \\ z \\ u \end{pmatrix} + \begin{pmatrix} 0 \\ -t^2 \\ -t^2 \\ e^t \end{pmatrix} .$$

We hope that this example will suffice to convey the general procedure to follow for the reduction of the order.

4.2 THE LINEAR EQUATION OF ORDER n

We have already mentioned in 4.1 the general form of the scalar linear equation of order n. Let us write it with variable t (time) and unknown function $x = x(t)$

$$x^{(n)} + a_{n-1}(t)\ x^{(n-1)} + \ldots + a_o(t)\ x = b(t).$$

The corresponding first order equation for

$$X = {}^t(x, x', \ldots ,x^{(n-1)})$$

is

$$X' = A(t)X + \mathbf{b}(t)$$

where

$$A(t) = \begin{pmatrix} 0 & 1 & 0 & \dots & 0 \\ 0 & 0 & 1 & \dots & 0 \\ \vdots & & 0 & & \vdots \\ \vdots & \vdots & \vdots & & 1 \\ -a_0 & -a_1 & -a_2 & \dots & -a_{n-1} \end{pmatrix}, \quad \mathbf{b}(t) = \begin{pmatrix} 0 \\ 0 \\ 0 \\ \vdots \\ b(t) \end{pmatrix}.$$

The initial condition consists in prescribing

$$X(0) = \begin{pmatrix} x(0) \\ x'(0) \\ \vdots \\ x^{(n-1)}(0) \end{pmatrix} = X_o ,$$

i.e. the values of the derivatives $x^{(i)}$ at the origin for $i < n$. This is the usual form adopted for initial conditions (e.g.initial position and speed for a movement submitted to the second order Newton's law).

Let us consider in particular the *homogeneous* case $\mathbf{b} = 0$. The map

$$x \longmapsto X = {}^t(x, x', \dots, x^{(n-1)})$$

defines an *isomorphism between the space of solutions of the linear homogeneous equation of order* n *and the space of solutions of the corresponding first order vector equation* $X' = A(t)X$. This mapping is indeed obviously linear and one-to-one, with inverse

$$X \longmapsto \text{first component } x_1 \text{ of } X.$$

In particular, linearly independent vector solutions of $X' = A(t)X$ automatically have linearly independent first components.

Proposition. Let $L(x) = x^{(n)} + a_{n-1}(t)\, x^{(n-1)} + \dots + a_o(t)\, x = 0$ be a linear homogeneous differential equation of order n and let $X' = A(t)X$ be the corresponding first order vector equation. For a system φ_i of solutions of $L(x) = 0$, we denote by

$$\Phi_i = {}^t(\varphi_i, \varphi_i', \dots, \varphi_i^{(n-1)})$$

the corresponding solutions of $X' = A(t)X$. Then, the following properties are equivalent :

i) the φ_i are linearly independent functions,

ii) the Φ_i are linearly independent (vector) functions,

iii) the $\Phi_i(0)$ are linearly independent vectors. ∎

Observe that an initial data $\Phi(0)$ for a vector solution Φ is precisely a data for all $\varphi(0)$, $\varphi'(0)$, ... ,$\varphi^{(n-1)}(0)$ as usual.

Let us now turn to the case of *constant coefficients*. Thus we are examining the order n linear homogeneous equation

$$L(x) = p(D)x = 0$$

where

$$D = d/dt,\ p(D) = D^n + a_{n-1}D^{n-1} + \dots + a_o \quad (a_i \in \mathbb{R} \text{ or } \mathbb{C}).$$

A classical method (due to Euler) is already known for finding the solutions of $L(x) = 0$: one looks for special solutions of the form $t \longmapsto x(t) = e^{\lambda t}$. In this case,

$$x'(t) = \lambda e^{\lambda t}, \dots, x^{(i)}(t) = \lambda^i e^{\lambda t}, \dots$$

and in order to obtain a solution, we must have $p(\lambda) = 0$, namely, λ must be a root of the polynomial p. Since reduction of the order leads to a new method, we compare them.

Proposition. Up to the sign, the characteristic polynomial of the matrix A (given above) coïncides with p. □

Proof. The characteristic polynomial of A is the determinant of $A - \lambda I$. To compute it, we expand this determinant

$$\det(A - \lambda I) = \begin{vmatrix} -\lambda & 1 & 0 & \dots & 0 \\ 0 & -\lambda & 1 & \dots & 0 \\ 0 & 0 & -\lambda & \ddots & \vdots \\ & \dots & & \ddots & 1 \\ -a_0 & -a_1 & -a_2 & \dots\ -a_{n-1} & -\lambda \end{vmatrix}$$

according to its last row

$$\det(A - \lambda I) = (-\lambda)^n + \sum_{i=0}^{n-1} (-1)^{n+i+1}(-a_i)\, M_i.$$

The *minors* M_i are obtained by erasing the row and column containing a_i. Thus

$$M_i = \left|\begin{array}{ccc|ccc} -\lambda & \dots & 0 & & & \\ & -\lambda & 1 & & 0 & \\ 0 & & -\lambda & & & \\ \hline & & & 1 & & 0 \\ & 0 & & -\lambda & 1 & \\ & & & 0 & \ddots & \\ & & & & -\lambda & 1 \end{array}\right| = (-\lambda)^i$$

i columns

For this computation, we have used

a) the determinant of a matrix given with diagonal blocks is the product of the determinants of the blocks

$$\left|\begin{array}{c|c} A & O \\ \hline O & B \end{array}\right| = \det A \cdot \det B$$

(in our case, the upper left block A had size $i \times i$ whereas B had size $(n-1-i)\times(n-1-i)$),

b) for an upper triangular matrix

$$\begin{vmatrix} a_1 & \cdots & * \\ \vdots & a_2 & \vdots \\ 0 & \ddots & \\ 0\ 0 & \cdots & a_i \end{vmatrix} = a_1 a_2 \ldots a_i$$

(and similarly for a lower triangular matrix B).

Summing up, we have proved

$$\det(A - \lambda I) = (-\lambda)^n + \sum_{i=0}^{n-1} (-1)^{n+i+1}(-a_i)(-\lambda)^i =$$

$$= (-1)^n \left(\lambda^n + \sum_{i=0}^{n-1} a_i \lambda^i\right) = (-1)^n p(\lambda). \qquad \blacksquare$$

Both methods thus lead to the roots of the polynomial p .

The theoretical discussion still has to account for multiple roots. Assume for example that the polynomial p has a root μ of multiplicity m

$$p(\lambda) = (\lambda - \mu)^m\ q(\lambda) \quad \text{with} \quad q(\mu) \neq 0.$$

Equivalently, we can write

$$p(\mu) = p'(\mu) = \ldots = p^{(m-1)}(\mu) = 0,\ p^{(m)}(\mu) \neq 0.$$

The classical method of Euler shows that $P(t)e^{\mu t}$ is a solution of the equation for *any polynomial* P *of degree* $\leq$ m-1.

Theorem. Let $L(x) = p(D)x = 0$ be a linear homogeneous equation of order n with constant coefficients. Then its space $W_{\mathbb{C}}$ of complex solutions has dimension n and admits a basis of the form $(x_{j\mu})$,

$$x_{j\mu}(t) = t^j e^{\mu t},$$

where μ ranges over the roots of p and $0 \leq j < m = m_\mu$ (= multiplicity of μ). □

Proof. Although the preceding result is well known, let us give an ad hoc proof for it. To check that the functions $x_{j\mu}$ are solutions of $L(x) = 0$, and to simplify computations, let us observe that

$$t^j e^{\mu t} = (\partial^j/\partial\lambda^j)e^{\lambda t}\Big|_{\lambda=\mu}.$$

This shows that

$$L(t^j e^{\mu t}) = p(D)t^j e^{\mu t} = p(D)(\partial^j/\partial\lambda^j)e^{\lambda t}\Big|_{\lambda=\mu}.$$

Using $D(\partial/\partial\lambda) = (\partial/\partial\lambda)D$ (recall $D = d/dt = \partial/\partial t$) the preceding expression can be evaluated quite simply using the Leibniz rule for the successive derivatives of a product

$$p(D)(\partial^j/\partial\lambda^j)e^{\lambda t} = (\partial^j/\partial\lambda^j)p(D)e^{\lambda t} = (\partial^j/\partial\lambda^j)(p(\lambda)e^{\lambda t}) =$$

$$= \sum_{0\leq k\leq j} \binom{j}{k} p^{(k)}(\lambda)\ t^{j-k}\ e^{\lambda t}.$$

But μ is a root of multiplicity m of p, namely

$$p(\mu) = p'(\mu) = \ldots = p^{(m-1)}(\mu) = 0 \quad \text{whereas} \quad p^{(m)}(\mu) \neq 0.$$

In the preceding expression, all terms vanish at $\lambda = \mu$ since

$$k \leq j \leq m-1.$$

This proves $L(t^j e^{\mu t}) = 0$. For a fixed root μ, we get an m-dimensional subpace of solutions consisting of the functions $p_{m-1} e^{\mu t}$ (p_{m-1} ranging over all polynomials of degree $< m$). Let us show that all these solutions correspond to a vector initial data which is a characteristic vector of A relative to μ. More precisely, if $\varphi = \varphi_q = q\, e^{\mu t}$ with a polynomial q of degree $< k \leq m$, I claim that the initial vector $\Phi_q(0)$ satisfies $(A - \mu I)^k\, \Phi_q(0) = 0$. As in 3.5,

$(A - \mu I)^k\, \Phi_q$ is a solution *and* is equal to

$(D - \mu I)^k \Phi_q = \Phi^\sim$ corresponding to $\varphi^\sim = q^{(k)} e^{\mu t}$

(examination of the first line is sufficient to identify a solution!). This is the trivial solution since

$$\deg(q) < k \quad \text{implies} \quad q^{(k)} = 0$$

and my statement is proved. Finally, we know that the vector space $V_{\mathbb{C}}$ of complex solutions of $X' = AX$ is the direct sum of the

$$V_\mu = \{\text{solutions X with } (A - \mu I)^m\, X(0) = 0\} \subset V_{\mathbb{C}}\ .$$

Correspondingly $W_{\mathbb{C}}$ is a direct sum of the $W_\mu \cong V_\mu$. ■

More explicitely, if q is a polynomial of degree $< m$, the solution $\varphi = q\, e^{\mu t}$ of $L(x) = 0$ gives rise to the vector solution

$$\Phi = {}^t(\varphi, \varphi', \ldots, \varphi^{(n-1)})$$

with

$$\varphi = q(t)e^{\mu t},$$
$$\varphi' = q'(t) + \mu q(t))e^{\mu t}\ ,$$
etc.

In particular,

$$\Phi = \mathbf{P}_{m-1}(t)e^{\mu t}$$

with a vector polynomial $\mathbf{P}_{m-1}$ of degree at most m-1 having arbitrary coefficients *in its first line only*. In this case, we see *a posteriori* that

$$\dim \operatorname{Ker}(A - \mu I)^k = k \quad \text{for } 1 \leq k \leq m = \text{multiplicity of } \mu.$$

4.3 THE WRONSKI DETERMINANT

Let us start with the computation of the determinant $\Delta(t)$ of the evolution operator $C(t)$ corresponding to the homogeneous linear equation X' = A(t)X (in dimension n).

When A(t) commutes with its primitive $B(t) = \int_0^t A(\tau)d\tau$, this computation is easy since we have C(t) = exp B(t) in this case :

$$\det C(t) = \det \exp B(t) = e^{TrB(t)}.$$

Hence

$$\Delta(t) = \exp \int_0^t TrA(\tau)\, d\tau.$$

It is remarkable that this formula holds in the general case also.

Proposition. The determinant of the evolution operator of the general homogeneous linear equation X' = A(t)X is

$$\Delta(t) = \det C(t) = \exp \int_0^t TrA(\tau)\, d\tau. \qquad \square$$

Proof. Let us recall that the columns of the evolution operator are the particular solutions Ξ_i with $\Xi_i(0) = \mathbf{e}_i$. Hence in particular $\Delta(0) = \det C(0) = \det(\mathbf{e}_i) = 1$. On the other hand, let us compute the derivative of the determinant of C(t) by successive derivation of the *lines* of this matrix (this may seem unnatural at first sight since the *columns* of C(t) are solutions of X' = A(t)X; however, the identities $\Xi_i' = A(t)\Xi_i$ will precisely have a nice *line* interpretation). The equation

$$C' = A(t)C \quad \text{(or equivalently } \Xi_i' = A(t)\Xi_i \text{ for } 1 \le i \le n)$$

with $A(t) = (a^i_j(t))$ implies that the derivative of the ith component Ξ^i_j of Ξ_j is given by

$$(\Xi^i_j)' = \textstyle\sum_k a^i_k(t)\Xi^k_j .$$

The derivative of the ith line Ξ^i of C(t) is

$$(\Xi^i)' = a^i_1(t)\, \Xi^1 + \ \ldots\ + a^i_n(t)\, \Xi^n.$$

The determinant containing the derivative of the ith line is

$$\begin{vmatrix} \Xi^1 \\ \vdots \\ (\Xi^i)' \\ \vdots \\ \Xi^n \end{vmatrix} = a^i_i(t) \begin{vmatrix} \Xi^1 \\ \vdots \\ \Xi^i \\ \vdots \\ \Xi^n \end{vmatrix} = a^i_i(t)\, \Delta(t).$$

Summing up these expressions, we get

$$\Delta(t)' = \textstyle\sum_i a^i_i(t)\, \Delta(t) = TrA(t)\cdot\Delta(t).$$

In other words, the scalar function Δ is a solution of the differential equation $\Delta' = \Delta\cdot TrA(t)$ with initial condition Δ(0) = 1. The solution of this problem is $\Delta(t) = \exp \alpha(t)$ where α is the primitive of TrA(t) vanishing for t = 0. ■

Observe that the preceding proof is also a consequence of the lemma of 2.3 : the determinant of the matrix curve $t \longmapsto C(t)$ must satisfy the differential equation

$$\Delta'(t) = \Delta(t)\cdot \mathrm{Tr}(C'C^{-1}).$$

Recalling $C' = AC$, we indeed get $C'C^{-1} = A$.

We can come back to the linear equation of order n

$$L(x) = p(D)x = x^{(n)} + a_{n-1}(t)x^{(n-1)} + \dots + a_o(t)x = 0.$$

We have already studied the first order vector equation corresponding to it. It is given by $X' = A(t)X$ with matrix

$$A(t) = \begin{pmatrix} 0 & 1 & 0 & \dots & 0 \\ 0 & 0 & 1 & \dots & 0 \\ 0 & 0 & 0 & \ddots & \vdots \\ \vdots & \vdots & \vdots & \ddots & 1 \\ -a_0 & -a_1 & -a_2 & \dots & -a_{n-1} \end{pmatrix}.$$

In particular, $\mathrm{Tr}A(t) = -a_{n-1}(t)$.

Definition. The **Wronski determinant** of a system of solutions $\varphi_1,\dots,\varphi_n$ of $L(x) = 0$ is the determinant

$$W(\varphi_1,\dots,\varphi_n) = \begin{vmatrix} \varphi_1 & \dots & \varphi_n \\ \varphi_1' & \dots & \varphi_n' \\ \vdots & & \vdots \\ \varphi_1^{(n-1)} & \dots & \varphi_n^{(n-1)} \end{vmatrix}$$

of the system of corresponding vector solutions

$$\Phi_i = {}^t(\varphi_i,\dots,\varphi_i^{(n-1)}) \quad \text{of } X' = A(t)X. \quad \blacksquare$$

According to the preceding computations, the derivative of the Wronski determinant satisfies

$$W'(\varphi_1,\dots,\varphi_n) = -a_{n-1}(t)W(\varphi_1,\dots,\varphi_n).$$

(This is also obvious if we directly compute the derivative of the determinant W by taking the derivatives of the lines : only the derivative of the last line produces a non trivial determinant.) This shows that —at least when $W \neq 0$— the logarithmic derivative of the Wronski determinant is *independent from the system of particular solutions chosen* and is $W'/W = -a_{n-1}$. The original equation can thus be written

$$L(x) = x^{(n)} -(W'/W)x^{(n-1)} + \dots + a_o x = 0$$

or equivalently

$$Wx^{(n)} - W'x^{(n-1)} + \dots + a_o Wx = 0.$$

An immediate consequence of the preceding facts is that if $a_{n-1} = 0$, then the Wronski determinant is *constant*. For example, the two functions $\sin t$ and $\cos t$ are solutions of $x'' + x = 0$ in which the coefficient of x' vanishes. The Wronski determinant of these solutions is

$$\begin{vmatrix} \sin t & \cos t \\ \cos t & -\sin t \end{vmatrix} = -\sin^2 t - \cos^2 t = -1.$$

Let us summarize.

Theorem. The Wronski determinant of n solutions $\varphi_1, \ldots, \varphi_n$ of the nth order scalar equation

$$L(x) = p(D)x = x^{(n)} + a_{n-1}(t)x^{(n-1)} + \ldots + a_o(t)x = 0$$

is given by the formula

$$W(\varphi_1,\ldots,\varphi_n)(t) = W(\varphi_1,\ldots,\varphi_n)(0)\cdot\exp -\int_0^t a_{n-1}(\tau)d\tau.$$

Its logarithmic derivative is independent from the particular system of solutions chosen : $W'/W = -a_{n-1}$, and when this coefficient of the linear equation L vanishes, the Wronski determinant is constant. ■

Let us emphasize that linear independence concerns the corresponding vector solutions Φ_i of $X' = A(t)X$. As was explained in 3.2 and 4.2, it is equivalent to linear independence of initial values $\Phi_i(0)$ and hence also to $W(0) \neq 0$ (on the above formula, we verify that $W(0) \neq 0 \implies W(t) \neq 0$ for all t).

There is an interesting converse result.

Proposition. Let $\varphi_1, \ldots, \varphi_n$ be functions of class $\mathcal{C}^n$ with non vanishing Wronski determinant. Then there is a linear homogeneous equation of order n admitting the φ_i as solutions. □

Proof. Consider the differential operator of order n defined by

$$L(x) = \begin{vmatrix} \varphi_1 & \ldots & \varphi_n & x \\ \varphi_1' & \ldots & \varphi_n' & x' \\ \vdots & & \vdots & \vdots \\ \varphi_1^{(n)} & \ldots & \varphi_n^{(n)} & x^{(n)} \end{vmatrix}.$$

Clearly $L(\varphi_i) = 0$ (a determinant with two equal columns vanishes). Moreover, with $W = W(\varphi_1, \ldots, \varphi_n)$ and $W^{\sim} = W(\varphi_1', \ldots, \varphi_n')$

$$L(x) = Wx^{(n)} + \ldots \pm W^{\sim}x.$$

If a normalized equation is required, we can take $(1/W)L$ instead of L. There can be *only one* such normalized equation (admitting the φ_i as solutions) for the following reason. If L and M are two normalized (linear homogeneous) differential operators of order n admitting all φ_i as solutions, the differential equation

$$(L - M)(x) = L(x) - M(x) = 0$$

of order n-1 admits n linearly independent solutions. This proves that $L - M = 0$ or $L = M$. ■

4.4 USE OF PARTICULAR SOLUTIONS

When a particular solution of the general linear homogeneous equation $X' = A(t)X$ is known, it is possible to use it to lower the vector dimension of the system. Call this a priori known solution $\Phi = {}^t(\varphi_1, \ldots, \varphi_n)$. If $\varphi_n(t) \neq 0$, make the *variable change of basis* (in the space of values of the solutions)

$$\varepsilon_1 = \mathbf{e}_1, \ldots, \varepsilon_{n-1} = \mathbf{e}_{n-1}, \varepsilon_n = \Phi(t).$$

The matrix corresponding to this change of coordinates is

$$S(t) = (s^i_j(t)) = \begin{pmatrix} 1 & 0 & \ldots & \varphi_1(t) \\ 0 & 1 & & \varphi_2(t) \\ \vdots & & \ddots & \vdots \\ 0 & 1 & \ldots & \varphi_n(t) \end{pmatrix}.$$

Let us write

$$X = \sum x_i \mathbf{e}_i = \sum y_j \varepsilon_j$$

with a new column vector $Y = (y_j)$. Since $\varepsilon_j = \sum_i s^i_j \mathbf{e}_i$ we find

$$x_i = \sum_j s^i_j y_j \quad \text{or in matrix form} \quad X = S(t)Y.$$

For X to be a solution of $X' = A(t)X$, we must have

$$SY' + S'Y = (SY)' = X' = AX = ASY$$

whence $SY' = ASY - S'Y$. In other words, we have to solve the new system

$$Y' = S^{-1}(AS - S')Y = BY.$$

The advantage here is that we know *a priori* that $\varepsilon_n = {}^t(0, \ldots, 1)$ is a solution. Substituting this value for Y leads to $0 = B\varepsilon_n$. This means that the last column of B consists only of 0's. The first equations of this new system are *decoupled* from y_n. They constitute a new n-1 dimensional system which can be solved independently. Finally, the last component will be determined by a quadrature from

$$y_n' = \sum_{i=1}^{n-1} b^n_i(t)\, y_i .$$

Thus we are reduced to solving the n-1 dimensional system

$$(Y^{\sim})' = B^{\sim}(t)Y^{\sim}$$

where

$$\tilde{Y} = {}^t(y_1, \ldots, y_{n-1}), \quad B(t) = \left(\begin{array}{ccc|c} & \tilde{B}(t) & & \\ \hline b_1^n & \ldots & b_{n-1}^n & 0 \end{array}\right). \qquad \blacksquare$$

This method also shows how the *a priori* knowledge of m (< n) solutions of X' = A(t)X can be used to lower the dimension of the system to n−m.

CHAPTER 5
SCALAR FIELDS : THE GRADIENT

5.1 DIRECTIONAL DERIVATIVES

To study the variation of a scalar function $f : \mathbb{R}^n \longrightarrow \mathbb{R}$ in the neighborhood of a point $\mathbf{a} \in \mathbb{R}^n$, it is quite natural to consider the restriction of f on lines going through the point **a**. From this point of view, we fix a direction given by a non vanishing vector $\mathbf{s} \in \mathbb{R}^n$ and consider the scalar function of one variable $t \longmapsto f(\mathbf{a}+t\mathbf{s})$ near $t = 0$.

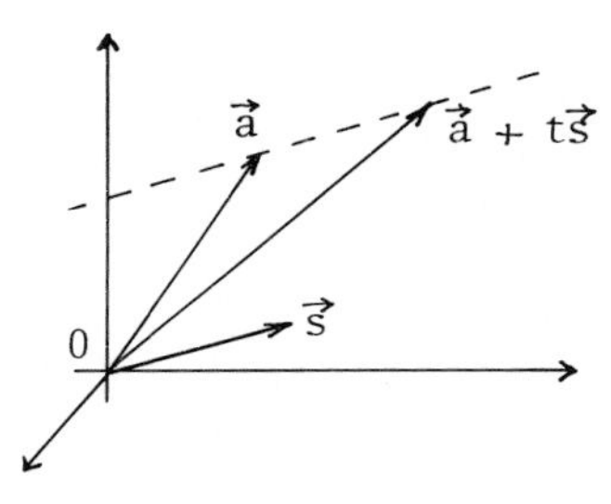

Fig.5.1

In particular, if f is continuous, we can see if this function is differentiable at $t = 0$. When it exists, the derivative at $t = 0$ of

$$t \longmapsto f(\mathbf{a}+t\mathbf{s})$$

is by definition the **directional derivative** of f at **a** in the direction **s**.

Let us denote by $D_{\mathbf{s}}f(\mathbf{a})$ this derivative. Alternatively, we can say that

$$D_{\mathbf{s}}f(\mathbf{a}) = \lim_{t\to 0} [f(\mathbf{a}+t\mathbf{s}) - f(\mathbf{a})]/t$$

or better

$$f(\mathbf{a}+t\mathbf{s}) = f(\mathbf{a}) + D_{\mathbf{s}}f(\mathbf{a})\cdot t + o(t) \quad \text{for } t \longrightarrow 0.$$

This definition depends on the length of **s**, and not only on the direction given by **s**. It will be enough to consider normed directions **s**, and in this case, we obtain a good notion of a directional derivative.

In particular, we can consider the directional derivatives with respect to the basic vectors $\mathbf{e}_i \in \mathbb{R}^n$. The corresponding directional derivatives are called **partial derivatives** of f and still denoted by

$$D_{\mathbf{e}_i}f(\mathbf{a}) = D_i f(\mathbf{a}) = \frac{\partial f}{\partial x_i}(\mathbf{a}) .$$

Proposition. When $f : \mathbb{R}^n \longrightarrow \mathbb{R}$ is differentiable at $\mathbf{a} \in \mathbb{R}^n$, the partial derivatives $\frac{\partial f}{\partial x_i}(\mathbf{a})$ exist and the differential $df(\mathbf{a})$ of f at **a** is represented by the line vector having the $\frac{\partial f}{\partial x_i}(\mathbf{a})$ as components. In other words, the derivative $f'(\mathbf{a})$ is the line vector

$$\left(\frac{\partial f}{\partial x_1}(\mathbf{a}), \dots , \frac{\partial f}{\partial x_n}(\mathbf{a}) \right) . \qquad \square$$

This line vector is by definition the **gradient** of f at **a** and denoted by $\mathbf{grad} f(\mathbf{a})$ or $\overrightarrow{\text{grad}}\, f\, (\mathbf{a})$ if emphasis on the vector character of this quantity is to be given. Thus we have

$$f'(\mathbf{a}) = \mathbf{grad} f(\mathbf{a}) = \overrightarrow{\text{grad}}\, f\, (\mathbf{a}) = \left(\frac{\partial f}{\partial x_1}(\mathbf{a}), \dots , \frac{\partial f}{\partial x_n}(\mathbf{a}) \right).$$

Proof. If f is differentiable at the point **a**, we have a limited expansion

$$f(\mathbf{a} + \mathbf{h}) = f(\mathbf{a}) + L(\mathbf{h}) + o(\|\mathbf{h}\|)$$

with a linear form $L : \mathbb{R}^n \longrightarrow \mathbb{R}$. Let us express L in the canonical basis $(\mathbf{e}_i) \subset \mathbb{R}^n$ by the line matrix $(l_1, \dots , l_n)$. This means that the entries l_i are defined by $l_i = L(\mathbf{e}_i)$. This implies that for an arbitrary vector $\mathbf{h} = {}^t(h_1, \dots , h_n) = \sum h_i\mathbf{e}_i$

$$L(\mathbf{h}) = \sum h_iL(\mathbf{e}_i) = \sum l_ih_i = (l_1, \dots , l_n)\,{}^t(h_1, \dots , h_n).$$

We can thus write

$$f(\mathbf{a} + \mathbf{h}) = f(\mathbf{a}) + l_1h_1 + \dots + l_nh_n + o(\|\mathbf{h}\|).$$

Taking $\mathbf{h} = l_1\mathbf{e}_1 = t\mathbf{e}_1$ (and successively of the form $l_i\mathbf{e}_i$), we get

$$f(\mathbf{a} + t\mathbf{e}_1) = f(\mathbf{a}) + l_1t + o(t)$$

since $t_2 = \dots = t_n = 0$ and $\|\mathbf{h}\| = t$ in this particular case. This shows that $D_1f(\mathbf{a})$ exists and

$$D_1f(\mathbf{a}) = l_1 = \frac{\partial f}{\partial x_1}(\mathbf{a})$$

and successively that $D_if(\mathbf{a})$ exists and

$$D_if(\mathbf{a}) = l_i = \frac{\partial f}{\partial x_i}(\mathbf{a}). \qquad \blacksquare$$

Corollary. Let $f : \mathbb{R}^n \longrightarrow \mathbb{R}$ be a scalar function. If f has a maximum (or more generally an extremum) at a point $\mathbf{a} \in \mathbb{R}^n$ where it is differentiable, then $\mathbf{grad} f(\mathbf{a}) = 0$. $\square$

Proof. This follows immediately from the fact that the restriction of f to all lines going through the point **a** have a maximum (resp. an extremum) at that point. Hence all directional derivatives—and in particular all partial derivatives— vanish at that point. $\blacksquare$

Observe that the differential of a scalar field is a linear form, thus its matrix representation is a *row vector*

$$f'(\mathbf{a}) = \mathbf{grad}\, f\, (\mathbf{a}) \in \mathbb{R}_n.$$

However, since we have a canonical scalar product in $\mathbb{R}^n$, we may also occasionally write this derivative as a column vector using this scalar product. Thus we would write

$$f(\mathbf{a} + \mathbf{h}) = f(\mathbf{a}) + \mathbf{gradf(a)} \cdot \mathbf{h} + o(\|\mathbf{h}\|).$$

General directional derivatives are easily linked with such a limited expansion. Take indeed a normed vector **s** and choose particular increments **h** = t**s**. Thus we have

$$f(\mathbf{a} + t\mathbf{s}) = f(\mathbf{a}) + \mathbf{gradf(a)} \cdot t\mathbf{s} + o(t).$$

This is a limited expansion with linear term

$$t \longmapsto f(\mathbf{a}) + t\, \mathbf{gradf(a)} \cdot \mathbf{s}.$$

The coefficient of t is by definition the directional derivative $D_{\mathbf{s}}f(a)$ of f in the direction given by **s** :

$$D_{\mathbf{s}}f(\mathbf{a}) = \mathbf{gradf(a)} \cdot \mathbf{s}.$$

The linear part in the limited expansion will be maximal when the vector **s** is parallel to **gradf(a)**. If this gradient vector does not vanish, we can take a normed vector **n** in the direction of **gradf(a)**, namely

$$\mathbf{n} = \mathbf{gradf(a)} / \|\mathbf{gradf(a)}\|$$

or in other words

$$\mathbf{gradf(a)} = \|\mathbf{gradf(a)}\| \cdot \mathbf{n}.$$

The corresponding directional derivative will be

$$D_{\mathbf{n}}f(\mathbf{a}) = \mathbf{gradf(a)} \cdot \mathbf{n} = \mathbf{gradf(a)} \cdot \mathbf{gradf(a)} / \|\mathbf{gradf(a)}\| = \|\mathbf{gradf(a)}\|.$$

This result gives a first interpretation of the meaning of the gradient vector **gradf(a)** :

- ▷ the direction of the vector **gradf(a)** indicates the direction of maximal growth of the function f (maximal directional derivative with respect to normed directions **s**)
- ▷ the norm (or length) of **gradf(a)** is the maximal directional derivative of f at **a** (with respect to normed directions **s**).

Warning. Although it is customary (since Legendre and Jacobi!) to denote partial derivatives of a function $f : \mathbb{R}^2 \longrightarrow \mathbb{R}$ by $\frac{\partial f}{\partial x}(x,y)$ and $\frac{\partial f}{\partial y}(x,y)$, these notations can lead to confusion. For example, when we restrict f to the diagonal y = x, what should we mean by $\frac{\partial f}{\partial x}(x,x)$? This could mean that the first partial derivative has to be evaluated at the point (x,x) on the diagonal and an unambiguous notation would then be $D_1f(x,x)$. But perhaps somebody might think that $\frac{\partial f}{\partial x}(x,x)$ represents the derivative of the function $x \longmapsto f(x,x)$. The situation would be even more confusing with notations like $\frac{\partial f}{\partial x}(y,x)$ which could equally well represent

$$D_1f(y,x) \quad \text{or} \quad D_2f(y,x) \ !$$

It will certainly be useful to reformulate the definition of differentiability in terms of partial derivatives.

Criterium. Let f be a scalar function defined in a neighborhood of the point $\mathbf{a} \in \mathbb{R}^n$. Then f is differentiable at the point **a** precisely when the following conditions are both satisfied

▷ the partial derivatives $D_i f(\mathbf{a})$ exist for all $1 \le i \le n$,

▷ $\dfrac{f(\mathbf{a}+\mathbf{h}) - f(\mathbf{a}) - \sum D_i f(\mathbf{a}) h_i}{\|\mathbf{h}\|} \longrightarrow 0$ when $\|\mathbf{h}\| \longrightarrow 0$.

5.2 SMOOTHNESS CONDITION

Let $f : \mathbb{R}^n \longrightarrow \mathbb{R}$ denote a scalar function which is differentiable at all points (or only of an open set $U \subset \mathbb{R}^n$...). Then each partial derivative $D_i f = \dfrac{\partial f}{\partial x_i} : \mathbb{R}^n \longrightarrow \mathbb{R}$ is still a function of the same type.

Definition. Let f be a scalar function. We say that f is **continuously differentiable** and we write $f \in \mathcal{C}^1$ when

▷ f is **differentiable** at all points

▷ the partial derivatives $D_i f = \dfrac{\partial f}{\partial x_i}$ of f are **continuous.**

Let us mention the following technical result.

Theorem. Let $f : \mathbb{R}^n \longrightarrow \mathbb{R}$ be a scalar function which admits continuous partial derivatives $\dfrac{\partial f}{\partial x_i}$. Then $f \in \mathcal{C}^1$. ■

In other words, when the partial derivatives **exist and are continuous**, then f is (continuously) differentiable. To understand the basic fact about this theorem, we give a proof in the two variables case (notations are indeed easier to master in this simpler case). More precisely, we show how differentiability follows from an even weaker condition.

Lemma. Let f be a scalar function defined in a neighborhood U of the origin of $\mathbb{R}^2$. Assume that

▷ $\dfrac{\partial f}{\partial x}(0,0)$ exists (i.e. $x \longmapsto f(x,0)$ differentiable at $x = 0$),

▷ $\dfrac{\partial f}{\partial y}(x,y)$ exists and is continuous at all points of a neighbourhood of the origin.

Then f is differentiable at the origin. □

Proof of the lemma. For small x and y we define

$$g_x(y) = f(x,y) - \frac{\partial f}{\partial y}(0,0) \cdot y$$

so that g_x represents a continuous function of y defined in a neighborhood of the origin . We have

$$g_x(y) - g_x(0) = f(x,y) - f(x,0) - \frac{\partial f}{\partial y}(0,0)\cdot y.$$

It is also obvious from the definition that $y \longmapsto g_x(y)$ is continuously differentiable and we let g'_x denote the derivative of this function (it is well defined in a neighborhood of the origin). Moreover, we have

$$g'_x(y) = \frac{\partial f}{\partial y}(x,y) - \frac{\partial f}{\partial y}(0,0).$$

For given small $\varepsilon > 0$, there exists a $\delta > 0$ with

$$|g'_x(y)| < \varepsilon \quad \text{as soon as} \quad |x| < \delta \text{ and } |y| < \delta$$

(this expresses the continuity of $\frac{\partial f}{\partial y}$ at the origin). Thus we have

$$|g_x(y) - g_x(0)| \le |y| \operatorname{Max}|g'_x(y)| \le \varepsilon|y|$$

for all small values of y, say $|y| \le \delta_1 < \delta$. This shows that

$$|f(x,y) - f(x,0) - \frac{\partial f}{\partial y}(0,0)\cdot y| \le \varepsilon|y|$$

for $|x| < \delta$ and $|y| < \delta_1$. On the other hand,

$$|f(x,0) - f(0,0) - \frac{\partial f}{\partial x}(0,0)\cdot x| < \varepsilon|x| \quad \text{for } |x| \le \delta_o < \delta$$

since $\frac{\partial f}{\partial x}(0,0)$ exists. Adding the two inequalities just found

$$|f(x,y) - f(0,0) - \frac{\partial f}{\partial x}(0,0)\cdot x - \frac{\partial f}{\partial y}(0,0)\cdot y| \le \varepsilon(|x| + |y|)$$

for $|x| \le \delta_o$ and $|y| \le \delta_1$. Observing that

$$|x| \text{ and } |y| \le r = \|{}^t(x,y)\| = (x^2+y^2)^{1/2}$$

we have $\varepsilon(|x| + |y|) \le 2\varepsilon r$ and this shows

$$f(x,y) - f(0,0) - \frac{\partial f}{\partial x}(0,0)\cdot x - \frac{\partial f}{\partial y}(0,0)\cdot y = o(r).$$

This proves differentiability of f at the origin. ∎

This is probably a good time to insist on the fact that existence of partial derivatives at one (or several) point(s) has no regularity consequence for f (apart from the fact that f must then be separately continuous). For example, any function $f : \mathbb{R}^n \longrightarrow \mathbb{R}$ vanishing identically on all coordinate axes has partial derivatives at the origin. But such a function is not necessarily continuous at the origin (in fact we could define such a function by requiring $f(x_1,\ldots,x_n) = 0$ on all coordinate axes, $= +1$ elsewhere!).

Morality. To avoid delicate existence questions, we can assume our functions f to be $\mathcal{C}^1$: to check this, it is enough to check that all partial derivatives $D_i f$ *exist and are continuous.* This condition implies differentiability at all points.

5.3 LEVEL CURVES AND SURFACES

Let $f : \mathbb{R}^n \longrightarrow \mathbb{R}$ denote a $\mathcal{C}^1$ scalar function. As in the one variable case, for $\mathbf{a} \in \mathbb{R}^n$ we define

▷ $\mathbf{a}$: **regular point** (of f)

whenever $\mathbf{grad}f(\mathbf{a}) \neq 0$,

▷ $\mathbf{a}$: **critical point** (of f)

whenever $\mathbf{grad}f(\mathbf{a}) = 0$.

On the other hand if $c \in \mathbb{R}$ is a value of f, we say

▷ c : **regular value** (of f) whenever

all points $\mathbf{a} \in f^{-1}(c)$ are regular points,

▷ c : **critical value** (of f) whenever

there is a critical point $\mathbf{a} \in f^{-1}(c)$.

We shall denote by S_c the level set $f^{-1}(c)$ of f and examine the structure of these subsets of $\mathbb{R}^n$. Here is the main result.

Proposition. If $\mathbf{a} \in \mathbb{R}^n$ is a regular point for $f \in \mathcal{C}^1$ and $c = f(\mathbf{a})$, then there is a neighborhood $V = \{\mathbf{x} \in \mathbb{R}^n : \|\mathbf{x} - \mathbf{a}\| < \varepsilon\}$ of $\mathbf{a} \in \mathbb{R}^n$ and a parametrization

$$\varphi : U \xrightarrow{\cong} V \cap S_c$$

of the corresponding neighborhood of $\mathbf{a}$ in S_c where U is an open subset of $\mathbb{R}^{n-1}$ and $\varphi'(\mathbf{a})$ is one-to-one into (i.e. rank $\varphi'(\mathbf{a})$ is n−1). ■

Corollary. The gradient of f at $\mathbf{a}$ is orthogonal to the level set S_c containing $\mathbf{a}$:

$$\operatorname{Im} \varphi'(a) = \mathbf{grad}f(\mathbf{a})^{\perp} = \{\mathbf{x} \in \mathbb{R}^n : \mathbf{x} \perp \mathbf{grad}f(\mathbf{a})\}. \quad ■$$

This proposition shows that regular points $\mathbf{a} \in \mathbb{R}^n$ are *smooth points of the level set* S_c. In particular, if c is a regular value of f, S_c will be a *codimension one submanifold* of $\mathbb{R}^n$, whereas in general, S_c will only be a *codimension one subvariety* of $\mathbb{R}^n$ (i.e. may have some singularities).

In the case n = 2, a level set S_c is a *curve* given by the *implicit equation* $f(x,y) - c = 0$.

In the case n = 3, a level set S_c is a *surface* given by the *implicit equation* $f(x,y,z) - c = 0$.

By analogy in general, the level sets S_c are often called **level hypersurfaces**.

Fix a point $\mathbf{a} \in f^{-1}(c)$ and consider a (differentiable) curve γ starting at the point $\mathbf{a}$ and situated on this hypersurface. Thus

$$\gamma : [0,1] \longrightarrow f^{-1}(c) \quad \text{with} \quad \gamma(0) = \mathbf{a}.$$

More explicitely, $\gamma(t) = {}^t(\gamma_1(t), \dots, \gamma_n(t))$ satisfies $f(\gamma_1(t), \dots, \gamma_n(t)) = c$ identically in $t \in [0,1]$. The derivative of the function $t \longmapsto f(\gamma_1(t), \dots, \gamma_n(t))$ vanishes identically and thus

$$D_1f(\mathbf{a})\cdot v_1 + \dots + D_nf(\mathbf{a})\cdot v_n = 0$$

where $\mathbf{v} = \gamma'(0)$, initial velocity vector of the curve γ, has components v_i. This proves

$$\mathbf{grad}f(\mathbf{a}) \perp \mathbf{v} \quad \text{for all } \mathbf{v} = \gamma'(0), \ (\gamma \text{ as explained})$$

i.e. $\mathbf{grad}f(\mathbf{a})$ is orthogonal to all tangent vectors $\mathbf{v}$ to the hypersurface $S_c = f^{-1}(c)$ at the point $\mathbf{a}$. Indeed, every tangent vector $\mathbf{v}$ to S_c at $\mathbf{a}$ is the initial velocity of a parameterized curve starting at $\mathbf{a}$ and contained in S_c : this results from the existence theorem for differential equations. We summarize this situation by saying that the *gradient is orthogonal to the level hypersurfaces.*

5.4 FUNDAMENTAL THEOREM OF CALCULUS

We study here a crucial case of differentiability for the composition of mappings (the general case will be considered in Chap.6). Namely, let $f : \mathbb{R}^n \longrightarrow \mathbb{R}$ denote a scalar function and $\gamma: \mathbb{R} \longrightarrow \mathbb{R}^n$ a parameterized curve. When both γ and f are differentiable, we would like to compute the derivative of the composite $f\circ\gamma : \mathbb{R} \longrightarrow \mathbb{R}$. Here is the result.

Theorem. Assume that the parameterized curve γ is differentiable at the origin and f is differentiable at $\mathbf{a} = \gamma(0) \in \mathbb{R}^n$. Then the composite $f\circ\gamma$ is differentiable at the origin with

$$(f\circ\gamma)'(0) = \mathbf{grad}f(\mathbf{a})\cdot\mathbf{v} = \mathbf{grad}f(\gamma(0))\cdot\gamma'(0). \qquad \square$$

Using partial derivatives, this result can be expressed in components as follows. Let $t \longmapsto \gamma_i(t)$ denote the components of γ. Then we are trying to differentiate the composite function

$$t \longmapsto f(\gamma_1(t), \dots, \gamma_n(t)).$$

The result of the theorem can then be written

$$(d/dt)_{t=0}\, f(\gamma_1(t), \dots, \gamma_n(t)) =$$
$$= D_1f(\mathbf{a})\cdot\dot\gamma_1(0) + \dots + D_nf(\mathbf{a})\cdot\dot\gamma_n(0) =$$
$$= \frac{\partial f}{\partial x_1}(\mathbf{a})\cdot\dot\gamma_1(0) + \dots + \frac{\partial f}{\partial x_n}(\mathbf{a})\cdot\dot\gamma_n(0).$$

When γ and $f \in \mathcal{C}^1$ are smooth we thus have

$$(d/dt)f(\gamma_1(t), \dots ,\gamma_n(t)) =$$

$$= \frac{\partial f}{\partial x_1}(\gamma(t))\cdot\gamma_1'(t) + \dots + \frac{\partial f}{\partial x_n}(\gamma(t))\cdot\gamma_n'(t).$$

The preceding formula is often called the **fundamental formula of differential calculus.**

Proof. The differentiability of the curve γ allows us to write a limited expansion

$$\gamma(t) = \mathbf{a} + t\mathbf{v} + o(t) \quad \text{where } \mathbf{a} = \gamma(0),\ \mathbf{v} = \gamma'(0).$$

In the similar expansion for f at the point **a**, we take

$$\mathbf{h} = \gamma(t) - \mathbf{a} = t\mathbf{v} + o(t)$$

obtaining

$$f(\gamma(t)) = f(\mathbf{a} + \mathbf{h}) = f(\mathbf{a}) + \mathbf{grad}f(\mathbf{a})\cdot(t\mathbf{v} + o(t)) + o(t) =$$

$$= f(\gamma(0)) + t\ \mathbf{grad}f(\gamma(0))\cdot\mathbf{v} + o(t).$$

In such an expression, the three $o(t)$ occurring represent *different* functions! Each of them should be written more explicitely in the form $t\phi(t)$. For example, we should write

$$\gamma(t) = \mathbf{a} + t\mathbf{v} + t\phi_1(t) \quad \text{with } \phi_1(t) \longrightarrow 0 \text{ (when } t \longrightarrow 0),$$

$$\mathbf{h} = t\mathbf{v} + t\phi_1(t),$$

$$f(\mathbf{a} + \mathbf{h}) = f(\mathbf{a}) + \mathbf{grad}f(\mathbf{a})\cdot\mathbf{h} + t\varphi(t) \quad \text{with } \varphi(t) \longrightarrow 0$$

hence

$$f(\gamma(t)) = f(\mathbf{a}) + \mathbf{grad}f(\mathbf{a})\cdot(t\mathbf{v} + t\phi_1(t)) + t\varphi(t)$$

$$= f(\mathbf{a}) + t\ \mathbf{grad}f(\mathbf{a})\cdot\mathbf{v} + t[\mathbf{grad}f(\mathbf{a})\cdot\phi\ (t) + \varphi(t)].$$

Obviously

$$\phi(t) = \mathbf{grad}f(\mathbf{a})\cdot\phi_1(t) + \varphi(t) \longrightarrow 0 \quad \text{when } t \longrightarrow 0$$

whence $t\phi(t) = o(t)$ as we claimed. Let us not repeat this kind of computation, being content to add different quantities $o(t)$ as we did at the beginning of this proof. ■

Observe that directional derivatives are particular cases of derivatives along curves. They correspond to the affine linear parameterized curves $\gamma(t) = \mathbf{a} + t\mathbf{s}$. Their treatment was possible even before the fundamental theorem of calculus was proved and, in particular, a necessary condition for an extremum of an $f \in \mathcal{C}^1$ could already be given earlier (sec. 5.1).

The result of this theorem shows that the derivative of the scalar function f *along a curve* is the same as the *directional derivative* of f relative to the direction $\mathbf{v} = \gamma'(0)$

$$(d/dt)_{t=0}\ f(\gamma(t)) = D_{\mathbf{v}}f\ (\mathbf{a}) \quad \text{if } \mathbf{v} = \gamma'(0).$$

Thus, at least for differentiable functions, derivatives along curves coïncide with directional derivatives and carry no additional information. For example, when **gradf(a)** = **0**, the restriction of f to any differentiable curve γ through **a** also has zero derivative.

5.5 CONDITIONAL EXTREMA

Instead of striving for the most general case with the weakest assumptions, let us try to explain how conditional extrema can be found when an extra-condition is imposed.

Let $f : \mathbb{R}^n \longrightarrow \mathbb{R}$ be a scalar $\mathcal{C}^1$-function and assume that we are interested in finding extrema for f restricted to a hypersurface of implicit equation

$$g(x_1, \dots, x_n) = 0.$$

We also assume that $g : \mathbb{R}^n \longrightarrow \mathbb{R}$ is a $\mathcal{C}^1$-function so that the hypersurface in question is a typical level hypersurface S of g. If the restriction of f to S presents an extremum at a (regular) point **a** of S, the restriction of f to any differentiable curve contained in S and going through the point **a** will also present an extremum. Thus, let us consider a differentiable parameterized curve

$$\gamma : \mathbb{R} \longrightarrow S \subset \mathbb{R}^n \quad \text{with } \gamma(0) = \mathbf{a} .$$

The derivative of $f \circ \gamma$ will vanish at the origin :

$$\mathbf{gradf(a)} \cdot \gamma'(0) = 0.$$

Thus we shall have

$$\mathbf{gradf(a)} \perp \mathbf{v}$$

for all vectors $\mathbf{v} = \gamma'(0)$ which are initial velocities of curves in S starting at **a**. These vectors $\mathbf{v} = \gamma'(0)$ make up the tangent space of S at **a** (we have already mentioned that for each tangent vector to S at the point **a**, there is a differentiable curve on S starting at **a** and having initial velocity **v**). We are assuming that **gradg(a)** ≠ **0**. Then these tangent vectors are also characterized by the property of being orthogonal to the gradient of g at the point **a** (S is a level hypersurface of g). Thus

$$\mathbf{gradf(a)} \perp \mathbf{v} \quad \text{for all } \mathbf{v} \perp \mathbf{gradg(a)} \neq \mathbf{0}.$$

But the orthogonal of the non zero vector **gradg(a)** is an n-1 dimensional subspace of $\mathbb{R}^n$ and

gradf(a) must be parallel to **gradg(a)**.

In other words, there exists a scalar λ with

$$\mathbf{grad}f(\mathbf{a}) = \lambda\ \mathbf{grad}g(\mathbf{a})$$

or more concisely, there exists $\lambda \in \mathbb{R}$ with $\mathbf{grad}(f - \lambda g)(\mathbf{a}) = \mathbf{0}$. We have obtained the following result.

Theorem (Lagrange parameters). Let f and g be two $\mathcal{C}^1$ scalar functions $\mathbb{R}^n \longrightarrow \mathbb{R}$. Let $S = g^{-1}(0)$ denote the hypersurface defined by $g(\mathbf{x}) = 0$ (implicit equation for S). If the restriction of f to S has an extremum at a point $\mathbf{a} \in S$ where the gradient of g does not vanish, then there exists $\lambda \in \mathbb{R}$ with $\mathbf{grad}(f - \lambda g)(\mathbf{a}) = \mathbf{0}$. ∎

Thus the search for extrema of the restriction of f to S should start with the search for vanishing points of $\mathbf{grad}(f - \lambda g)$ *situated in the hypersurface* S. The method thus consists in finding the *unconditional* extrema of $f - \lambda g$ *situated on* S. For this, one tries to find common solutions $(\lambda,\mathbf{a})$ of

$$\mathbf{grad}(f - \lambda g)(\mathbf{a}) = \mathbf{0} \quad \text{and} \quad g(\mathbf{a}) = 0.$$

The first vector equation is equivalent to n scalar equations, and the second is a single scalar equation. Together, they furnish n+1 scalar equations for the n+1 unknown scalars λ, $a_1, \dots, a_n$.

Before giving an example of the method, let us indicate how it can be generalized. Suppose that we are looking for an extremum of the restriction of a $\mathcal{C}^1$ function $f : \mathbb{R}^n \longrightarrow \mathbb{R}$ to a subset defined by $p < n$ conditions given in the form

$$g_j(\mathbf{x}) = 0 \qquad 1 \le j \le p$$

where the g_j also denote $\mathcal{C}^1$ functions $\mathbb{R}^n \longrightarrow \mathbb{R}$. Let S denote the intersection of the level hypersurfaces $S_j = g_j^{-1}(0)$, so that we are looking for an extremum of the restriction of f to S. Assume that the point $\mathbf{a} \in S$ is such an extremum for the restriction $f|_S$ of f to S. Let us also denote by V the vector subspace of $\mathbb{R}^n$ generated by the gradients of the conditions g_j at the point $\mathbf{a}$:

$$V = \mathcal{L}\left\langle \mathbf{grad}g_j(\mathbf{a}) : 1 \le j \le p \right\rangle$$

(the gothic $\mathcal{L}$ means that we are to take the linear span of the indicated set of vectors). For any differentiable curve $\gamma \subset S$ starting at the point $\mathbf{a}$ with velocity $\mathbf{v}$ (necessarily tangent to S), the composite $f \circ \gamma$ will have a vanishing derivative at the origin. As before, this implies

$$\mathbf{grad}f(\mathbf{a}) \perp \mathbf{v} = \gamma'(0)$$

and thus

$$\mathbf{grad}f(\mathbf{a}) \text{ orthogonal to the tangent space of } S \text{ at } \mathbf{a}.$$

But this tangent space is the orthogonal of V, hence

$$\mathbf{grad}f(\mathbf{a}) \in V$$

and $\mathbf{grad}f(\mathbf{a})$ is a linear combination of the $\mathbf{grad}g_j(\mathbf{a})$. There are scalars $\lambda_j \in \mathbb{R}$ with

$$\mathbf{grad}f(\mathbf{a}) = \sum \lambda_j \, \mathbf{grad}g_j(\mathbf{a}).$$

Equivalently,

$\mathbf{grad}(f - \sum \lambda_j g_j)$ vanishes at the point $\mathbf{a}$.

Again, the search for extrema of $f|_S$ would be reduced to the search of common solutions $(\lambda_1, \ldots, \lambda_p, \mathbf{a})$ of

$$\mathbf{grad}(f - \sum \lambda_j g_j)(\mathbf{a}) = \mathbf{0} \quad \text{and} \quad g_j(\mathbf{a}) = 0 \ (1 \le j \le p).$$

The first vector condition amounts to n scalar equations so that we are faced with n+p scalar equations for the n+p unknown scalar quantities a_i $(1 \le i \le n)$ and λ_j $(1 \le j \le p)$.

5.6 AN APPLICATION

The preceding formulation of conditional extrema introduces extra parameters λ_j that are —a priori— irrelevant to the problem in question. However, it turns out in most applications that these natural parameters have an interesting interpretation for the problem. Instead of getting rid of them as soon as possible, one should rather try to understand their role and keep them.

We treat the case of extrema of quadratic forms as an example. Let $A = (a_{ij})$ be a square matrix $A \in M_n(\mathbb{R})$ and let

$$Q(x) = \mathbf{x} \cdot A\mathbf{x} \ (= {}^t\mathbf{x} \circ A\mathbf{x} \text{ with matrix products!})$$

be the corresponding quadratic form

$$Q(x_1, \ldots, x_n) = \sum a_{ij} x_i x_j.$$

We are looking for the extrema of this quadratic form restricted to normed vectors $\|\mathbf{x}\| = 1$. As a preliminary, let us compute the derivative of the scalar mapping $\mathbf{x} \longmapsto Q(\mathbf{x})$ at a vector $\mathbf{x} = \mathbf{a}$. For this, we take $\mathbf{x} = \mathbf{a} + \mathbf{h}$ and write

$$Q(\mathbf{a} + \mathbf{h}) = (\mathbf{a} + \mathbf{h}) \cdot A(\mathbf{a} + \mathbf{h}) = \mathbf{a} \cdot A\mathbf{a} + \mathbf{a} \cdot A\mathbf{h} + \mathbf{h} \cdot A\mathbf{a} + \mathbf{h} \cdot A\mathbf{h}.$$

Thus

$$Q(\mathbf{a} + \mathbf{h}) = Q(\mathbf{a}) + {}^tA\mathbf{a} \cdot \mathbf{h} + A\mathbf{a} \cdot \mathbf{h} + Q(\mathbf{h}).$$

Since $Q(\mathbf{h}) = o(\|\mathbf{h}\|)$, this is a limited expansion of the first order and the derivative $Q'(\mathbf{a})$ is identified with the linear form obtained by scalar product against the vector $({}^tA + A)\mathbf{a}$. (This linear form is represented by the row vector

$$\mathbf{grad}Q(\mathbf{a}) = {}^t\mathbf{a}\circ(A + {}^tA).)$$

We come back to the problem of determining the extrema of the restriction of the quadratic form Q to normed vectors. The condition can be represented by $g(\mathbf{x}) = \|\mathbf{x}\|^2 - 1 = \mathbf{x}\cdot\mathbf{x} - 1 = 0$. If $\mathbf{x} = \mathbf{v}$ is an extremum of our problem, the derivative —gradient— of $Q - \lambda g$ must vanish. The preceding result with $A = I$ (identity matrix) furnishes the derivative of g and allows us to write the condition

$$A\mathbf{v} + {}^tA\mathbf{v} - 2\lambda\mathbf{v} = 0.$$

But $A_s = (A + {}^tA)/2$ is the *symmetric part* of A and the found condition is

$$A_s\mathbf{v} = \lambda\mathbf{v}.$$

The extrema of the restriction to the unit sphere $\|\mathbf{x}\| = 1$ of the quadratic form $Q(\mathbf{x}) = \mathbf{x}\cdot A\mathbf{x}$ are the eigenvectors of the symmetric part A_s of A. Observe that the quadratic form also only depends on the symmetric part of A (and we could have assumed A to be symmetric to start with).

Similar considerations would apply to the search of extrema of $\|\mathbf{x}\|$ under the condition $\mathbf{x}\cdot A\mathbf{x} = 1$. In dimension 3, this would correspond to the search of main axes of the quadric of equation $\mathbf{x}\cdot A\mathbf{x} = 1$ (e.g. the axes of an ellipsoid if A is symmetric and positive definite).

CHAPTER 6
JACOBIANS

6.1 MATRIX FORM OF DERIVATIVES

Let us express a vector map $F : \mathbb{R}^n \longrightarrow \mathbb{R}^m$ in components :

$$F = \begin{pmatrix} f_1 \\ \cdots \\ f_m \end{pmatrix} = {}^t(f_1, \ldots, f_m)$$

with some scalar functions $f_i : \mathbb{R}^n \longrightarrow \mathbb{R}$. I claim that if each f_i is differentiable at the point $\mathbf{a} \in \mathbb{R}^n$, then F is also differentiable at that point. Let us indeed write limited expansions of the first order around the point $\mathbf{a}$:

$$f_i(\mathbf{a} + \mathbf{h}) = f_i(\mathbf{a}) + \mathbf{grad} f_i(\mathbf{a}) \cdot \mathbf{h} + o(\|\mathbf{h}\|)$$

(each limited expansion has its own remainder $o(\|\mathbf{h}\|) = \varphi_i(\|\mathbf{h}\|)$ but we only need to know that these quantities are negligible in front of $\|\mathbf{h}\|$ when this length goes to 0!). In the canonical basis of $\mathbb{R}^n$, $f'_i(\mathbf{a}) = \mathbf{grad} f_i(\mathbf{a})$ is represented by the row vector of partial derivatives of f_i at $\mathbf{a}$

$$f'_i(\mathbf{a}) = \mathbf{grad} f_i(\mathbf{a}) = (\partial f_i/\partial x_1(\mathbf{a}), \ldots, \partial f_i/\partial x_n(\mathbf{a})).$$

Grouping these scalar limited expansions in vector form, we obtain a limited expansion for F at $\mathbf{a}$

$$F(\mathbf{a} + \mathbf{h}) = F(\mathbf{a}) + \begin{pmatrix} \partial f_1/\partial x_1(\mathbf{a})h_1 + \ldots + \partial f_1/\partial x_n(\mathbf{a})h_n \\ \cdots \quad \cdots \quad \cdots \\ \partial f_m/\partial x_1(\mathbf{a})h_1 + \ldots + \partial f_m/\partial x_n(\mathbf{a})h_n \end{pmatrix} + o(\|\mathbf{h}\|).$$

The linear part of this limited expansion is easily identified to the matrix product

$$\begin{pmatrix} \partial f_1/\partial x_1(\mathbf{a}) & \ldots & \partial f_1/\partial x_n(\mathbf{a}) \\ \cdots & \cdots & \cdots \\ \partial f_m/\partial x_1(\mathbf{a}) & \ldots & \partial f_m/\partial x_n(\mathbf{a}) \end{pmatrix} \begin{pmatrix} h_1 \\ \cdots \\ h_n \end{pmatrix}.$$

This shows simultaneously that F is differentiable at $\mathbf{a}$ and that the matrix of $dF(\mathbf{a})$ (or of $F'(\mathbf{a})$) with respect to the canonical bases of $\mathbb{R}^n$ and $\mathbb{R}^m$ is the matrix of partial derivatives

$$\begin{pmatrix} \partial f_1/\partial x_1(\mathbf{a}) & \ldots & \partial f_1/\partial x_n(\mathbf{a}) \\ \cdots & \cdots & \cdots \\ \partial f_m/\partial x_1(\mathbf{a}) & \ldots & \partial f_m/\partial x_n(\mathbf{a}) \end{pmatrix}$$

also called **Jacobian matrix of** F at **a** and denoted by

$$\mathrm{Jac}(F)(\mathbf{a}) \quad \text{or} \quad \mathrm{Jac}_F(\mathbf{a}).$$

For memory :

the lines of the Jacobian matrix of F are the gradient of the components f_i *of* F.

We also say that F is $\mathcal{C}^1$ (in some open subset U of $\mathbb{R}^n$) when its components f_i are $\mathcal{C}^1$ (in U), namely when these components are continuously partially differentiable (in U). When this is the case, the derivative F' of F (in U) is represented by the matrix function

$$\mathrm{Jac}_F : \mathbb{R}^n \longrightarrow M_n^m(\mathbb{R}) \text{ (space of n by m matrices).}$$

This emphasizes once more the fact that the derivative of a vector function $F : \mathbb{R}^n \longrightarrow \mathbb{R}^m$ is no more a function of the same type, but rather a function

$$F' : \mathbb{R}^n \longrightarrow \mathcal{L}(\mathbb{R}^n,\mathbb{R}^m) \text{ (space of linear maps } \mathbb{R}^n \longrightarrow \mathbb{R}^m).$$

6.2 DERIVATIVES OF COMPOSITE VECTOR MAPS

Let $F : \mathbb{R}^n \longrightarrow \mathbb{R}^m$ and $G : \mathbb{R}^m \longrightarrow \mathbb{R}^k$ be two vector maps such that the composite $G \circ F : \mathbb{R}^n \longrightarrow \mathbb{R}^k$ can be defined. Assume that F is differentiable at the point $\mathbf{a} \in \mathbb{R}^n$ and G differentiable at the image point $\mathbf{b} = F(\mathbf{a}) \in \mathbb{R}^m$. We claim that $G \circ F$ is then differentiable at **a** with

$$(G \circ F)'(\mathbf{a}) = G'(\mathbf{b}) \circ F'(\mathbf{a}) = G'(F(\mathbf{a})) \circ F'(\mathbf{a}).$$

In words,

the tangent linear map to the composite $G \circ F$ *at* **a** *is simply the composite ot the two tangent linear maps* $F'(\mathbf{a})$ *and* $G'(\mathbf{b})$.

The following picture well describes the situation

$$\mathbb{R}^n \xrightarrow{F'(\mathbf{a})} \mathbb{R}^m \xrightarrow{G'(\mathbf{b})} \mathbb{R}^k \qquad \underset{(G \circ F)'(\mathbf{a})}{\longrightarrow} \quad .$$

As it is written, the formula $(G \circ F)'(\mathbf{a}) = G'(F(\mathbf{a})) \circ F'(\mathbf{a})$ also reminds us of the chain rule for the derivative of usual scalar functions $\mathbb{R} \longrightarrow \mathbb{R}$. But of course here, if we work in the canonical bases of the vector spaces in question, this chain rule for the derivative of a composition will give rise to a matrix multiplication for the Jacobians

$$\mathrm{Jac}_{G \circ F}(\mathbf{a}) = \mathrm{Jac}_G(\mathbf{b}) \circ \mathrm{Jac}_F(\mathbf{a}) = \mathrm{Jac}_G(F(\mathbf{a})) \circ \mathrm{Jac}_F(\mathbf{a}).$$

There is no advantage (but a great discomfort for the typesetter!)

in writing the matrix multiplication with all partial derivatives.

Let us **indicate a proof of our claim.**

For this, we write limited expansions of F at the point **a** (resp. G at the point **b**)

$$F(\mathbf{a} + \mathbf{h}) = F(\mathbf{a}) + F'(\mathbf{a})\mathbf{h} + o(\|\mathbf{h}\|),$$

$$G(\mathbf{b} + \mathbf{k}) = G(\mathbf{b}) + G'(\mathbf{b})\mathbf{k} + o(\|\mathbf{k}\|).$$

Take now the increment **k** of **b** in the form

$$\mathbf{k} = F(\mathbf{a} + \mathbf{h}) - F(\mathbf{a}) = F'(\mathbf{a})\mathbf{h} + o(\|\mathbf{h}\|).$$

We have

$$\mathbf{b} + \mathbf{k} = F(\mathbf{a}) + \mathbf{k} = F(\mathbf{a} + \mathbf{h})$$

and

$$G(F(\mathbf{a} + \mathbf{h})) = G(\mathbf{b} + \mathbf{k}) =$$

$$= G(\mathbf{b}) + G'(\mathbf{b})\Big(F'(\mathbf{a})\mathbf{h} + o(\|\mathbf{h}\|)\Big) + o(\|\mathbf{k}\|) =$$

$$= G(F(\mathbf{a})) + G'(\mathbf{b})(F'(\mathbf{a})\mathbf{h}) + G'(\mathbf{b})o(\|\mathbf{h}\|) + o(\|\mathbf{k}\|).$$

All we have to do to finish the proof is to prove that

$$G'(\mathbf{b})o(\|\mathbf{h}\|) + o(\|\mathbf{k}\|) = o(\|\mathbf{h}\|).$$

If we divide by $\|\mathbf{h}\|$ the *remainder* $G'(\mathbf{b})o(\|\mathbf{h}\|) + o(\|\mathbf{k}\|)$ we find a sum of two terms

$G'(\mathbf{b})o(\|\mathbf{h}\|)/\|\mathbf{h}\|$ which tends to zero when $\|\mathbf{h}\| \longrightarrow 0$

(simply since the linear mapping $G'(\mathbf{b})$ is continuous),

$o(\|\mathbf{k}\|)/\|\mathbf{h}\| = \|\mathbf{k}\|\varphi(\mathbf{k})/\|\mathbf{h}\|$ where $\varphi(\mathbf{k}) \longrightarrow 0$ when $\|\mathbf{k}\| \longrightarrow 0$.

Recall that by definition $\mathbf{k} = F'(\mathbf{a})\mathbf{h} + o(\|\mathbf{h}\|)$ so that

$$\|\mathbf{h}\| \longrightarrow 0 \implies \|\mathbf{k}\| \longrightarrow 0 \implies \varphi(\mathbf{k}) \longrightarrow 0$$

whereas $\|\mathbf{k}\|/\|\mathbf{h}\|$ remains bounded. This concludes the proof. ■

6.3 JACOBIAN DETERMINANTS

When a vector mapping $F : \mathbb{R}^n \longrightarrow \mathbb{R}^n$ (observe that we take $m = n$ here) is differentiable at the point $\mathbf{a} \in \mathbb{R}^n$, its differential is represented by a square matrix

$$dF(\mathbf{a}) = F'(\mathbf{a}) : \mathrm{Jac}_F(\mathbf{a}) = \left(\frac{\partial f_i}{\partial x_j}\right)_{1\le i,j\le n}$$

The determinant of this matrix is called the **Jacobian determinant of** F (at **a**), or the **functional determinant of** F (at **a**) and denoted by

$$\det(F'(\mathbf{a})) = \Delta_F(\mathbf{a}) = \frac{\partial(f_1,\ldots,f_n)}{\partial(x_1,\ldots,x_n)}(\mathbf{a}) .$$

Observe that this determinant is independent of the basis chosen to express the derivative $F'(\mathbf{a})$. The chain rule for the derivative of a composite vector map immediately shows that —with the same notations as in the last section, but with k = m = n here—

$$\Delta_{G\circ F}(\mathbf{a}) = \Delta_G(\mathbf{b})\cdot\Delta_F(\mathbf{a}) = \Delta_G(F(\mathbf{a}))\cdot\Delta_F(\mathbf{a})$$

with an ordinary *multiplication* of determinants (another nightmare to write explicitely using partial derivatives!).

6.4 INVERSION : NECESSARY CONDITION

A vector map $F : \mathbb{R}^n \longrightarrow \mathbb{R}^n$ is **invertible** when there exists an inverse $G = F^{-1} : \mathbb{R}^n \longrightarrow \mathbb{R}^n$, namely $G\circ F$ and $F\circ G$ are the identity on $\mathbb{R}^n$. Assume that F is invertible with

F differentiable at $\mathbf{a} \in \mathbb{R}^n$,

$G = F^{-1}$ differentiable at $\mathbf{b} = F(\mathbf{a}) \in \mathbb{R}^n$.

Then the chain rule for the derivative of the identity map

$$G\circ F = I \text{ (identity map of } \mathbb{R}^n\text{)}$$

gives

$$G'(\mathbf{b})\circ F'(\mathbf{a}) = (G\circ F)'(\mathbf{a}) = I'(\mathbf{a}) = I$$

since I —being linear— is differentiable at all points, with a tangent linear map equal to I itself! This shows that the tangent linear map $F'(\mathbf{a})$ is invertible with

$$F'(\mathbf{a})^{-1} = G'(\mathbf{b}).$$

We have found a *necessary condition* for a vector map $F : \mathbb{R}^n \longrightarrow \mathbb{R}^n$ to be differentiably invertible at a point $\mathbf{a} \in \mathbb{R}^n$: its differential $dF(\mathbf{a})$ —or derivative $F'(\mathbf{a})$— must be an invertible linear map. This condition is known to be equivalent to the non vanishing of the determinant of this map. Thus the necessary condition just found is the non vanishing of the Jacobian determinant

$$\Delta_F(\mathbf{a}) = \det F'(\mathbf{a}) \neq 0.$$

But even if F is $\mathcal{C}^1$ in the whole $\mathbb{R}^n$ with non vanishing Jacobian determinant, F may not be invertible.

For **example**, let $F : \mathbb{R}^2 \longrightarrow \mathbb{R}^2$ be defined in components by

$$F(x,y) = \begin{pmatrix} e^x\cos y \\ e^x\sin y \end{pmatrix}.$$

The matrix of the derivative $F'(a,b)$ has the gradients for lines, hence is given by

$$F(a,b) = \begin{pmatrix} e^a\cos b & -e^a\sin b \\ e^a\sin b & e^a\cos b \end{pmatrix}$$

with Jacobian determinant

$$\Delta_F(a,b) = e^a(\cos^2 b + \sin^2 b) = e^a.$$

In this case $\Delta_F > 0$ in $\mathbb{R}^2$ but F is not invertible : F is periodic of period 2π in y and is thus not one to one.

6.5 LOCAL INVERSION THEOREM

For a $\mathcal{C}^1$-function $F : \mathbb{R}^n \longrightarrow \mathbb{R}^n$, the non vanishing of the functional determinant $\Delta_F(\mathbf{a}) = \det F'(\mathbf{a}) \neq 0$ at one point $\mathbf{a} \in \mathbb{R}^n$ only implies local invertibility of F near **a**. More precisely

Theorem (local inversion). Let $F : \mathbb{R}^n \longrightarrow \mathbb{R}^n$ be a $\mathcal{C}^1$-function with $\det F'(\mathbf{a}) \neq 0$ at some point $\mathbf{a} \in \mathbb{R}^n$. Then there exist open neighborhoods

$$U \text{ of } \mathbf{a},\ V \text{ of } \mathbf{b} = F(\mathbf{a})$$

such that

$$F \text{ is one to one on } U \text{ with } F(U) = V.$$

The inverse G of F on V is $\mathcal{C}^1$. ■

A most useful *global* $\mathcal{C}^1$ inversion result is immediately obtained from the local one.

Corollary. Let $F : \mathbb{R}^n \longrightarrow \mathbb{R}^n$ be $\mathcal{C}^1$ with non vanishing functional determinant. If $G : W \longrightarrow \mathbb{R}^n$ is a *continuous* inverse of F defined in some open set $W \subset \mathbb{R}^n$ i.e. $G(F(\mathbf{x})) = \mathbf{x}$ for all $\mathbf{x} \in G(W)$, then G is $\mathcal{C}^1$ in W. □

Proof. It is enough to prove that G is $\mathcal{C}^1$ in the neighborhood of each point $\mathbf{v} \in W$. Take $\mathbf{b} \in W$ and let $\mathbf{a} = G(\mathbf{b})$ so that $F(\mathbf{a}) = \mathbf{b}$ and let the open neighborhoods U and V be as in the local inversion theorem. By continuity of G, it is possible to take U small enough to ensure $G^{-1}(U) \subset V$. This forces $G|_V$ to coïncide with the unique $\mathcal{C}^1$ inverse of F in V■

Observe that the identity $G \circ F|_U = id_U$ immediately leads to $G'(\mathbf{v}) = F'(\mathbf{u})^{-1}$ (inverse linear operator) for all $\mathbf{u} \in U$ and $\mathbf{v} \in V$ related by $\mathbf{v} = F(\mathbf{u})$.

Definition. A mapping F between two open sets U and $V \subset \mathbb{R}^n$ is called a **diffeomorphism** when

a) $F : U \longrightarrow V$ is one to one onto,

b) both F and its inverse G are $\mathcal{C}^1$.

When D is an open set in $\mathbb{R}^n$, a mapping $F : D \longrightarrow \mathbb{R}^n$ is a **local diffeomorphism** if each point $\mathbf{a} \in D$ possesses an open neighborhood U such that $F|_U$ is a diffeomorphism between U and $V = F(U)$. ■

The example of a function given at the end of the preceding section is thus a *local* diffeomorphism (in particular, we see that local diffeomorphisms are not one to one). The above theorem shows that if a function $F : D \longrightarrow \mathbb{R}^n$ (D open in $\mathbb{R}^n$) has a nowhere vanishing functional determinant

$$\Delta_F(\mathbf{x}) \neq 0 \quad \text{for} \quad \text{all } \mathbf{x} \in D,$$

then F is a local diffeomorphism (but may fail to be one to one).

Another quite useful result that we have to mention here is the **implicit function theorem**. This theorem is a generalization of the following statement. When f is a $\mathcal{C}^1$ scalar field in $\mathbb{R}^2$ with

$$D_2f(a,b) = \frac{\partial f}{\partial y}(a,b) \neq 0,$$

the *implicit equation* $f(x,y) = 0$ is equivalent to $y = \varphi(x)$ for some $\mathcal{C}^1$ function φ provided

(x,y) belongs to a sufficiently small neighborhood of (a,b).

In other words —under this assumption— the implicitly defined curve coïncides with the graph of a $\mathcal{C}^1$ function (at least in a neighborhood of the point in question). A similar statement would hold if the partial derivative with respect to x does not vanish at the point (a,b). Thus we see that an implicit equation $f(x,y) = 0$ defines a *smooth curve* in the neighborhood of each point (a,b) where

▷ $f(a,b) = 0$,

▷ $\mathbf{grad} f(a,b) \neq 0$.

For the generalization of the preceding statement, let us consider the situation where

$$F : \mathbb{R}^{n+m} = \mathbb{R}^n \times \mathbb{R}^m \longrightarrow \mathbb{R}^m,\ (\mathbf{x},\mathbf{y}) \longmapsto F(\mathbf{x},\mathbf{y}) \quad (\mathbf{x} \in \mathbb{R}^n,\ \mathbf{y} \in \mathbb{R}^m)$$

is a fixed $\mathcal{C}^1$ function and

▷ $F(\mathbf{a},\mathbf{b}) = 0 \quad (\mathbf{a} \in \mathbb{R}^n,\ \mathbf{b} \in \mathbb{R}^m)$,

▷ $\det D_2F(\mathbf{a},\mathbf{b}) \neq 0$.

The notation $D_2F(\mathbf{a},\mathbf{b})$ represents the derivative of

$$\mathbf{y} \longmapsto F(\mathbf{a},\mathbf{y}) = F_{\mathbf{a}}(\mathbf{y})$$

computed at the point $\mathbf{y} = \mathbf{b}$. In particular, this derivative is a linear map $\mathbb{R}^m \longrightarrow \mathbb{R}^m$ and the second assumption means that this linear map is *invertible*.

Under these conditions, there are open neighborhoods U of **a**, V of **b** such that for each $\mathbf{x} \in U$ there is a unique $\mathbf{y} = f(\mathbf{x}) \in V$ with $F(\mathbf{x},\mathbf{y}) = 0$. Moreover, the function f thus defined is $\mathcal{C}^1$. Thus $f(\mathbf{a}) = \mathbf{b}$ and f is *implicitely* defined by $F(\mathbf{x},f(\mathbf{x})) = 0$. ■

For proofs of both the local inversion and the implicit function theorem, we refer to the book by Rudin [Principles of...].

Example. Let $F : \mathbb{R}^2 \times \mathbb{R}^2 \longrightarrow \mathbb{R}^2$ be given in components by

$$F_1(x_1,x_2,y_1,y_2) = \cos(x_1+y_1) - \exp(x_2+y_2),$$
$$F_2(x_1,x_2,y_1,y_2) = \sin(x_1+x_2+y_1+y_2).$$

Then $F(0,0) = \mathbf{0}$ and

$$D_2F = \partial_y F = \begin{pmatrix} -\sin(x_1+y_1) & -\exp(x_2+y_2) \\ \cos(x_1+x_2+y_1+y_2) & \cos(x_1+x_2+y_1+y_2) \end{pmatrix}$$

so that

$$D_2F(0,0) = \begin{pmatrix} 0 & -1 \\ 1 & 1 \end{pmatrix}$$

is invertible. There is consequently one $\mathcal{C}^1$-mapping

$$f = \begin{pmatrix} f_1 \\ f_2 \end{pmatrix} : U \longrightarrow \mathbb{R}^2$$

defined in a neighborhood U of $\mathbf{0} \in \mathbb{R}^2$, say

$$y_1 = f_1(x_1,x_2)\ ,\ y_2 = f_2(x_1,x_2)$$

with

$$f_i(0,0) = 0 \text{ and } F(x_1,x_2,f_1(x_1,x_2),f_2(x_1,x_2)) \equiv 0.$$

CHAPTER 7
SECOND ORDER DERIVATIVES

7.1 ITERATION OF PARTIAL DERIVATIVES

For a scalar function $f : \mathbb{R}^n \longrightarrow \mathbb{R}$, the partial derivatives $D_i f$ are still scalar functions of the same type, i.e.

$$D_i f : \mathbb{R}^n \longrightarrow \mathbb{R}.$$

Thus, it is again possible to compute partial derivatives of these and speak of $D_j D_i f : \mathbb{R}^n \longrightarrow \mathbb{R}$ $(1 \le i,j \le n)$. Higher partial derivatives can be introduced similarly.

Definition. A scalar function $f : \mathbb{R}^n \longrightarrow \mathbb{R}$ is called k **times continuously differentiable** when all partial derivatives

$$D_{i_m} \ldots D_{i_2} D_{i_1} f \qquad (1 \le i_j \le n)$$

exist and are continuous for $m \le k$. We abbreviate this condition by the notation $f \in \mathcal{C}^k$.

Here is a basic regularity property.

Theorem (Schwarz). When $f \in \mathcal{C}^2$, we have $D_j D_i f = D_i D_j f$ for all couples of indices $1 \le i,j \le n$. □

Proof. It is enough to consider the case $n = 2$, in which case f is a function of x and y and prove $D_1 D_2 f = D_2 D_1 f$ at a point (a,b). We shall have to consider positive increments h of a, resp. k of b and thus introduce an auxiliary function of one variable

$$\varphi(t) = f(t,b+k) - f(t,b)$$

measuring the vertical increase of f between y = b and b+k. The proof will now be divided into two steps.

First step. The existence of $D_1 f$ and $D_2 D_1 f$ at all points of an open set containing the rectangle R with corners

$$(a,b),\ (a+h,b),\ (a+h,b+k),\ (a,b+k)$$

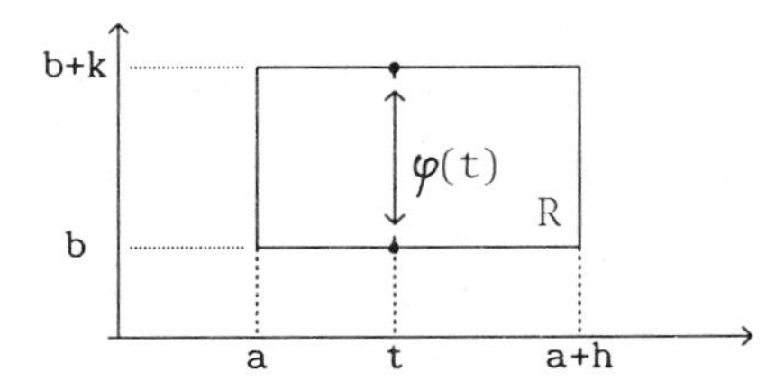

Fig.7.1

allows us to write the difference $\Delta(f;a,b;h,k) = \varphi(a+h) - \varphi(a)$ between two vertical increases of f by means of two applications of the mean value theorem as follows. This difference is

$$\Delta = \varphi(a+h) - \varphi(a) = h\varphi'(x)$$

for some intermediate point $a < x < a+h$. By definition of the function φ, we have

$$\Delta = h\ \varphi'(x) = h\{D_1f(x,b+k) - D_1f(x,b)\}.$$

Another application of the mean value theorem gives

$$\Delta = hkD_2D_1f(x,y)$$

for some intermediate point $b < y < b+k$ (depending on the choice of x...).

Second step. Assume now that D_1f, D_2f, and D_2D_1f exist (at all points... as before) *and* D_2D_1f *is continuous at* the point (a,b). Then we can show $D_1D_2f(a,b)$ exists and equals $D_2D_1f(a,b)$. Let

$$D_2D_1f(a,b) = \lambda.$$

For small $\varepsilon > 0$, (and sufficiently small h and $k > 0$) we shall have

$$|D_2D_1f(x,y) - \lambda| < \varepsilon$$

at all points of the rectangle R (by continuity of D_2D_1f at (a,b)). In particular, this proves $|\Delta/hk - \lambda| < \varepsilon$ because the intermediate point found in the first step lies in the rectangle R. Recall the definition of Δ :

$$\Delta = \varphi(a+h) - \varphi(a) =$$
$$= f(a+h,b+k) - f(a+h,b) - [f(a,b+k) - f(a,b)],$$

fix $h > 0$ and let $k \longrightarrow 0$ in $|\Delta/hk - \lambda| < \varepsilon$. Since D_2f is assumed to exist at all points, the limit will still satisfy

$$\left|\frac{D_2f(a+h,b) - D_2f(a,b)}{h} - \lambda\right| \le \varepsilon.$$

But ε is arbitrarily small : this proves that D_1D_2f exists at (a,b) and is equal to $\lambda = D_2D_1f(a,b)$. ∎

Without the smoothness assumption of the theorem, nearly anything can occur! Here is an example of Peano. Consider the scalar function of two variables defined by

$$f(x,y) = xy(x^2-y^2)/(x^2+y^2) \quad \text{(and } f(0,0) = 0).$$

This function is $\mathcal{C}^1$ in $\mathbb{R}^2$. Moreover D_2D_1f and D_1D_2f exist at every point (and are continuous except at the origin). For example, one checks that

$$D_1f(x,y) = (x^4y + 2x^2y^3 + 2x^2y^2 - y^5)/r^4 \text{ (and } = 0 \text{ for } x = y = 0)$$

where $r^2 = x^2 + y^2$. In particular,

$$D_1f(0,y) = -y^5/r^4 = -y.$$

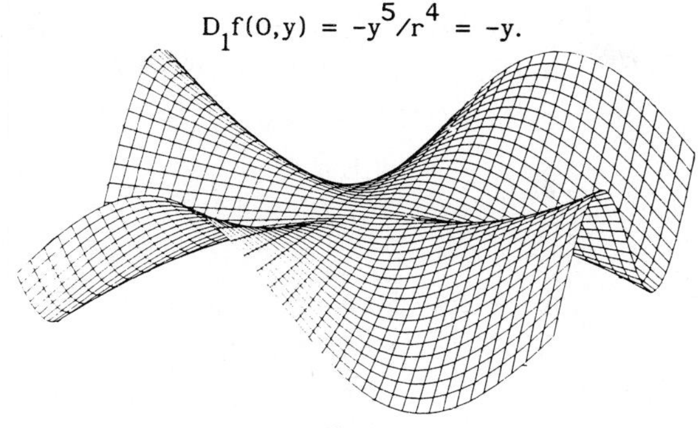

Fig.7.2

From this, it immediately follows that

$$D_2D_1f(0,0) \text{ exists and is equal to } -1.$$

One can check similarly that $D_1D_2f(0,0) = +1$.

7.2 THE THEOREM OF FROBENIUS

Let $\mathbf{u}$ be a $\mathcal{C}^1$ map : $\mathbb{R}^n \longrightarrow \mathbb{R}^m$. For each $\mathbf{x} \in \mathbb{R}^n$, $\mathbf{u}'(\mathbf{x})$ is a linear map : $\mathbb{R}^n \longrightarrow \mathbb{R}^m$, namely $\mathbf{u}'(\mathbf{x}) \in \mathcal{L}(\mathbb{R}^n,\mathbb{R}^m)$. A first order differential equation —in solved form— satisfied by $\mathbf{u}$ would thus have the form

$$\mathbf{u}'(\mathbf{x}) = F(\mathbf{x},\mathbf{u}(\mathbf{x}))$$

where $F(\mathbf{x},\mathbf{y}) \in \mathcal{L}(\mathbb{R}^n,\mathbb{R}^m)$ is given for each $\mathbf{x} \in \mathbb{R}^n$, $\mathbf{y} \in \mathbb{R}^m$ and

$$F : \mathbb{R}^n \times \mathbb{R}^m \longrightarrow \mathcal{L}(\mathbb{R}^n,\mathbb{R}^m)$$

is a continuous map. Conversely, let F be a continuous map as before and consider the differential equation

$$\mathbf{y}' = F(\mathbf{x},\mathbf{y}).$$

Can we always find a solution $\mathbf{y} = \mathbf{u}(\mathbf{x})$ with a prescribed initial condition $\mathbf{u}(\mathbf{a}) = \mathbf{b}$? The answer is NO in general if $n > 1$ and we can easily see what happens by considering the particular case where F is $\mathcal{C}^1$. In this case, it is indeed easy to give a *necessary condition* for the existence of a (local) solution. Let $\mathbf{u}$ be a solution and apply the matrix equation $\mathbf{u}'(\mathbf{x}) = F(\mathbf{x},\mathbf{u}(\mathbf{x}))$ to the basic vectors $\mathbf{e}_j$

$$\partial\mathbf{u}/\partial x_j(\mathbf{x}) = \mathbf{u}'(\mathbf{x})\mathbf{e}_j = F_j(\mathbf{x},\mathbf{u}(\mathbf{x})) \in \mathbb{R}^m .$$

The symmetry relations for second order derivatives (Schwarz' theorem 7.1) implies non trivial relations for the partial derivatives of the F_j's. By the fundamental theorem of calculus (5.4)

$$\partial^2\mathbf{u}/\partial x_i\partial x_j = (\partial/\partial x_i)F_j(\mathbf{x},\mathbf{u}(\mathbf{x})) =$$
$$= (D_1F_j)_i + D_2F_j\cdot\partial\mathbf{u}/\partial x_i \text{ evaluated at } \mathbf{x}\ .$$

More explicitely, if we list the variables of F_j as

$$x_1, \ldots, x_n, x_{n+1} = y_1, \ldots, x_{n+m} = y_m ,$$

the preceding equality means

$$\partial^2\mathbf{u}/\partial x_i\partial x_j = \partial_i F_j + \sum_{1\le k\le m} \partial_{n+k}F_j\ \partial u_k/\partial x_i$$

when evaluated at the point $(\mathbf{x},\mathbf{u}(\mathbf{x}))$. The symmetry relations are

$$\partial_i F_j + \sum_{1\le k\le m} \partial_{n+k}F_j\ \partial u_k/\partial x_i = \partial_j F_i + \sum_{1\le k\le m} \partial_{n+k}F_i\ \partial u_k/\partial x_j$$
$$(\text{for all } 1 \le i,j \le n).$$

These are restrictive as soon as $n > 1$, even if $m = 1$ (cf.infra).

Theorem (Frobenius). Let $F : \mathbb{R}^n \times \mathbb{R}^m \longrightarrow \mathcal{L}(\mathbb{R}^n,\mathbb{R}^m)$ be a $\mathcal{C}^1$ map. Then for each initial data $\mathbf{u}(\mathbf{a}) = \mathbf{b} \in \mathbb{R}^m$ $(\mathbf{a} \in \mathbb{R}^n)$, there is a $\mathcal{C}^1$ solution $\mathbf{u} : U \longrightarrow \mathbb{R}^m$, $\mathbf{x} \longmapsto \mathbf{u}(\mathbf{x})$ (defined in an open neighborhood U of $\mathbf{a}$ in $\mathbb{R}^n$) of the differential equation $\mathbf{y}' = F(\mathbf{x},\mathbf{y})$ precisely when the symmetry relations

$$\partial_i F_j + \sum_{1\le k\le m} \partial_{n+k}F_j\ \partial u_k/\partial x_i = \partial_j F_i + \sum_{1\le k\le m} \partial_{n+k}F_i\ \partial u_k/\partial x_j$$
$$(\text{for all } i \ne j,\ 1 \le i,j \le n)$$

are satisfied. ■

Let us write the Frobenius symmetry condition in the simplest case ($n = 2$, $m = 1$), namely for a scalar field $u : \mathbb{R}^2 \longrightarrow \mathbb{R}$. Changing notations slightly, the problem is now to find a local function $z = u(x,y)$ satisfying a differential equation

$$\mathbf{grad}\ u = u' = F(x,y,z),$$

i.e. $(\mathbf{grad}\ u)(x,y) = F(x,y,u(x,y))$, or in extenso

$$(\partial u/\partial x)(x,y) = F_1(x,y,u(x,y)),\ (\partial u/\partial y)(x,y) = F_2(x,y,u(x,y)),$$

where F is a given $\mathcal{C}^1$ function in $\mathbb{R}^3$. The symmetry condition is

$$\partial_2F_1 + F_2\cdot\partial_3F_1 = \partial_1F_2 + F_1\cdot\partial_3F_2\ .$$

Geometrically, we would represent the function $z = u(x,y)$ by a piece of surface in $\mathbb{R}^3$, the initial condition $u(a_1,a_2) = b$ simply meaning that the surface contains the point (a_1,a_2,b). The given partial derivatives $\partial u/\partial x = F_1$, $\partial u/\partial y = F_2$ represent slopes of the tangent planes to this surface. Since we are going to use it later (10.3), let us formulate this special case of the Frobenius theorem in geometrical terms.

Corollary. Let $\mathcal{F}$ be a $\mathcal{C}^1$ field of non vertical planes in an open subset U of $\mathbb{R}^3$. We can assume that for $P \in U$, the equation of the plane $\Pi = \Pi_P \in \mathcal{F}$ going through P is given by the equation

$$z = F_1(P)(x - x_P) + F_2(P)(y - y_P)$$

where $F_i : U \longrightarrow \mathbb{R}$ are $\mathcal{C}^1$ functions. Then, for each $P \in U$, it will be possible to find a $\mathcal{C}^1$ surface $\Sigma_P \ni P$ having its tangent planes in the family $\mathcal{F}$ *precisely when*

$$\partial_2 F_1 + F_2 \cdot \partial_3 F_1 = \partial_1 F_2 + F_1 \cdot \partial_3 F_2 \quad \text{in } U\ . \qquad ■$$

In the preceding statement, it is implicitely assumed that Σ_P is defined in a neighborhood of P and consequently, can be described by an equation $z = u(x,y)$ where u is a $\mathcal{C}^1$ function in an open neighborhood of (x_P, y_P).

7.3 HESSIAN OF A SCALAR FIELD

Let $f \in \mathcal{C}^2$ be a scalar field $\mathbb{R}^n \longrightarrow \mathbb{R}$ and $F \in \mathcal{C}^1$ its derivative

$$F = \mathbf{grad} f = f' : \mathbb{R}^n \longrightarrow \mathbb{R}_n = \mathcal{L}(\mathbb{R}^n, \mathbb{R}).$$

We intend to compute $F'(\mathbf{a})$ and explain its meaning. For this, we establish the representative matrix of this linear map (in the canonical bases). In other words, we are looking for the Jacobian matrix of the gradient. Let us find the images of the basis elements $\mathbf{e}_j$ of $\mathbb{R}^n$

$$F'(\mathbf{a})\mathbf{e}_j = \frac{\partial F}{\partial x_j}(\mathbf{a}) = \frac{\partial}{\partial x_j}\mathbf{grad} f(\mathbf{a}) = \left(\frac{\partial^2 f}{\partial x_j \partial x_i}(\mathbf{a})\right)_{1 \le i \le n}.$$

This means that

$$F'(\mathbf{a})\mathbf{e}_j \in \mathbb{R}_n$$

is the linear form represented by the row vector with coefficients equal to the second partial derivatives. The evaluation of this linear form on a vector $\mathbf{h} \in \mathbb{R}^n$ yields

$$F'(\mathbf{a})\mathbf{e}_j(\mathbf{h}) = \sum_i \frac{\partial^2 f}{\partial x_j \partial x_i}(\mathbf{a}) \cdot h_i \qquad \text{if } \mathbf{h} = \sum h_i \mathbf{e}_i.$$

For $\mathbf{k} = \sum k_j \mathbf{e}_j$ we shall get the corresponding linear combination of values

$$F'(\mathbf{a})(\mathbf{k})(\mathbf{h}) = \sum_{i,j} \frac{\partial^2 f}{\partial x_j \partial x_i}(\mathbf{a}) \cdot h_i k_j.$$

As we have proved in 7.1 that these second partial derivatives are symmetric (we are indeed assuming $f \in \mathcal{C}^2$)

$$\frac{\partial^2 f}{\partial x_j \partial x_i}(\mathbf{a}) = \frac{\partial^2 f}{\partial x_i \partial x_j}(\mathbf{a}),$$

we see that $F'(\mathbf{a}) = d(df)(\mathbf{a}) = f''(\mathbf{a})$ is a symmetric bilinear form represented by the symmetric matrix

$$\left(\frac{\partial^2 f}{\partial x_i \partial x_j}(\mathbf{a})\right).$$

This symmetric matrix is also called the **Hessian matrix** of f (at the point **a**). The quadratic form obtained by taking the diagonal values of this bilinear form

$$H_f(\mathbf{a})(\mathbf{h}) = F'(\mathbf{a})(\mathbf{h},\mathbf{h}) = f''(\mathbf{a})(\mathbf{h},\mathbf{h})$$

is the **Hessian quadratic form** of f (at the point **a**). Thus, for a scalar field $f \in \mathcal{C}^2$ on $\mathbb{R}^n$, the second derivative of f is interpreted as a continuous *field of symmetric bilinear forms* in $\mathbb{R}^n$ and the corresponding *field of quadratic forms* in $\mathbb{R}^n$ is the Hessian of f. The importance of these concepts will be seen presently.

7.4 LIMITED EXPANSIONS OF THE SECOND ORDER

Let $f : \mathbb{R}^n \longrightarrow \mathbb{R}$ be a scalar field and $\mathbf{a} \in \mathbb{R}^n$. We are interested in a limited expansion of f near the point **a** of the form

$$f(\mathbf{a} + \mathbf{h}) = f(\mathbf{a}) + f'(\mathbf{a})\mathbf{h} + Q_{\mathbf{a}}(\mathbf{h}) + o(\|\mathbf{h}\|^2)$$

where $Q_{\mathbf{a}}$ is a *quadratic form*. Let us recall that by definition,

$$Q_{\mathbf{a}}(\lambda\mathbf{h}) = \lambda^2 Q_{\mathbf{a}}(\mathbf{h}),$$

$$(\mathbf{h},\mathbf{k}) \longmapsto Q_{\mathbf{a}}(\mathbf{h} + \mathbf{k}) - Q_{\mathbf{a}}(\mathbf{h}) - Q_{\mathbf{a}}(\mathbf{k}) \quad \text{is bilinear.}$$

Moreover, the notation $o(\|\mathbf{h}\|^2)$ represents a remainder $R(\mathbf{h})$ which has the property

$$R(\mathbf{h})/\|\mathbf{h}\|^2 \longrightarrow 0 \text{ for } \|\mathbf{h}\| \longrightarrow 0$$

or equivalently, $R(\mathbf{h}) = \|\mathbf{h}\|^2\phi(\mathbf{h})$ with $\phi(\mathbf{h}) \longrightarrow 0$ for $\|\mathbf{h}\| \longrightarrow 0$.

Proposition. Let $f : \mathbb{R}^n \longrightarrow \mathbb{R}$ be a $\mathcal{C}^1$ scalar field. Assume moreover that $f' = \mathbf{grad} f : \mathbb{R}^n \longrightarrow \mathbb{R}_n$ is differentiable at $\mathbf{a} \in \mathbb{R}^n$. Then f admits a second order limited expansion of the above type with quadratic form $Q_{\mathbf{a}} = H_f(\mathbf{a})/2$. In other words, we have

$$f(\mathbf{a} + \mathbf{h}) = f(\mathbf{a}) + f'(\mathbf{a})\mathbf{h} + \frac{1}{2} H_f(\mathbf{a})(\mathbf{h}) + o(\|\mathbf{h}\|^2)$$

$$= f(\mathbf{a}) + f'(\mathbf{a})\mathbf{h} + \frac{1}{2} f''(\mathbf{a})(\mathbf{h},\mathbf{h}) + o(\|\mathbf{h}\|^2). \qquad \square$$

Proof. Consider the auxiliary function

$$\varphi(\mathbf{h}) = f(\mathbf{a} + \mathbf{h}) - f(\mathbf{a}) - f'(\mathbf{a})\mathbf{h} - \frac{1}{2} f''(\mathbf{a})(\mathbf{h},\mathbf{h}).$$

We shall compute the derivative $\varphi'(\mathbf{h}) = \mathbf{grad}\varphi(\mathbf{h})$. As we have already seen in 5.6, the derivative of the term $\frac{1}{2} f''(\mathbf{a})(\mathbf{h},\mathbf{h})$ is the linear map

$$\mathbf{k} \longmapsto f''(\mathbf{a})(\mathbf{h},\mathbf{k}) \ (= f''(\mathbf{a})(\mathbf{h})(\mathbf{k}) = f''(\mathbf{a})(\mathbf{h})\cdot\mathbf{k} \ ).$$

Hence we have

$$\begin{aligned}\varphi'(\mathbf{h})\mathbf{k} &= \mathbf{grad}\varphi(\mathbf{h})\cdot\mathbf{k} = \\ &= f'(\mathbf{a} + \mathbf{h})\mathbf{k} - f'(\mathbf{a})\mathbf{k} - f''(\mathbf{a})(\mathbf{h},\mathbf{k}) \\ &= [f'(\mathbf{a} + \mathbf{h}) - f'(\mathbf{a}) - f''(\mathbf{a})(\mathbf{h})](\mathbf{k}).\end{aligned}$$

But f' is assumed to be differentiable at the point **a** :

$$f'(\mathbf{a} + \mathbf{h}) - f'(\mathbf{a}) - f''(\mathbf{a})(\mathbf{h}) = o(\|\mathbf{h}\|) = \|\mathbf{h}\|\phi(\mathbf{h})$$

with $\phi(\mathbf{h}) \longrightarrow 0$ when $\|\mathbf{h}\| \longrightarrow 0$ (recall that $\phi(\mathbf{h}) \in \mathbb{R}_n$). This implies

$$\varphi'(\mathbf{h})\mathbf{k} = \|\mathbf{h}\|\phi(\mathbf{h})\mathbf{k} \ , \ |\varphi'(\mathbf{h})\mathbf{k}| \le \|\mathbf{h}\|\|\mathbf{k}\|\cdot\|\phi(\mathbf{h})\|$$

and

$$\|\varphi'(\mathbf{h})\| \le \|\mathbf{h}\|\|\phi(\mathbf{h})\|.$$

If $\varepsilon > 0$ is given, there will exist $\eta > 0$ with

$$\|\varphi'(\mathbf{h})\| \le \varepsilon\|\mathbf{h}\| \quad \text{for all } \mathbf{h} \text{ satisfying } \|\mathbf{h}\| \le \eta.$$

the basic inequality of 2.1 gives then

$$\|\varphi(\mathbf{h}) - \varphi(0)\| \le \|\mathbf{h}\| \ \sup_{0\le t\le 1} \|\varphi'(t\mathbf{h})\| \le \varepsilon\|\mathbf{h}\|^2.$$

But $\varphi(0) = 0$ by definition and this proves

$$\|\varphi(\mathbf{h})\| \le \varepsilon\|\mathbf{h}\|^2 \ , \ \text{i.e. } \varphi(\mathbf{h}) = o(\|\mathbf{h}\|^2).$$

■

7.5 SECOND ORDER DISCUSSION OF EXTREMA

The extrema of a $\mathcal{C}^1$ scalar field $f : \mathbb{R}^n \longrightarrow \mathbb{R}$ satisfy the condition $f'(\mathbf{a}) = \mathbf{grad}f(\mathbf{a}) = 0$. This shows that extrema are found among critical points of f. Conversely, it is easy to give sufficient conditions for a critical point to be an extremum of f.

Theorem. Let f be a $\mathcal{C}^1$ scalar field in $\mathbb{R}^n$ and $\mathbf{a} \in \mathbb{R}^n$ a point where

- ▷ $f'(\mathbf{a}) = \mathbf{grad}f(\mathbf{a}) = 0$,
- ▷ f' is differentiable at **a**,
- ▷ $H_f(\mathbf{a})$ is a positive definite quadratic form.

Then **a** is a local minimum of f. □

Proof. We shall indeed prove that for some positive $\varepsilon > 0$

$$f(\mathbf{a} + \mathbf{h}) > f(\mathbf{a}) \quad \text{for all increments } \mathbf{h} \ne 0 \text{ with } \|\mathbf{h}\| < \varepsilon.$$

Let Q denote the positive definite quadratic form $H_f(\mathbf{a})/2$. We have $\min_{\|\mathbf{h}\|=1} Q(\mathbf{h}) = c > 0$ and if $\varepsilon > 0$ is suitably small, the remainder in the second order expansion of f at **a** will have norm $\le \frac{c}{2}\|\mathbf{h}\|^2$ for all increments **h** with $\|\mathbf{h}\| \le \varepsilon$. In particular, for all normed increments **h**, and all $0 < t < \varepsilon$

$$f(\mathbf{a} + t\mathbf{h}) - f(\mathbf{a}) = t^2Q(\mathbf{h}) + o(t^2)$$

(recall that —by assumption— $f'(\mathbf{a}) = 0$) will lead to an inequality

$$f(\mathbf{a} + t\mathbf{h}) - f(\mathbf{a}) > t^2 c - \frac{c}{2} t^2 = \frac{c}{2} t^2 > 0. \qquad \blacksquare$$

Of course, if the Hessian quadratic form is negative definite at an extremum $\mathbf{a} \in \mathbb{R}^n$ this point will be a relative maximum for f. But if the Hessian quadratic form takes both positive and negative values for normed vectors $\mathbf{h}$, the point $\mathbf{a}$ is neither a local maximum, nor a local minimum of f. Such points are called *stationary points*. Examples of stationary points are easily given : the function $f : \mathbb{R}^2 \longrightarrow \mathbb{R}$ defined by $f(x,y) = xy$ has such a stationary point at the origin. The graph of f in $\mathbb{R}^3$ is a surface with a *saddle point* at the origin. This surface contains the Ox-axis (where $y = 0 \Longrightarrow z = f(x,y) = 0$) and the Oy-axis (where $x = 0 \Longrightarrow z = f(x,y) = 0$).

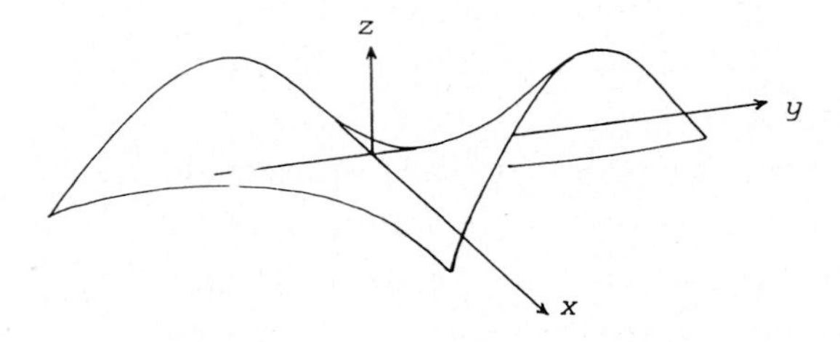

Fig.7.3

Quadratic forms Q with $Q(\mathbf{h}) \geq 0$ for all normed vectors $\mathbf{h}$ are commonly called **positive quadratic forms** whereas quadratic forms Q with $Q(\mathbf{h}) \geq c > 0$ (for all normed vectors $\mathbf{h}$) are said to be **positive definite**. Positive forms are characterized by $\operatorname{Inf}_{\|\mathbf{h}\|=1} Q(\mathbf{h}) \geq 0$ and positive definite quadratic forms are characterized by

$$\operatorname{Inf}_{\|\mathbf{h}\|=1} Q(\mathbf{h}) > 0.$$

In general however, when the quadratic form $H_f(\mathbf{a})$ is positive (or negative) without being definite, i.e.

$$\operatorname{Inf}_{\|\mathbf{h}\|=1} H_f(\mathbf{a})(\mathbf{h}) = 0,$$

the second order discussion does not allow one to recognize the precise nature of the critical point.

7.6 SECOND ORDER DISCUSSION FOR CONDITIONAL EXTREMA

We only consider a particular —but typical— case. Namely, we are looking for the extrema of a $\mathcal{C}^2$ scalar field f in the plane $\mathbb{R}^2$ when one restriction $\varphi(x,y) = 0$ is given. We also assume that the scalar function φ is smooth, say $\varphi \in \mathcal{C}^2$. Critical points for the restriction $f|_C$ of f to the curve C defined implicitely by the equation $\varphi(x,y) = 0$ are characterized by the *first order conditions*

▷ $\mathbf{grad}(f - \lambda\varphi)(\mathbf{a}) = 0$,

▷ $\varphi(\mathbf{a}) = 0$.

The *second order* condition is best written for independent variables first. So let us assume that $x = \xi(t)$ and $y = \vartheta(t)$ constitute a parametrization of the curve $\varphi = 0$:

$$\varphi(\xi(t),\vartheta(t)) \equiv 0 \quad \text{(identically in t)}.$$

Also assume for simplification that the parametrization is chosen with origin at the point $\mathbf{a}$: $\mathbf{a} = (\xi(0),\vartheta(0))$. We want to express the second order condition for a minimum of

$$t \longmapsto F(t) = f(\xi(t),\vartheta(t))$$

at $t = 0$. This condition is simply $F''(t) > 0$. Using the fundamental theorem of calculus (5.4) twice, we get

$$F'(t) = D_1f(\xi(t),\vartheta(t))\cdot\xi'(t) + D_2f(\xi(t),\vartheta(t))\cdot\vartheta'(t)$$

and

$$\begin{aligned}F''(t) = {} & D_1D_1f(\xi(t),\vartheta(t))\cdot(\xi'(t))^2 + 2D_1D_2f(\xi(t),\vartheta(t))\cdot\xi'(t)\vartheta'(t) + \\ & + D_2D_2f(\xi(t),\vartheta(t))\cdot(\vartheta'(t))^2 + \\ & + D_1f(\xi(t),\vartheta(t))\cdot\xi''(t) + D_2f(\xi(t),\vartheta(t))\cdot\vartheta''(t).\end{aligned}$$

We recognize

$$\begin{aligned}F''(t) = {} & f''(\xi(t),\vartheta(t))^t(\xi'(t),\vartheta'(t)) + \\ & \mathbf{grad}f(\xi(t),\vartheta(t))^t(\xi''(t),\vartheta''(t)).\end{aligned}$$

The same computation for $\Phi(t) = \varphi(\xi(t),\vartheta(t)) \equiv 0$ leads to

$$\begin{aligned}\Phi''(t) = {} & \varphi''(\xi(t),\vartheta(t))^t(\xi'(t),\vartheta'(t)) + \\ & \mathbf{grad}\varphi(\xi(t),\vartheta(t))^t(\xi''(t),\vartheta''(t)) \equiv 0.\end{aligned}$$

With the particular value λ of the Lagrange parameter for which the first order condition is satisfied at the point $\mathbf{a}$, we deduce

$$\begin{aligned}F''(0) = (F - \lambda\Phi)''(0) &= (f - \lambda\varphi)''(\mathbf{a})(\mathbf{h},\mathbf{h}) + \mathbf{grad}(f - \lambda\varphi)(\mathbf{a})(\mathbf{h}') = \\ &= (f - \lambda\varphi)''(\mathbf{a})(\mathbf{h},\mathbf{h})\end{aligned}$$

where $\mathbf{h} = {}^t(\xi'(0),\vartheta'(0)) \perp \mathbf{grad}\varphi(\mathbf{a})$ and $\mathbf{h}' = {}^t(\xi''(0),\vartheta''(0))$.

The second order condition for a minimum $\mathbf{a}$ of f restricted to

$\varphi = 0$ (or of F at $t = 0$) is thus

$$H_{f-\lambda\varphi}(\mathbf{a})(\mathbf{h}) > 0 \quad \text{whenever } 0 \neq \mathbf{h} \perp \mathbf{grad}\varphi(\mathbf{a}).$$

In words : *the Hessian of* $f - \lambda\varphi$ *—computed as in the case of independent variables, with the value of* λ *determined by the first order condition at* **a**— *should have a positive definite restriction to the tangent subspace of* $\varphi = 0$ *at* **a** (this tangent subspace is precisely the orthogonal to $\mathbf{grad}\varphi(\mathbf{a})$).

7.7 THE LAPLACE OPERATOR

Let f be a $\mathcal{C}^2$ scalar field in $\mathbb{R}^n$. We have defined the Hessian matrix f" as a field of symmetric matrices in $\mathbb{R}^n$. The *trace* of this field of matrices is a continuous scalar field called **Laplacian** of f and denoted by Δf. By definition, we thus have

$$\Delta f = \sum_{i=1}^{n} \frac{\partial^2 f}{\partial x_i^2} .$$

Due to the importance of this differential operator, we shall have to come back to some of its properties in more detail, especially when $n = 3$ (cf.14.2). Already observe that since the trace is an invariant under (linear) change of coordinates, the Laplace operator Δ has an intrinsic meaning, independently from the choice of basis in $\mathbb{R}^n$ (we have worked in the canonical basis, but any other basis would lead to the same operator on scalar fields).

7.8 NON DEGENERATE CRITICAL POINTS

A **non degenerate critical point** is by definition a critical point for which the **determinant of the Hessian matrix** does not vanish. Non degenerate critical points are then classified according to the number of negative eigenvalues of the Hessian matrix. Since the Hessian matrix is symmetric, it is diagonalizable over the reals, and since its determinant is not zero, no eigenvalue is 0 and there is a partition of the spectrum in positive, resp. negative, eigenvalues.

A typical example of a non degenerate critical point is the following. Let $0 \leq \lambda \leq n$ and decompose $\mathbb{R}^n$ as product $\mathbb{R}^\lambda \times \mathbb{R}^{n-\lambda}$ by grouping the first λ components of an $\mathbf{x} \in \mathbb{R}^n$ in a vector $\mathbf{u} \in \mathbb{R}^\lambda$ (resp. the last $n-\lambda$ components in a vector $\mathbf{v} \in \mathbb{R}^{n-\lambda}$). Thus we simply write $\mathbf{x} = (\mathbf{u},\mathbf{v}) \in \mathbb{R}^n = \mathbb{R}^\lambda \times \mathbb{R}^{n-\lambda}$. The scalar function

$$f_\lambda : \mathbb{R}^n \longrightarrow \mathbb{R} \quad \text{defined by} \quad f_\lambda(\mathbf{u},\mathbf{v}) = -\|\mathbf{u}\|^2 + \|\mathbf{v}\|^2$$

has the origin as only critical point with $f_\lambda = H_\lambda/2$ for second order limited expansion. This shows that the Hessian of f_λ is $2f_\lambda$. The Hessian matrix of f_λ has λ eigenvalues equal to -2 and $n-\lambda$ eigenvalues equal to $+2$.

In fact, every non degenerate critical point of a scalar function $f : \mathbb{R}^n \longrightarrow \mathbb{R}$ looks like one f_λ (at the origin).

Theorem. Let $f : \mathbb{R}^n \longrightarrow \mathbb{R}$ be a $\mathcal{C}^2$ scalar field having the point $\mathbf{a}$ as a non degenerate critical point with, say, $f(\mathbf{a}) = 0$. There is a unique $0 \le \lambda \le n$ with the following property

> there exists an open neighborhood U of $\mathbf{a} \in \mathbb{R}^n$ and a diffeomorphism $\varphi : U \longrightarrow \mathbb{R}^\lambda \times \mathbb{R}^{n-\lambda}$ with $\varphi(\mathbf{a}) = 0$ such that the restriction $f|_U$ of f to U is the composite of f_λ and φ

$$\begin{array}{ccc} \mathbf{x} \in \mathbb{R}^n \supset U & \xrightarrow{\;f|_U\;} & \mathbb{R} \ni f(\mathbf{x}) = -\|\mathbf{u}\|^2 + \|\mathbf{v}\|^2 \\ & \searrow^{\varphi} \quad \nearrow_{f_\lambda} & \\ \varphi(\mathbf{x}) = (\mathbf{u},\mathbf{v}) \in \mathbb{R}^\lambda \times \mathbb{R}^{n-\lambda} & & \end{array}$$

□

Corollary. Non degenerate critical points are isolated. □

Proof of the corollary. If U is as in the statement of the theorem

$$f'(\mathbf{x}) = f_\lambda'(\varphi(\mathbf{x}))\circ\varphi'(\mathbf{x}) \quad \text{for all } \mathbf{x} \in U.$$

Since $\varphi'(\mathbf{x})$ is an isomorphism at each $\mathbf{x} \in U$, the statement of the corollary follows from the fact that the only critical point of f_λ is the origin. ■

Proof of the theorem. For $x \in \mathbb{R}^n$ with small enough norm we can write a limited expansion of $f(t\mathbf{x})$ valid for $0 \le t \le 1$. For this, we integrate the derivative of $t \longmapsto f(t\mathbf{x}) = g(t)$. Using the fundamental theorem of calculus (5.4) we find

$$g'(t) = \sum D_i f(t\mathbf{x})\cdot x_i \quad (\text{where } D_i f = \partial f/\partial x_i \in \mathcal{C}^1)$$

and since $g(0) = f(0) = 0$

$$f(\mathbf{x}) = g(1) = \int_0^1 g'(t)dt = \sum x_i \int_0^1 D_i f(t\mathbf{x})dt = \sum x_i g_i(\mathbf{x}).$$

We have used the abbreviation $g_i(\mathbf{x}) = \int_0^1 D_i f(t\mathbf{x})dt$. Starting afresh with each g_i, the same procedure will lead to

$$g_i(\mathbf{x}) = \sum_j x_j h_{ij}(\mathbf{x}) \qquad (\text{for each } 1 \le i \le n).$$

Consequently

$$f(\mathbf{x}) = \sum_{i,j} x_i x_j h_{ij}(\mathbf{x}).$$

Observe that by the Schwarz theorem (7.1)

$$h_{ij}(0) = (\partial g_i/\partial x_j)(0) = \frac{1}{2}\frac{\partial^2 f}{\partial x_j \partial x_i}(0) = h_{ji}(0)$$

but the $h_{ij}(\mathbf{x})$ are not necessarily symmetrical for $\mathbf{x} \neq 0$! We can always define

$$\tilde{h}_{ij}(\mathbf{x}) = (h_{ij}(\mathbf{x}) + h_{ji}(\mathbf{x}))/2$$

and still have

$$f(\mathbf{x}) = \sum_{i,j} x_i x_j \tilde{h}_{ij}(\mathbf{x}) \text{ with symmetric } \tilde{h}_{ij}(\mathbf{x}) = \tilde{h}_{ji}(\mathbf{x}).$$

In a suitable basis, this field of quadratic forms can be reduced to a difference of squares and the implicit function theorem will take care of the existence of the diffeomorphism φ... ∎

VECTOR FIELDS IN THE USUAL PHYSICAL SPACE $\mathbb{R}^3$

8.1 JACOBIAN OF A VECTOR FIELD IN $\mathbb{R}^3$

The general theory developed in the preceding chapters will now be applied to the case of the usual physical space $\mathbb{R}^3$. A vector map $F : \mathbb{R}^3 \longrightarrow \mathbb{R}^3$ will be viewed here as a vector field and we shall write $F = \mathbf{u}$ such a vector map. With

$$\mathbf{r} = \vec{r} = {}^t(x_1,x_2,x_3) = {}^t(x,y,z),$$

we can assume that the vector field $\mathbf{u}$ is given in components by three scalar functions of x, y and z, say $\mathbf{u} = {}^t(u_1,u_2,u_3)$ where u_i depends on x, y and z. When $\mathbf{u} \in \mathcal{C}^1$, the Jacobian matrix of $\mathbf{u}$ is a 3 by 3 matrix whose lines are the usual gradient functions of the u_i. We shall also call Gradient matrix this Jacobian :

$$\text{Grad } \mathbf{u} = \text{Jac}_{\mathbf{u}} = \begin{pmatrix} \mathbf{grad}\ u_1 \\ \mathbf{grad}\ u_2 \\ \mathbf{grad}\ u_3 \end{pmatrix} \quad (= \mathbf{u}' = \mathbf{du}\ !).$$

To get a better *feeling* for this matrix derivative, we shall extract some invariants from it. Each of them will have a simpler geometrical meaning.

First recall that any square matrix can be uniquely written as a sum of a symmetric and an antisymmetric matrix. This decomposition is given by :

$$M_s = (M + {}^tM)/2 \quad : \text{symmetric matrix}$$
$$\text{(average between M and } {}^tM\text{)},$$
$$M_a = (M - {}^tM)/2 \quad : \text{antisymmetric matrix.}$$

Obviously

$$M = M_s + M_a.$$

In order to prove uniqueness, let S be a symmetric matrix and A an antisymmetric matrix with $M = S + A$. Thus we shall have

$$M = M_s + M_a = S + A$$

and

$$M_s - S = A - M_a \text{ is both symmetric and antisymmetric.}$$

Obviously, a matrix which is both symmetric and antisymmetric is the 0 matrix (its transpose N satisfies $N = -N$, $2N = 0$). This shows that $M_s - S = A - M_a = 0$ proving uniqueness $M_s = S$, $M_a = A$.

Let us examine more closely a 3 by 3 antisymmetric matrix :

it has the form

$$\begin{pmatrix} 0 & a & b \\ -a & 0 & c \\ -b & -c & 0 \end{pmatrix}$$

with three arbitrary parameters a, b and c. Such a matrix corresponds to a vector product operator $M_\omega = \vec{\omega} \wedge\ : \mathbf{r} \longmapsto \vec{\omega} \wedge \mathbf{r}$. If the vector ω (let us delete the arrow on it for typographical simplicity) has components ω_1, ω_2 and ω_3 one checks that $M_\omega(\mathbf{e}_1) = \omega \wedge \mathbf{e}_1$ has components 0, ω_3 and $-\omega_2$. These three scalars constitute the first column of the matrix of M_ω (in the canonical basis of $\mathbb{R}^3$). Eventually, the representative matrix of the operator M_ω is seen to be

$$M_\omega : \begin{pmatrix} 0 & -\omega_3 & \omega_2 \\ \omega_3 & 0 & -\omega_1 \\ -\omega_2 & \omega_1 & 0 \end{pmatrix}.$$

In particular, we recognize that the three parameters a, b and c above are resp. given by $-\omega_3$, ω_2 and $-\omega_1$. Conversely, we find the components of the vector ω in the antisymmetric matrix in the places $-c$, b and $-a$.

Since the antisymmetric part of the Grad matrix is

$$\frac{1}{2}\begin{pmatrix} 0 & *** & \partial u_1/\partial z - \partial u_3/\partial x \\ \partial u_2/\partial x - \partial u_1/\partial y & 0 & *** \\ *** & \partial u_3/\partial y - \partial u_2/\partial z & 0 \end{pmatrix}$$

the associated vector has components

$$\frac{1}{2}\begin{pmatrix} \partial u_3/\partial y - \partial u_2/\partial z \\ \partial u_1/\partial z - \partial u_3/\partial x \\ \partial u_2/\partial x - \partial u_1/\partial y \end{pmatrix}.$$

We define **curl u** to be the double of this vector

$$\mathbf{curl\ u} = \begin{pmatrix} \partial u_3/\partial y - \partial u_2/\partial z \\ \partial u_1/\partial z - \partial u_3/\partial x \\ \partial u_2/\partial x - \partial u_1/\partial y \end{pmatrix}.$$

The symmetric part of the the Grad matrix is more difficult to interpret. It has six arbitrary parameters (the diagonal values are arbitrary). The easiest invariant which can be extracted from it is the trace (equal to the trace of the initial matrix since the antisymmetric part has 0 diagonal entries). Thus we also define

$$\text{div } \mathbf{u} = \partial u_1/\partial x + \partial u_2/\partial y + \partial u_3/\partial z.$$

This is a scalar field and the scalar matrix $(\frac{1}{3}$ div $\mathbf{u})\cdot I$ has the same trace as Grad $\mathbf{u}$. Summarizing, we have obtained

$$\text{Grad } \mathbf{u} = \frac{1}{2} M_{\mathbf{curl\ u}} + (\frac{1}{3} \text{ div } \mathbf{u})\cdot I + S_o$$

with a symmetric matrix S_o having zero trace. Although this component also plays an important role in elasticity —it is called **shear** of the vector field $\mathbf{u}$— we shall not deal with it any further (just observe that it depends on five extra parameters).

Roughly speaking, the divergence of a field gives an idea of the local inflation rate of the field. For example, the field

$$\mathbf{u}(\mathbf{r}) = a\mathbf{r} \quad (\text{in } \mathbb{R}^3)$$

has divergence equal to 3a (constant scalar field). The value of the divergence at a point is a number representing the scalar inflation of the vector field at that point. The following Fig.

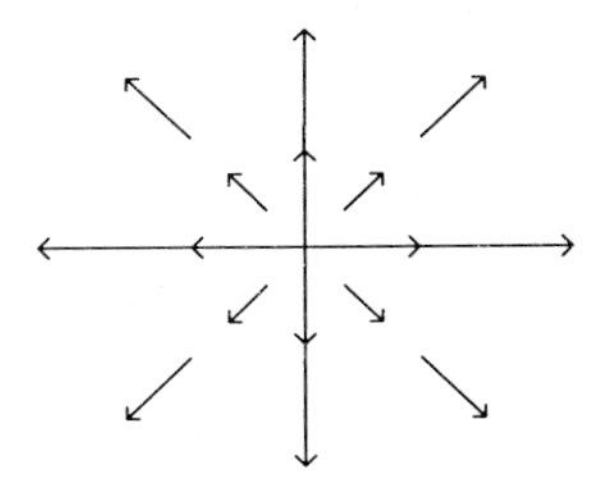

Fig.8.1

illustrates the situation at the origin. Observe that the divergence of the fields $\mathbf{r} \longmapsto a(\mathbf{r} - \mathbf{r}_o)$ are all equal to 3a and the situation of the origin is irrelevant for the inflation rate of this field. The divergence also measures an *expansion* produced by the vector field $\mathbf{u}$: it will vanish for the velocity field of an incompressible fluid in movement. Provided that there are no sources, an electrical field will have zero divergence (more on this in the chapter on applications!). One can check that **curl u** vanishes for this field.

Similarly, it is easy to get an idea of the rotational of a vector field $\mathbf{u}$. For this, one should first find the Jacobian (or Grad) matrix of the field and plot the vector field associated to its antisymmetric part. The *whirlwind* of the field is measured by the vector **curl** of the field. For example, if $\mathbf{u}(\mathbf{r}) = \omega \wedge \mathbf{r}$, the derivative of **u** is the constant matrix mapping given by $\omega \wedge \ldots$(because **u** is *linear* hence coïncides with its derivative!) and thus, **curl u** = 2ω in this case. A picture of the vector field $\omega \wedge \mathbf{r}$ will reveal the presence of a whirl around the axis $\mathbb{R}\omega$, and as before, the whirl of the field $\mathbf{r} \longmapsto \omega \wedge (\mathbf{r}-\mathbf{r}_o)$ being the same as the preceding one, we get a constant whirl in the whole space (cf. Fig.).

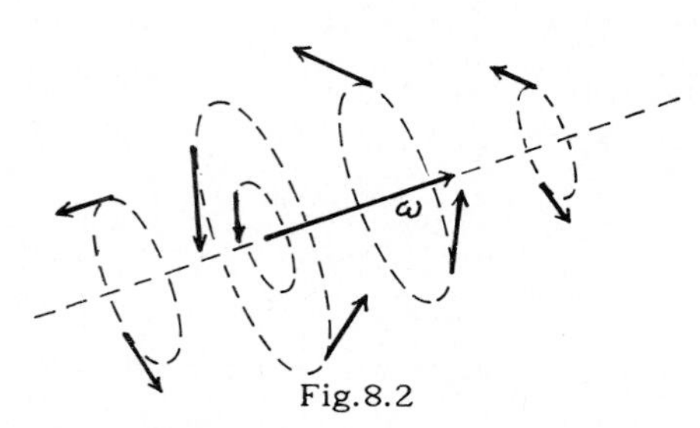

Fig.8.2

It is known that the distribution of velocities for a solid rotating around the axis $\mathbb{R}\omega$ with angular velocity $\|\omega\|$ is given by the the field $\mathbf{v}(\mathbf{r}) = \omega \wedge \mathbf{r}$ which is thus the prototype of a vector field with a constant **curl** equal to 2ω. Here, div **v** vanishes.

The above geometrical interpretations do not pretend to have a precise mathematical meaning, but we shall see —once the Stokes formula has been given— that their mathematical content can be considerably sharpened.

8.2 THE DERIVATION OPERATOR DEL (NABLA)

It would be difficult to memorize the definition of the **curl** of a vector field **u** without the mnemotechnical method that we are going to explain now. Let us introduce a *formal line vector* ∇ having for components the three partial derivation operators

$$\partial_x = \partial/\partial x,\ \partial_y = \partial/\partial y,\ \partial_z = \partial/\partial z.$$

We shall call **del operator** —or **nabla operator**— this formal expression

$$\nabla = (\partial_x, \partial_y, \partial_z).$$

Recall that this del operator is meant to be a "line vector" and that its components are differential operators. The gradient of a scalar field f in $\mathbb{R}^3$ can be abbreviated by

$$\mathbf{grad} f = (\partial_x f, \partial_y f, \partial_z f) = \nabla f.$$

Similarly, if **u** is a vector field in $\mathbb{R}^3$, its divergence is abbreviated by

$$\operatorname{div} \mathbf{u} = \partial_x u_1 + \partial_y u_2 + \partial_z u_3 = \nabla \cdot \mathbf{u}.$$

Finally, the curl of the vector field **u** in $\mathbb{R}^3$ is also found to be given by

$$\mathbf{curl}\ \mathbf{u} = {}^t(\partial_y u_3 - \partial_z u_2, \ldots) = \nabla \wedge \mathbf{u}.$$

Recall here that a cross product in $\mathbb{R}^3$ is computed by means of a symbolic determinant (which turns out to be *doubly* symbolic here since the line corresponding to ∇ has *operator* components...)

$$\nabla \wedge \mathbf{u} = \begin{vmatrix} \mathbf{e}_1 & \mathbf{e}_2 & \mathbf{e}_3 \\ \partial_x & \partial_y & \partial_z \\ u_1 & u_2 & u_3 \end{vmatrix} =$$

$$= \mathbf{e}_1(\partial_y u_3 - \partial_z u_2) + \mathbf{e}_2(\partial_z u_1 - \partial_x u_3) + \mathbf{e}_3(\partial_x u_2 - \partial_y u_1).$$

Another important mnemotechnical observation about the curl vector field is the following. Once its first component

$$(\mathbf{curl}\ \mathbf{u})_1 = \partial_y u_3 - \partial_z u_2$$

has been memorized (or re-computed), the other components can be obtained by a *cyclic permutation of the indices*

$$1 \longmapsto 2 \longmapsto 3 \longmapsto 1 \text{ and resp. } x \longmapsto y \longmapsto z \longmapsto x.$$

This observation can economize a lot of writing for theoretical considerations by reducing component proofs to consideration of the first component only.

To achieve an even more general use of the del operator, we indicate that it is still possible to use it for the Jacobian itself. For this, a preliminary notion of linear algebra is now reviewed.

Let $\mathbf{a} \in \mathbb{R}_n$ and $\mathbf{b} \in \mathbb{R}^m$ so that $\mathbf{a} = (a_1,\ldots,a_n)$ is a row vector identified to a linear form on $\mathbb{R}^n$ and $b = {}^t(b_1,\ldots,b_m)$ is a column vector (with m components). By definition, the **tensor product** of **a** and **b**, denoted by $\mathbf{a} \otimes \mathbf{b}$ represents the *operator*

$$\mathbb{R}^n \longrightarrow \mathbb{R}^m : \mathbf{x} \longmapsto (\mathbf{ax})\mathbf{b}.$$

The expression $\mathbf{ax} = a_1x_1 + \ldots + a_nx_n$ can also be computed as the matrix product $\mathbf{a}\circ\mathbf{x}$ of the row matrix **a** with the column matrix **x** (they indeed have the same number n of components so that the product is possible!). By definition, the operator $\mathbf{a} \otimes \mathbf{b}$ sends all vectors $\mathbf{x} \in \mathbb{R}^n$ on multiples of the vector $\mathbf{b} \in \mathbb{R}^m$. The image of the operator $\mathbf{a} \otimes \mathbf{b}$ is contained in the subspace generated by $\mathbf{b}$ in $\mathbb{R}^m$, and at least if $\mathbf{a}$ and $\mathbf{b} \neq 0$, $\mathbf{a} \otimes \mathbf{b}$ is an operator of rank 1. In general, we can only say that

$$\operatorname{rank}(\mathbf{a} \otimes \mathbf{b}) \leq 1.$$

Let us look for the matrix representation (in the canonical bases) for such an operator $\mathbf{a} \otimes \mathbf{b}$. The image of the jth basis vector is

$$(\mathbf{a} \otimes \mathbf{b})(\mathbf{e}_j) = (\mathbf{a}\mathbf{e}_j)\mathbf{b} = a_j\mathbf{b} = \sum_i a_j b_i \mathbf{e}_i .$$

The jth column of the representative matrix is proportional to $\mathbf{b}$ (all columns are thus proportional to $\mathbf{b}$ and thus furnishing a matrix of rank ≤ 1) and the matrix is

$$\mathbf{a} \otimes \mathbf{b} : \begin{pmatrix} a_1b_1 & \dots & a_nb_1 \\ \dots & & \dots \\ a_1b_m & \dots & a_nb_m \end{pmatrix} = \begin{pmatrix} b_1 \\ .. \\ b_m \end{pmatrix} \circ (a_1 \dots a_n) .$$

The matrix representation of $\mathbf{a} \otimes \mathbf{b}$ *is given by the product of the two matrices* $\mathbf{b}$ *and* $\mathbf{a}$ (in reversed order, thus furnishing an n by m matrix).

Coming back to the $\mathbb{R}^3$ context, with the line vector ∇ (called "del") in place of $\mathbf{a}$, we obtain

$$\nabla \otimes \mathbf{u} = \begin{pmatrix} u_1 \\ u_2 \\ u_3 \end{pmatrix} \circ (\partial_x, \partial_y, \partial_z) = \begin{pmatrix} \partial_x u_1 & \partial_y u_1 & \partial_z u_1 \\ \partial_x u_2 & \partial_y u_2 & \partial_z u_2 \\ \partial_x u_3 & \partial_y u_3 & \partial_z u_3 \end{pmatrix}$$

provided we remember that the differential operators ∂_i *operate on* $\mathbf{u}$ *and are thus to be written first.* This shows that we can also write

$$\operatorname{Grad} \mathbf{u} = \nabla \otimes \mathbf{u}.$$

We hope that the reader who practises exercises will be convinced of the usefulness of these notations.

8.3 PROPERTIES OF DIFFERENTIAL OPERATORS

The operator del ∇ has some important properties. First of all, it is a *linear operator*. This contains in condensed form the following properties. For a and b $\in \mathbb{R}$,

$\nabla(af + bg) = a\nabla f + b\nabla g$ (f and g $\mathcal{C}^1$ scalar functions),

$\nabla(a\mathbf{u} + b\mathbf{v}) = a\nabla\mathbf{u} + b\nabla\mathbf{v}$ ($\mathbf{u}$ and $\mathbf{v}$ $\mathcal{C}^1$ vector fields),

$\nabla\wedge(a\mathbf{u} + b\mathbf{v}) = a\nabla\wedge\mathbf{u} + b\nabla\wedge\mathbf{v}$ ($\mathbf{u}$ and $\mathbf{v}$ $\mathcal{C}^1$ vector fields),

$\nabla\otimes(a\mathbf{u} + b\mathbf{v}) = a\nabla\otimes\mathbf{u} + b\nabla\otimes\mathbf{v}$ ($\mathbf{u}$ and $\mathbf{v}$ $\mathcal{C}^1$ vector fields).

These respectively mean

▷ $\mathbf{grad}(af + bg) = a\,\mathbf{grad}f + b\,\mathbf{grad}g$

(f and g $\mathcal{C}^1$ scalar functions),

▷ $\mathrm{div}(a\mathbf{u} + b\mathbf{v}) = a\,\mathrm{div}\mathbf{u} + b\,\mathrm{div}\mathbf{v}$,

▷ $\mathbf{curl}(a\mathbf{u} + b\mathbf{v}) = a\,\mathbf{curl}\,\mathbf{u} + b\,\mathbf{curl}\,\mathbf{v}$,

▷ $\mathrm{Grad}(a\mathbf{u} + b\mathbf{v}) = a\,\mathrm{Grad}\mathbf{u} + b\,\mathrm{Grad}\mathbf{v}$

(**u** and **v** $\mathcal{C}^1$ vector fields).

Secondly, the del operator ∇ is a *derivation operator*. Recall that *derivations* are linear operators on $\mathcal{C}^1$ functions having the characteristic property

$$D(fg) = (Df)g + f(Dg) \qquad (f \text{ and } g \in \mathcal{C}^1).$$

(To be more precise, *punctual derivations* are defined as linear mappings $:\mathcal{C}^1 \longrightarrow \mathbb{R}$ having the preceding property whereas *derivation operators* are linear operators $\mathcal{C}^1 \longrightarrow \mathcal{C}$ having the preceding property. Of the first type is $(\partial/\partial x)_{x=0}$, and of the second type is $\partial_x = \partial/\partial x$.) The del operator ∇ inherits this property, e.g. in the form

$$\nabla(fg) = (\nabla f)g + f(\nabla g),$$

i.e.

$$\mathbf{grad}(fg) = (\mathbf{grad}f)g + f(\mathbf{grad}g) = g\,\mathbf{grad}f + f\,\mathbf{grad}g.$$

In these formulas, everything looks as if the del operator acts first on f (treating g as a constant) and then acts on g (treating f as a constant). Borrowing a methodology from G. Juvet, I shall occasionally indicate a variable by an arrow under it, writing thus

$$\nabla(fg) = \nabla(\underset{\uparrow}{f}g) + \nabla(f\underset{\uparrow}{g}) = g\nabla f + f\nabla g,$$

understanding that the del operator only acts on the following functions *unless otherwise stated!* The formula just found reads

$$\mathbf{grad}(fg) = g\,\mathbf{grad}f + f\,\mathbf{grad}g.$$

Without striving towards completeness (the reader should complete our formulas as an exercise!), let us indicate a few more examples of the derivation property. For example, for $\mathcal{C}^1$ scalar functions f and vector fields **u**, we have

$$\nabla\wedge(f\mathbf{u}) = \nabla\wedge(\underset{\uparrow}{f}\mathbf{u}) + \nabla\wedge(f\underset{\uparrow}{\mathbf{u}}) = \underset{\uparrow}{\nabla f}\wedge(\mathbf{u}) + f\nabla\wedge(\underset{\uparrow}{\mathbf{u}}) = -\mathbf{u}\wedge\nabla f + f\nabla\wedge\mathbf{u}.$$

This formula reads

$$\mathbf{curl}(f\mathbf{u}) = -\mathbf{u}\wedge\mathbf{grad}f + f\,\mathbf{curl}\,\mathbf{u}.$$

It could of course be verified in components, expanding for example the first one

$$\partial_y(fu_3) - \partial_z(fu_2) = u_3\partial_y f + f\partial_y u_3 - u_2\partial_z f - f\partial_z u_2 =$$

$$= -(u_2 \partial_z f - u_3 \partial_y f) + f(\partial_y u_3 - \partial_z u_2)$$

which is indeed the first component of the announced expression. The advantage of using a unified formalism for the operators **grad**, div, **curl** and Grad should be obvious by now.

A last example will show how to proceed in a somewhat less trivial situation. Let us try to find a formula for the gradient of the scalar product of two $\mathcal{C}^1$ vector fields $\mathbf{u}$ and $\mathbf{v}$, namely for $\nabla(\mathbf{uv})$. We can certainly write

$$\nabla(\mathbf{uv}) = \nabla(\underset{\uparrow}{\mathbf{u}}\mathbf{v}) + \nabla(\mathbf{u}\underset{\uparrow}{\mathbf{v}}),$$

but now, we are stuck... until we think of using the **Gibbs formula** for double cross products

$$\mathbf{a}\wedge(\mathbf{b}\wedge\mathbf{c}) = \mathbf{b}(\mathbf{ac}) - \mathbf{c}(\mathbf{ab}).$$

The term $\nabla(\underset{\uparrow}{\mathbf{u}}\mathbf{v})$ can be identified to $\mathbf{b}(\mathbf{ca})$ and thus

$$\mathbf{b}(\mathbf{ca}) = \mathbf{c}(\mathbf{ab}) + \mathbf{a}\wedge(\mathbf{b}\wedge\mathbf{c}) \text{ gives } \nabla(\underset{\uparrow}{\mathbf{u}}\mathbf{v}) = \mathbf{v}\wedge(\nabla\wedge\underset{\uparrow}{\mathbf{u}}) + (\mathbf{v}\nabla)\underset{\uparrow}{\mathbf{u}}\ .$$

Now, the term $(\mathbf{v}\nabla)\underset{\uparrow}{\mathbf{u}} = (\nabla \otimes \underset{\uparrow}{\mathbf{u}})\mathbf{v}$ is recognized to be $(\mathrm{Grad}\underset{\uparrow}{\mathbf{u}})(\mathbf{v})$, where the derivations appearing in ∇ only act on the vector field $\mathbf{u}$. The second term is treated similarly. Summarizing, we have found the formula

$$\mathbf{grad}(\mathbf{uv}) = \mathbf{v}\wedge(\mathbf{curl}\ \mathbf{u}) + \mathrm{Grad}\mathbf{u}(\mathbf{v}) + \mathbf{u}\wedge(\mathbf{curl}\ \mathbf{v}) + \mathrm{Grad}\mathbf{v}(\mathbf{u}).$$

8.4 ITERATION OF THE DEL OPERATOR

The del operator can be iterated twice on $\mathcal{C}^2$ fields. For example, if f is a $\mathcal{C}^2$ scalar field, **grad**f $= \nabla f$ is $\mathcal{C}^1$ and we can compute both $\nabla\nabla f = \mathrm{div}\,\mathbf{grad} f = \Delta f$ (Laplacian of f already encountered in 7.7) and $\nabla\wedge(\nabla f) = \mathbf{curl}(\mathbf{grad} f)$. Similarly, if $\mathbf{u}$ is a $\mathcal{C}^2$ vector field, div$\mathbf{u}$ and **curl u** are $\mathcal{C}^1$ fields and we can compute

$$\nabla(\nabla\mathbf{u}) = \mathbf{grad}\ \mathrm{div}\mathbf{u} \quad \text{(vector field)},$$

$$\nabla(\nabla\wedge\mathbf{u}) = \mathrm{div}\ \mathbf{curl}\ \mathbf{u} \quad \text{(scalar field)},$$

$$\nabla\wedge(\nabla\wedge\mathbf{u}) = \mathbf{curl}\ \mathbf{curl}\ \mathbf{u} \quad \text{(vector field)}.$$

Lemma. a) For any $\mathcal{C}^2$ scalar field f we have

$$\mathbf{curl}(\mathbf{grad} f) = 0.$$

b) For any $\mathcal{C}^2$ vector field $\mathbf{u}$ we have

$$\mathrm{div}\ \mathbf{curl}\ \mathbf{u} = 0. \qquad \square$$

Proof. Observe that the assertion of a) can be written in the form

$$\nabla\wedge(\nabla f) = 0.$$

Formally, ∇f is parallel to ∇ and this implies that the cross product with ∇ vanishes. Completely explicitely,

$$\nabla\wedge(\nabla f) = \begin{vmatrix} \mathbf{e}_1 & \mathbf{e}_2 & \mathbf{e}_3 \\ \partial_x & \partial_y & \partial_z \\ \partial_x f & \partial_y f & \partial_z f \end{vmatrix}$$

has first component

$$\partial_y \partial_z f - \partial_z \partial_y f = 0$$

by Schwarz theorem (7.1). The other components are obtained by cyclic permutation : they also vanish. The parallelism invoked at the beginning of the proof was of course purely formal. But the vanishing of the determinant occurs for the same formal reason : components commute with each other and compensations occur as with genuine scalar components of vectors.

Let us now prove b) : div **curl u** = 0. As before, the formal reason is clear : the mixed product $\nabla(\nabla\wedge\mathbf{u})$ has two parallel terms and thus furnishes zero volume. Less formally, we would have to compute the following expression

$$\partial_x(\partial_y u_3 - \partial_z u_2) + \partial_y(\partial_z u_1 - \partial_x u_3) + \partial_z(\partial_x u_2 - \partial_y u_1)$$

(in which the three terms are obtained by a cyclic permutation). Three applications of the Schwarz theorem also show that it vanishes. ∎

Warning. One should emphasize that in spite of the simplicity of the parallelism (or orthogonality) arguments applied to the formal del vector, careless use of them can lead one into pitfalls...

Here is an interesting **example**. If **u** is a $\mathcal{C}^1$ vector field, one can form the scalar field **u**·(**curl u**) = **u**·($\nabla\wedge$**u**). One could argue that **u** ⊥ $\nabla\wedge$**u** implies the vanishing of **u**·(**curl u**)... In fact, the vector field $\nabla\wedge$**u** = **curl u** is not always orthogonal to **u** and thus **u**·(**curl u**) is often different from zero. What happens? The ∇ operator only acts on the components of the field **u** that follows and the components of the first field **u** do not commute with the components of ∇. Explicitely, the scalar product in question is given by

$$u_1(\partial_y u_3 - \partial_z u_2) + u_2(\partial_z u_1 - \partial_x u_3) + u_3(\partial_x u_2 - \partial_y u_1)$$

(an expression looking very much like the one occuring in the proof of part b) of the above lemma). But now

$$u_1 \partial_y u_3 - u_3 \partial_y u_1 \quad \text{does not vanish in general !}$$

Non trivial iteration of the del operator has already produced the Laplacian

$$\Delta f = \nabla(\nabla f) = \mathrm{div}(\mathbf{grad} f),$$

and we remind the reader that solutions f of $\Delta f = 0$ are called harmonic functions. Another interesting case of iteration is furnished by

$$\nabla \wedge (\nabla \wedge \mathbf{u}) = \mathbf{curl\ curl\ u}.$$

Using Gibbs formula, we can transform this as follows

$$\nabla \wedge (\nabla \wedge \mathbf{u}) = \nabla(\nabla \mathbf{u}) - (\nabla\nabla)\mathbf{u} = \mathbf{grad}(\mathrm{div}\mathbf{u}) - \Delta\mathbf{u}.$$

The meaning of the Laplacian of a vector field is the following

$$\Delta \mathbf{u} = \begin{pmatrix} \Delta u_1 \\ \Delta u_2 \\ \Delta u_3 \end{pmatrix}.$$

One can (and one should) verify this in components...

8.5 POTENTIALS : DEFINITIONS

Vector fields **u** of the form $\mathbf{u} = \mathbf{grad} f$ (for some $f \in \mathcal{C}^1$) are said to **derive from a potential**. Physicists prefer to write

$$\mathbf{u} = -\mathbf{grad}\varphi$$

and call this scalar field φ the potential producing **u**. Obviously, the choice of sign has no mathematical implication and we choose to ignore it. A *necessary condition* for a vector field **u** to derive from a potential is

$$\mathbf{curl\ u} = \mathbf{curl}(\mathbf{grad} f) = 0$$

(part a) of the lemma of 8.4).

Vector fields **v** of the form $\mathbf{u} = \mathbf{curl\ v}$ (for some $\mathcal{C}^1$ vector field **v**) are said to **derive from a vector potential**. Again, physicists would prefer to denote this situation by $\mathbf{v} = \mathbf{curl\ A}$ to remember the situation for the magnetic field (this application is treated in a subsequent chapter). A *necessary condition* for a vector field to derive from a vector potential is given by the part b) of the lemma of (8.4).

$$\mathrm{div}\ \mathbf{u} = \mathrm{div}(\mathbf{curl\ u}) = 0.$$

Potentials play the role of *primitive* for vector fields and the *necessary conditions* given are by no means *sufficient* for their existence. (In certain regions however, we shall see that they are.)

An example. Let us consider the horizontal vector field

$$\mathbf{u}(\mathbf{r}) = \rho^{-2}\begin{pmatrix} -y \\ x \\ 0 \end{pmatrix} \qquad \text{where } \rho^2 = x^2 + y^2 \text{ (independent of z)}$$

which is well defined outside the Oz-axis in $\mathbb{R}^3$. The **curl** of this vector field vanishes (outside the Oz-axis) since the first two components of the rotational of any horizontal field with components independent of z vanish (why?), whereas the third one is

$$\partial_x[x/(x^2+y^2)] - \partial_y[-y/(x^2+y^2)] =$$

$$= (x^2+y^2-2x^2 + x^2+y^2-2y^2)/(x^2+y^2)^2 = 0 .$$

One can write

$$\mathbf{u}(\mathbf{r}) = \mathbf{grad}(\arctan y/x)$$

in regions where it is possible to choose a continuous (uniform) determination of the angle $\varphi = \arctan y/x$. For example, this will be the case in the half space $x > 0$, or in the complement of $\{(x,0,z) : x \le 0\}$ in $\mathbb{R}^3$.

Remark. A vector field **u** can very well derive from both a scalar potential f and a vector potential **v**. It is easy to write a condition for this. Assume $\mathbf{u} = \mathbf{grad}f$ also derives from a vector potential so that $\mathrm{div}\mathbf{u} = 0$. We find

$$\Delta f = \mathrm{div}\mathbf{grad}f = 0,$$

namely, the potential f is a harmonic function. Symmetrically, if $\mathbf{u} = \mathbf{curl}\ \mathbf{v}$ also derives from a scalar potential, we shall have $\mathbf{curl}\ \mathbf{u} = 0$, namely

$$\mathbf{curl}\ \mathbf{curl}\ \mathbf{v} = 0$$

i.e. $\mathbf{grad}\ \mathrm{div}\mathbf{v} - \Delta\mathbf{v} = 0$ and the vector potential **v** will satisfy

$$\Delta\mathbf{v} = \mathbf{grad}\ \mathrm{div}\mathbf{v}.$$

8.6 POTENTIALS : EXISTENCE

The sufficiency of the condition $\mathbf{curl}\ \mathbf{u} = 0$ for **u** to derive from a scalar potential depends very much on the nature of the region in which the vector field **u** is defined. Thus, we have to depart from our tacit convention of only considering vector fields defined in the whole space.

A **star-shaped** region $S \subset \mathbb{R}^3$ is a union of segments with origin at $0 \in \mathbb{R}^3$. Such a region contains 0, and whenever it contains a point P, it must contain all the segment $\overline{OP}$ linking the origin to P. Balls, parallelepipeds, half spaces are examples of star-shaped domains (if they contain the origin). More generally, if a region R contains a point A, we shall say that R is **star-shaped with respect to** A if it is a union of segments having origin at the point A (cf.Fig.)

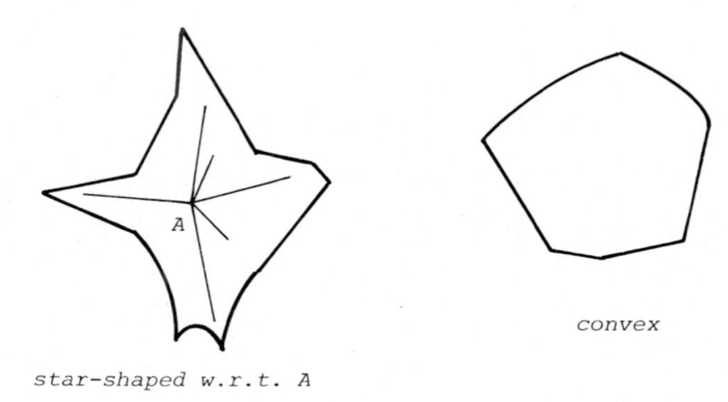

Fig.8.3

A non empty region R which is star-shaped with respect to all its points is called **convex**. Equivalently, convex regions K are characterized by the following property :

For each couple A, B *of* K, *the segment* $\overline{AB} = [A,B]$ *is in* K :

$$A \in K,\ B \in K \quad \Longrightarrow \quad \overline{AB} \subset K.$$

By segment $\overline{AB}$, we mean all points of the line going through A and B *which are situated between* A *and* B. Since the line passing through A and B can be parameterized by $t \longmapsto tA + (1-t)B$ (assume that $A \neq B$ and recall that A and B are represented by column vectors with 3 components so that $A = \overrightarrow{OA},\ldots$). The segment $\overline{AB}$ is now described by

$$\overline{AB} = [A,B] = \{tA + (1-t)B : 0 \le t \le 1\} =$$
$$= \{\alpha A + \beta B : \alpha+\beta = 1,\ \alpha\ge 0,\ \beta\ge 0\} .$$

It is obvious that any intersection of convex sets is still convex and that half spaces are convex. In particular, any intersection of half spaces is convex. Balls, cylinders, parallelepipeds, are examples of convex sets.

Theorem. Let **u** be a vector field defined in some open region $R \subset \mathbb{R}^3$ and let S be some open star shaped sub-region $S \subset R$. Then

a) If **curl u** = 0 in S, there exists a scalar function f in S with $\mathbf{u} = \mathbf{grad} f$ *in* S (i.e. $\mathbf{u}|_S = \mathbf{grad} f$),

b) If div **u** = 0 in S, there exists a vector potential **v** for **u** in S : **u** = **curl v** *in* S (i.e. $\mathbf{u}|_S = \mathbf{curl}\ \mathbf{v}$). □

Corollary. If R is open and convex, the necessary conditions

$$\mathbf{curl}\ \mathbf{u} = 0 \text{ in } R \text{ (resp. div } \mathbf{u} = 0 \text{ in } R)$$

are sufficient for the existence of a scalar (resp. vector) potential for **u** in R. □

Proof of part a) **of the theorem**. For $\mathbf{r} \in S$ we let

$$f(\mathbf{r}) = \int_0^1 \mathbf{r}\cdot\mathbf{u}(t\mathbf{r})dt = \int_0^1 [xu_1(t\mathbf{r}) + yu_2(t\mathbf{r}) + zu_3(t\mathbf{r})]dt\ .$$

This function f is well defined since the points $t\mathbf{r}$ belong to S. Compute now the gradient of this function f. It will be enough to compute the first partial derivative $\partial_x f = D_1 f$ of this function. Computing the derivative by differentiation under the integral sign, we find

$$D_1 f(x,y,z) = \int_0^1 u_1(t\mathbf{r})dt + \int_0^1 [xD_1u_1(t\mathbf{r}) + yD_1u_2(t\mathbf{r}) + zD_1u_3(t\mathbf{r})]tdt$$

and since **curl u** = 0 by assumption, $D_1u_2 = D_2u_1$, $D_1u_3 = D_3u_1$ lead to

$$D_1 f(\mathbf{r}) = \int_0^1 u_1(t\mathbf{r})dt + \int_0^1 [xD_1u_1(t\mathbf{r}) + yD_2u_1(t\mathbf{r}) + zD_3u_1(t\mathbf{r})]tdt.$$

We recognize now the derivative of the function

$$t \longmapsto u_1(tx,ty,tz)$$

which is indeed (by the fundamental theorem of calculus 5.4)

$$xD_1u_1(t\mathbf{r}) + yD_2u_1(t\mathbf{r}) + zD_3u_1(t\mathbf{r})\ .$$

Integrating by parts, we get

$$D_1 f(\mathbf{r}) = \int_0^1 u_1(t\mathbf{r})dt + \int_0^1 \frac{d}{dt}(u_1(t\mathbf{r}))\cdot tdt =$$

$$= \int_0^1 u_1(t\mathbf{r})dt + \Big[u_1(t\mathbf{r})\cdot t\Big]_0^1 - \int_0^1 u_1(t\mathbf{r})dt\ .$$

There only remains

$$D_1 f(\mathbf{r}) = \Big[u_1(t\mathbf{r})\cdot t\Big]_0^1 = u_1(\mathbf{r}).$$ ■

Instead of proving part b) of the theorem, we defer to an exercise giving a formula for vector potentials in certain domains (the case of star-shaped domains will also be treated in the part on differential forms).

8.7 POTENTIALS : UNIQUENESS

When a vector field $\mathbf{u}$ derives from a potential, say $\mathbf{u} = \mathbf{grad}f$ one can ask whether the function f is uniquely determined (up to a constant). This question can be formulated in the following way : when $\mathbf{grad}f_1 = \mathbf{grad}f_2$, can we infer that $f_1 = f_2 + c$? This will indeed be the case in connected domains since the difference $f_1 - f_2$ will have vanishing partial derivatives

$$\partial_x(f_1 - f_2) = 0,\ \partial_y(f_1 - f_2) = 0,\ \partial_z(f_1 - f_2) = 0,$$

hence $f_1 - f_2 = \text{const.}$ in each connected component of the region considered.

Similarly, if $\mathrm{Grad}\mathbf{u}_1 = \mathrm{Grad}\mathbf{u}_2$, we shall infer that the components of $\mathbf{u}_1$ and $\mathbf{u}_2$ will have same gradient and thus

$$(\mathbf{u}_1)_i = (\mathbf{u}_2)_i + c_i.$$

This proves that $\mathbf{u}_1 = \mathbf{u}_2 + \mathbf{c}$ (with a vector constant $\mathbf{c} = (c_i)$) in each connected component of the domain considered.

From an equality

$$\mathbf{curl}\ \mathbf{u}_1 = \mathbf{curl}\ \mathbf{u}_2$$

one can infer $\mathbf{curl}(\mathbf{u}_1 - \mathbf{u}_2) = 0$, hence $\mathbf{u}_1 - \mathbf{u}_2$ derives from a potential f in any convex (star-shaped, or contractible) region. In such a region we shall thus be able to write

$$\mathbf{u}_1 = \mathbf{u}_2 + \mathbf{grad}f$$

where the scalar potential f is determined up to addition of a constant (by the preceding discussion).

Finally

$$\mathrm{div}\ \mathbf{u}_1 = \mathrm{div}\ \mathbf{u}_2$$

implies $\mathrm{div}(\mathbf{u}_1 - \mathbf{u}_2) = 0$ and $\mathbf{u}_1 - \mathbf{u}_2$ derives from a vector potential in any convex (star-shaped, or contractible) region. In such a region, we shall have

$$\mathbf{u}_1 = \mathbf{u}_2 + \mathbf{curl}\ \mathbf{v}$$

for some vector potential $\mathbf{v}$ (determined up to addition of a field $\mathbf{grad}f$ by the preceding discussion!).

CHAPTER 9
CURVILINEAR COORDINATES

9.1 OBLIQUE BASES IN EUCLIDEAN SPACES

The space $\mathbb{R}^n$ is equipped with the canonical *scalar product*

$$\mathbf{x}\cdot\mathbf{y} = \sum x_i y_i \quad \text{if } \mathbf{x} = (x_i),\ \mathbf{y} = (y_i).$$

Any finite dimensional real vector space with a scalar product is called a **Euclidean space**, so that $\mathbb{R}^n$ is the *prototype* or *model* of a Euclidean space.

In this section, we review a few results valid in Euclidean spaces, and in particular, we deal with not necessarily orthonormal bases: let us call them **oblique bases** to emphasize the fact that the canonical basis is no longer the only one in circulation. But of course, results valid for oblique bases will also be valid for orthonormal bases !

Let us fix an oblique basis $(\mathbf{e}_i)_{1\le i\le n}$ of $\mathbb{R}^n$: each vector $\mathbf{e}_i$ is a column vector

$$\mathbf{e}_i = \begin{pmatrix} e_i^1 \\ \vdots \\ e_i^n \end{pmatrix} = {}^t(e_i^1, \dots, e_i^n) \in \mathbb{R}^n.$$

As the reader has observed, we put in upper position the indices of the components of these vectors, hoping that they will not be confused with powers (the reason for this is —at first— purely formal: we want to avoid double indices, but we shall see other advantages of this notation). To be consistent, we shall also denote by upper indices the components of any vector $\mathbf{x} \in \mathbb{R}^n$ *in the basis* $(\mathbf{e}_i)$

$$\mathbf{x} = x^1\mathbf{e}_1 + \dots + x^n\mathbf{e}_n = \sum x^i\mathbf{e}_i.$$

When the basis $(\mathbf{e}_i)$ is **orthonormal** i.e. when

$$\|\mathbf{e}_i\| = 1 \quad \text{and } \mathbf{e}_i\cdot\mathbf{e}_j = 0 \text{ for } i \neq j$$

—a situation which is best abbreviated by $\mathbf{e}_i\cdot\mathbf{e}_j = \delta_{ij}$ with the **Kronecker symbol** $\delta_{ii} = 1$ and $\delta_{ij} = 0$ for $i\neq j$ — it is easy to determine the value of the components x^i of $\mathbf{x}$ in the basis $(\mathbf{e}_i)$. A simple scalar product will furnish them according to

$$\mathbf{x}\cdot\mathbf{e}_j = \sum_i x^i\mathbf{e}_i\cdot\mathbf{e}_j = \sum_i x^i\delta_{ij} = x^j.$$

When, more generally, the vectors of the basis $(\mathbf{e}_i)$ are only normed without being orthogonal, the scalar products $\mathbf{x}\cdot\mathbf{e}_j$ are no more the components of $\mathbf{x}$: they represent the length of the projection of $\mathbf{x}$ on the line $\mathbb{R}\mathbf{e}_j$ spanned by the vector $\mathbf{e}_j$. These scalar products are quite generally called **covariant components** of $\mathbf{x}$. A justification of this terminology will appear later. These covariant components will be denoted by lower indices. If we define the n^2 numbers $g_{ij} = \mathbf{e}_i \cdot \mathbf{e}_j$, we shall be able to write

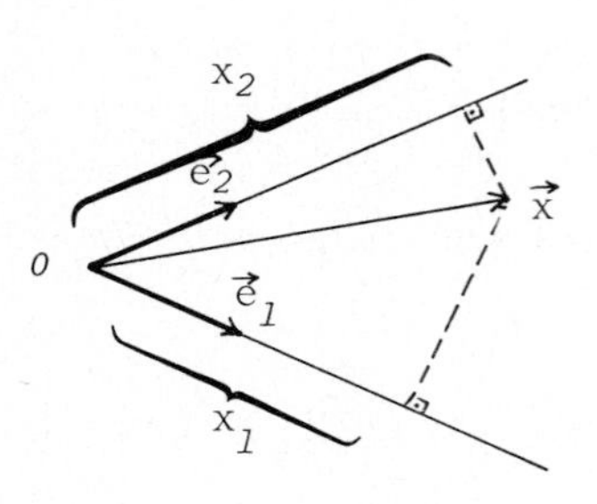

Fig.9.1

$$x_j = \mathbf{x} \cdot \mathbf{e}_j = \sum_i x^i \mathbf{e}_i \cdot \mathbf{e}_j = \sum_i x^i g_{ij}.$$

This is a linear system of n unknown quantities x^i when the g_{ij} and x_j have been computed. Let us show that this system has a unique solution.

Lemma. The determinant $g = \det(g_{ij})$ does not vanish. More precisely, $g > 0$. □

Proof. By definition,

$$g_{ij} = \mathbf{e}_i \cdot \mathbf{e}_j = \sum_{1 \le k \le n} e_i^k \, e_j^k \, .$$

This sum is the entry situated in the line i and column j of the matrix product ${}^t E \circ E$ of $E = (e^i_j)$ with its transpose ${}^t E = (e'^i_j)$, where $e'^i_j = e^j_i$. Consequently,

$$\det(g_{ij}) = \det({}^t E \circ E) = \det({}^t E) \cdot \det(E) = (\det E)^2 > 0.$$

Recall indeed that

- ▷ $\det({}^t E) = \det(E)$,
- ▷ $\det(E) \neq 0$ since $(\mathbf{e}_i)$ is a basis. ■

This lemma proves that the matrix (g_{ij}) is invertible. Let (g^{ij}) denote its inverse. It is computed according to the formula

$$g^{ij} = (-1)^{i+j} \, G^{ji}/g$$

where G^{ji} represents the determinant of the minor of g_{ji} in the matrix (g_{ij}). With this notation for the inverse matrix, the system of linear equations for the components of x is solved by

$$x^i = \sum_{1 \le j \le n} g^{ij} x_j \, .$$

The determinant $g = \det(g_{ij})$ is called Gram determinant of the basis (e_i).

This is a good place to introduce the **Einstein summation convention** :

any index which appears both in upper and lower position is to be summed over and the resulting expression is independent of this index (summation index).

For example, with this convention we shall write $g^{ij}x_j$ in place of $\sum_{1\le j\le n} g^{ij}x_j$. Similarly

$$g_{ij}x^j \quad \text{means} \quad \sum_{1\le j\le n} g_{ij}x^j,$$

$$a^i_i = a^j_j \quad \text{means} \quad \sum_{1\le i\le n} a^i_i = \sum_{1\le j\le n} a^j_j = \mathrm{Tr}(a^i_j).$$

With this convention, formulas acquire a greater simplicity and they are easily remembered (or recovered). Since any index appearing in upper and lower position is automatically summed over, the result is independent of the name of this index and formally, this index is cancelled —as in a fraction— in the result. For example in $g_{ij}x^i$, the index i cancels out and the result is simply x_j :

$$g_{ij}x^i = x_j .$$

A nice homogeneity of notations is attained. To preserve it, we would write the relation between (g_{ij}) and its inverse (g^{ij}) as

$$g_{ik}g^{kj} = \delta^j_i .$$

In particular, the Kronecker symbol will indifferently be written

$$\delta_{ij} = \delta^i_j = \delta^{ij} \quad (= 0 \text{ if } i\ne j, \quad = +1 \text{ if } i=j)$$

according to the situation (the summation convention will impose a certain position for the index to be summed over, and the position of the index in the result will suggest the position of the other!). The two relations

$$g_{ij}x^i = x_j , \quad g^{ij}x_i = x^j$$

formally mean that the matrix (g_{ij}) and its inverse (g^{ij}) allow respectively to *lower* or *raise* indices.

9.2 COVARIANT COMPONENTS AND THE GRADIENT

Let us show how covariant components appear very naturally in our context. Let f be a scalar field in $\mathbb{R}^2$: imagine that a few points where $f = 0$ have been plotted, some points where $f = 1$ similarly, and thus a few level curves of f have been drawn. Let P be a point where this field has to be examined so that we introduce a coordinate system $\mathbf{e}_1 = \mathbf{e}_u$, $\mathbf{e}_2 = \mathbf{e}_v$ with origin P in $\mathbb{R}^2$. The components of a vector $\mathbf{x}$ will be denoted by $u = x^1$ and $v = x^2$

$$\mathbf{x} = x^1\mathbf{e}_1 + x^2\mathbf{e}_2 = u\mathbf{e}_u + v\mathbf{e}_v .$$

Simultaneously, the function f will be represented by a certain *expression* in u and v

$$f(Q) = \varphi(u,v) \quad \text{if} \quad \overrightarrow{OQ} = \mathbf{x} = u\mathbf{e}_u + v\mathbf{e}_v .$$

(At this point, let us observe that if $\overrightarrow{OQ} = \mathbf{x} = \begin{pmatrix} x \\ y \end{pmatrix} \in \mathbb{R}^2$, we would write $f(Q) = f(x,y)$, but this function f is certainly different from the function φ introduced above in general...) The **question** is now

> *How can we compute the gradient of* f *in its representation in coordinates* u *and* v *given by* φ *?*

More abruptly, can we say that the components of this gradient are the *partial derivatives of* φ and use

$$\frac{\partial\varphi}{\partial u}\cdot\mathbf{e}_u + \frac{\partial\varphi}{\partial v}\cdot\mathbf{e}_v \quad ?$$

The answer is definitely **NO !** A geometrical reason is easily given (cf.Fig.). Assume that the oblique basis has a first $\mathbf{e}_u$ tangent to a level curve of f. The directional derivative of f (i.e.the partial derivative $\frac{\partial\varphi}{\partial u}$) will vanish whatever choice we make for second basis vector $\mathbf{e}_v$ and thus

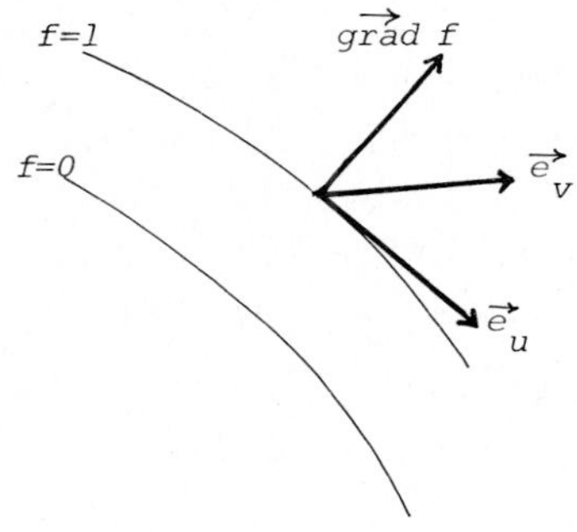

$$\frac{\partial\varphi}{\partial u}\cdot\mathbf{e}_u + \frac{\partial\varphi}{\partial v}\cdot\mathbf{e}_v = \frac{\partial\varphi}{\partial v}\cdot\mathbf{e}_v .$$

The direction of this vector varies with the choice of $\mathbf{e}_v$!

Fig.9.2

It cannot represent the gradient of f, which is well defined, orthogonal to the level curves of f ! Chemists and physicists (in alphabetical order!) are used to being more careful : when they speak of a partial derivative, they indicate the variable with respect to which it is computed *and the variable(s) which should be kept constant*. Thus, instead of $\frac{\partial\varphi}{\partial u}$ which *simply has no meaning before the other* v*-coordinate has been chosen*, they write more accurately

$$\left(\frac{\partial\varphi}{\partial u}\right)_v \quad \text{or} \quad \left(\frac{\partial\varphi}{\partial u}\right)_{v=\text{const.}}$$

The *directional derivative in the direction of* $\mathbf{e}_u$ however is well defined before the choice of the second basis vector $\mathbf{e}_v$. All these distinctions and relations are the cause of many misconceptions and we have even described them in a negative way to insist on the *necessity of using proper concepts*. More comments for users will be given in 9.6 below.

In particular, the functions f and φ have to be carefully distinguished (and here, chemists and physicists are less careful...). We can —and should— write $f(x,y) = \varphi(u,v)$ whenever

$$\begin{pmatrix} x \\ y \end{pmatrix} = \mathbf{x} = u\mathbf{e}_u + v\mathbf{e}_v = \overrightarrow{OQ} \in \mathbb{R}^2.$$

But now, the gradient of f at P is *a linear form* : it is the tangent linear form to f at P. This form gives the linear increase of f (first order approximation) in all directions. In particular, $\mathbf{gradf}\cdot\mathbf{e}_u$ represents the linear increase of f in the direction $\mathbf{e}_u$. This scalar product *is the covariant component of* **gradf**

$$\mathbf{gradf}\cdot\mathbf{e}_u = (\mathbf{gradf})_u = (\mathbf{gradf})_1$$

and similarly for the scalar product with the second basis vector $\mathbf{e}_v$. The *components* of **gradf** are linked with the *covariant components according to the theory that has just been developed* :

$$\mathbf{gradf} = (\mathbf{gradf})^i\mathbf{e}_i \quad \text{with} \quad (\mathbf{gradf})^i = g^{ij}(\mathbf{gradf})_j .$$

(Recall that these expressions represent *sums* over the index j : to retain the validity of the convention summation, we shall always interpret an upper index in a denominator as a *lower* index, thus considering that the index j in $\partial f/\partial x^j$ is a *lower* index !)

9.3 CHANGE OF BASIS

To justify the terminology, we consider a change of basis and look for the transformation formula of the *covariant* components of a given vector **x**. For this, let us come back to the general context of an oblique basis $(\mathbf{e}_i) \subset \mathbb{R}^n$ and let us choose a second arbitrary basis (ε_i) of $\mathbb{R}^n$. To distinguish quantities relating to the "new" basis (ε_i) from similar quantities relating to the "old" basis $(\mathbf{e}_i)$ we shall adopt similar Greek letters for the new basis.

For example, the components of a vector $\mathbf{x} \in \mathbb{R}^n$ in the old basis $(\mathbf{e}_i)$ are called x^i, and the components of the same vector $\mathbf{x}$ in the new basis are called ξ^i. By definition, we have

$$\mathbf{x} = x^i\mathbf{e}_i = \xi^i\varepsilon_i \quad (= \xi^j\varepsilon_j).$$

The same notations will be applied for covariant components resp. defined by

$$x_i = \mathbf{x}\cdot\mathbf{e}_i \ , \qquad \xi_i = \mathbf{x}\cdot\varepsilon_i \ .$$

The relation between new and old bases will be taken in the form

$$\varepsilon_j = s^i_j\mathbf{e}_i \quad \text{or conversely} \quad \mathbf{e}_j = t^i_j\varepsilon_i.$$

In other words, the matrix $S = (s^i_j)$ contains the information concerning the transformation $(\mathbf{e}_i) \longmapsto (\varepsilon_i)$ and its inverse

$$T = (t^i_j) = S^{-1}$$

contains the information concerning the reverse direction.

Proposition. The following relations hold

$$\xi_i = s^j_i x_j \qquad (\textit{co-variance}),$$

$$\xi^i = t^i_j x^j \qquad (\textit{contra-variance}). \qquad \square$$

Proof. Both relations follow from an easy computation once notations have been clearly defined. For example, we can write

$$\xi_i = \mathbf{x}\cdot\varepsilon_i = \mathbf{x}\cdot s^j_i\mathbf{e}_j = s^j_i(\mathbf{x}\cdot\mathbf{e}_j) = s^j_i x_j.$$

On the other hand, in

$$\mathbf{x} = \xi^i\varepsilon_i = x^j\mathbf{e}_j = x^j t^i_j\varepsilon_i$$

the coefficients of the basis vectors ε_i must be the same :

$$\xi^i = t^i_j x^j. \qquad \blacksquare$$

This proposition shows that the *covariant components* of the vector $\mathbf{x}$ are transformed with the matrix S expressing the change of basis $(\mathbf{e}_i) \longmapsto (\varepsilon_i)$ whereas the *true components —also called contravariant* components for this reason— transform according to the inverse matrix $T = S^{-1}$.

For practice, let us also derive the formula for the transformation of the matrix representation of a linear operator in $\mathbb{R}^n$. Let $f : \mathbb{R}^n \longrightarrow \mathbb{R}^n$ be linear and

$$A = (a^i_j) = \mathrm{Mat}_{(\mathbf{e}_i)}(f) \quad , \quad A' = (\alpha^i_j) = \mathrm{Mat}_{(\varepsilon_i)}(f).$$

By definition, the matrix of f in the new basis has in its j column the components of $f(\varepsilon_j)$ (in the basis (ε_i)) :

$$f(\varepsilon_j) = \alpha^i_j\varepsilon_i.$$

Let us compute them :

$$f(\varepsilon_j) = f(s^i_j \mathbf{e}_i) = s^i_j f(\mathbf{e}_i) = s^i_j (a^k_i \mathbf{e}_k) = s^i_j (a^k_i t^\ell_k \varepsilon_\ell).$$

We have used the definition of the matrix A of f in the basis $(\mathbf{e}_i)$ and in order to be able to identify the coefficient of ε_i in the second expression, we have to rename summation indices (the last expression is really a sum over i, k and ℓ). Let us write

$$f(\varepsilon_j) = s^\ell_j (a^k_\ell t^i_k \varepsilon_i)$$

and compare it to

$$f(\varepsilon_j) = \alpha^i_j \varepsilon_i.$$

We find

$$\alpha^i_j = t^i_k a^k_\ell s^\ell_j \quad \text{(summations over k and } \ell\text{)}.$$

In this expression, we see that the upper index of a is responsible for an occurence of the matrix T (contra-variance) and the lower index of a introduces the apparition of the matrix S (co-variance).

In matrix form, the relation just found is simply

$$A' = T \circ A \circ S = S^{-1} \circ A \circ S$$

i.e.

$$\mathrm{Mat}_{(\varepsilon_i)}(f) = S^{-1} \circ \mathrm{Mat}_{(\mathbf{e}_i)}(f) \circ S.$$

One should not exaggerate the importance of this formalism, but it provides very useful simplifications when properly understood !

9.4 CURVILINEAR COORDINATES : TERMINOLOGY

Curvilinear coordinates in some open domain D of $\mathbb{R}^n$ are introduced by means of a diffeomorphism $F : U \longrightarrow D$ (U also being an open set in $\mathbb{R}^n$). As the example of polar coordinates

$$\begin{cases} x = \rho \cos\varphi \\ y = \rho \sin\varphi \end{cases}$$

already suggests, we are more interested in the *target* D of the map F than in its source U. Indeed, to get an idea of the polar coordinates, we always make a picture of circles and radii in the (x,y)-space. We proceed similarly in general.

We shall adopt the following notations for the introduction of curvilinear coordinates in some open region $D \subset \mathbb{R}^n$

$$F : U \longrightarrow D \text{ , (where } U = f^{-1}(D))$$

is a mapping

$$\xi = {}^t(\xi^1, \dots ,\xi^n) \longmapsto \mathbf{x} = {}^t(x^1, \dots ,x^n)$$

where the $x^j = f^j(\xi^1, \dots ,\xi^n) = f^j(\xi)$ are some given functions.

The **coordinate curves** are by definition the parametric curves obtained by letting only one parameter ξ^j vary :

$$\mathcal{C}_j : t \longmapsto F(\xi^1,\ldots,\xi^j{=}t,\ldots,\xi^n).$$

These curves are the images by F of the parallel to the coordinate lines.

The **coordinate surfaces** are by definition the parametric surfaces obtained by letting only two parameters ξ^i and ξ^j $(i \neq j)$ vary. These surfaces

$$\mathcal{S}_{ij} : (u,v) \longmapsto F(\xi^1,\ldots,\xi^i{=}u,\ldots,\xi^j{=}v,\ldots,\xi^n)$$

are the images of the planes parallel to coordinate planes (generated by two canonical basis vectors).

Through each point $P \in D$, we should thus make a sketch of the coordinate curves $\mathcal{C}_j$ going through P, and of the coordinate surfaces $\mathcal{S}_{ij}$ through P.

The coordinate curves $\mathcal{C}_i$ going through a given point $P \in D$ will have initial tangent vectors $\mathbf{e}_i(P)$ making up a basis of $\mathbb{R}^n$: in components, calling $\xi_P = F^{-1}(P) \in U$ the *representation of* $P \in U$

$$\mathbf{e}_i(P) = (\partial \mathbf{x}/\partial \xi^i)(\xi_P) = \begin{pmatrix} (\partial f^1/\partial \xi^i)(\xi_P) \\ \cdots\cdots \\ (\partial f^n/\partial \xi^i)(\xi_P) \end{pmatrix}$$

is the ith column of the Jacobian matrix $\mathrm{Jac}_F(\xi_P)$ of F at ξ_P and this matrix is invertible since F is assumed to be a diffeomorphism. This basis $(\mathbf{e}_i(P))$ is called **local frame** or **local reference system** at P associated to the curvilinear coordinates given by F. The introduction of curvilinear coordinates in an open region D of $\mathbb{R}^n$ immediately furnishes a *field of local frames* in D.

The curvilinear coordinates F are called **orthogonal** when all local frames are orthogonal ones. In general, we have to compute the scalar products

$$g_{ij}(P) = \mathbf{e}_i(P) \cdot \mathbf{e}_j(P).$$

As we have seen in the proof of the lemma of 9.1, the determinant $g(P) = \det(g_{ij}(P))$ is the square of the determinant of the matrix obtained by ranging the components of the $\mathbf{e}_j(P)$ (in the canonical basis) in columns. This shows immediately

$$g(P) = \det(g_{ij}(P)) = \left|(\partial f^i/\partial \xi^j)(\xi_P)\right|^2 = \Delta_F(\xi_P)^2 ,$$

and conversely, the Jacobian determinant (or functional determinant) of F is given by $|\Delta_F(\xi_P)| = \sqrt{g(P)}$.

A suitable choice of function ψ will simplify computations : observe that

$$\mathrm{div}\mathbf{A}' + \frac{1}{c}\frac{\partial\varphi'}{\partial t} = \mathrm{div}\mathbf{A} + \frac{1}{c}\frac{\partial\varphi}{\partial t} + \mathrm{div}\ \mathbf{grad}\psi + \frac{1}{c}\frac{\partial}{\partial t}(-\frac{1}{c}\frac{\partial\psi}{\partial t}) =$$

$$= \mathrm{div}\mathbf{A} + \frac{1}{c}\frac{\partial\varphi}{\partial t} + \Delta\psi - \frac{1}{c^2}\frac{\partial^2\psi}{\partial t^2} .$$

Having made a first (arbitrary) choice of potentials $\mathbf{A}$ and φ, we can compute the function

$$f = \mathrm{div}\mathbf{A} + \frac{1}{c}\frac{\partial\varphi}{\partial t}$$

and then take for ψ any solution of

$$\Delta\psi - \frac{1}{c^2}\frac{\partial^2\psi}{\partial t^2} = -f$$

furnishing then

$$\mathrm{div}\mathbf{A}' + \frac{1}{c}\frac{\partial\varphi'}{\partial t} = 0. \tag{6}$$

Since the partial differential operator $\Delta - c^{-2}\partial^2/\partial t^2$ is going to reappear several times, let us introduce the abbreviation

$$\square = \Delta - c^{-2}\partial^2/\partial t^2.$$

This operator has been considered by Jean le Rond d'Alembert and is often called Dalembertian (or d'Alembertian ?!).

Provided we are able to solve a non homogeneous linear partial differential equation of the type $\square\psi = -f$, we shall be able to choose the potentials $\mathbf{A}'$, φ' in the **Lorenz gauge**, i.e. satisfying (6).

10.2 THE NON HOMOGENEOUS MAXWELL EQUATIONS

Let us now consider the Maxwell equations linked to sources : we introduce

- ρ : charge density (scalar function of position and time),
- $\mathbf{j}$: current density (vector function of position and time).

These fields are linked to the physical fields by the non homogeneous linear partial differential equations

$$\mathrm{div}\ \mathbf{E} = 4\pi\rho, \tag{1}$$

$$\mathbf{curl}\ \mathbf{H} - \frac{1}{c}\frac{\partial\mathbf{E}}{\partial t} = \frac{4\pi}{c}\mathbf{j}. \tag{2}$$

As we have seen in the preceding section, we can choose potentials $\mathbf{A}$ and φ with

$$\mathbf{H} = \mathbf{curl}\ \mathbf{A},$$

$$\mathbf{E} = -\frac{1}{c}\frac{\partial\mathbf{A}}{\partial t} - \mathbf{grad}\varphi.$$

Substitution of these expressions in the non homogeneous Maxwell equations (1) leads to the condition

$$-4\pi\rho = \mathrm{div}(-\mathbf{E}) = \mathrm{div}(\frac{1}{c}\frac{\partial \mathbf{A}}{\partial t} + \mathbf{grad}\varphi) = \Delta\varphi + \frac{1}{c}\frac{\partial}{\partial t}\,\mathrm{div}\mathbf{A}.$$

We are also assuming that the Lorenz gauge is satisfied, namely

$$\mathrm{div}\mathbf{A} + \frac{1}{c}\frac{\partial \varphi}{\partial t} = 0 \tag{3}$$

and this choice will permit us to write

$$-4\pi\rho = \Delta\varphi + \frac{1}{c}\frac{\partial}{\partial t}(-\frac{1}{c}\frac{\partial \varphi}{\partial t}) = \square\varphi.$$

If the density of charge ρ is known, the scalar potential will have to satisfy

$$\square\varphi = -4\pi\rho. \tag{4}$$

Quite similarly, the non homogeneous equation (2) leads to

$$\frac{-4\pi}{c}\,\mathbf{j} = -\mathbf{curl}\ \mathbf{H} + \frac{1}{c}\frac{\partial \mathbf{E}}{\partial t} = -\mathbf{curl}\ \mathbf{curl}\ \mathbf{A} + \frac{1}{c}\frac{\partial}{\partial t}(-\mathbf{grad}\varphi - \frac{1}{c}\frac{\partial \mathbf{A}}{\partial t})$$

in which it is easy to get rid of the potential φ by using (3)

$$\frac{-4\pi}{c}\,\mathbf{j} = -\mathbf{curl}\ \mathbf{curl}\ \mathbf{A} + \frac{1}{c}\frac{\partial}{\partial t}(-\frac{1}{c}\frac{\partial \mathbf{A}}{\partial t}) + \mathbf{grad}\ \mathrm{div}\mathbf{A}.$$

But we have seen in 8.4 that

$$\mathbf{curl}\ \mathbf{curl}\ \mathbf{A} = \mathbf{grad}\ \mathrm{div}\mathbf{A} - \Delta\mathbf{A}$$

so that we are left with

$$\frac{-4\pi}{c}\,\mathbf{j} = \Delta\mathbf{A} - c^{-2}\partial^2\mathbf{A}/\partial t^2 = \square\mathbf{A}.$$

Once more, we find that each component of the vector potential satisfies the typical equation that we have already encountered

$$\square A_i = \frac{-4\pi}{c}\,j_i \qquad (1 \le i \le 3) \tag{5}$$

The advantage of the four scalar equations given in (4) and (5) is —apart from their similarity— that they give four *uncoupled* equations for the scalar components of the vector potentials.

To summarize, we shall be able to solve Maxwell equations as soon as we can solve the typical scalar partial differential equation $\square u = f$. Even the special choice of potentials that we have made to achieve the Lorenz gauge (3) was based on the resolution of an equation of this type. For given f, this is a linear non homogeneous PDE in the unknown function u. Extra conditions have to be imposed if uniqueness of u is to be attained. Physical considerations will have to be invoked about

asymptotic behavior, limit conditions, etc.

on u to achieve this goal. This is another chapter of mathematics!

10.3 INTRODUCTION TO THE DYNAMICS OF FLUIDS

A vector field $\mathbf{v}$ in $\mathbb{R}^3$ can be interpreted as the velocity field in a *stationary* flow. (More generally, the velocity field in a moving fluid would be a variable vector field, hence described by a mapping $\mathbb{R} \times \mathbb{R}^3 \longrightarrow \mathbb{R}^3 : (t,\mathbf{r}) \longmapsto \mathbf{v}(t,\mathbf{r})$.) Such a description was considered by Euler. If we are interested in the trajectories of the particles which constitute the fluid in movement, we have to find the parameterized curves $t \longmapsto X(t)$ which satisfy $X' = \mathbf{v}(X)$. As in Chapter 3, the solutions of this differential equation will depend on initial conditions and we can denote by $t \longmapsto X(t,\mathbf{r})$ the solution which goes through the point $\mathbf{r}$ at $t = 0 : X(0,\mathbf{r}) = \mathbf{r}$. The following result will be taken for granted.

Theorem. Let $F : \mathbb{R} \times \mathbb{R}^3 \longrightarrow \mathbb{R}^3 : (t,\mathbf{r}) \longmapsto F(t,\mathbf{r})$ be a $\mathcal{C}^1$-mapping. Then, for each initial data $\mathbf{r}$, there is one, and only one, solution of the differential equation $X' = F(t,X)$, say $t \longmapsto X(t,\mathbf{r})$ defined in a neighborhood of the origin for which $X(0,\mathbf{r}) = \mathbf{r}$. Moreover, the mapping $(t,\mathbf{r}) \longmapsto X(t,\mathbf{r})$ is $\mathcal{C}^1$. ■

This existence theorem for non-linear differential equations is well-known, but the smoothness in the initial conditions is quite subtle (cf. the article by J.W. Robbin, Proc. of AMS, vol.10 (1968) pp.1005-1006).

Taking for F the stationary distribution of velocities in a fluid, we are going to obtain a nice interpretation of the divergence of $\mathbf{v}$. Denote as above by $X(t,\mathbf{r})$ the solution of $X' = \mathbf{v}(X)$ for which $X(0,\mathbf{r}) = \mathbf{r}$. We obtain a family of transformations

$$\Gamma_t : \mathbb{R}^3 \longrightarrow \mathbb{R}^3 , \ \mathbf{r} \longmapsto X(t,\mathbf{r})$$

describing the movement of the fluid (Lagrange description of the *flow*). These maps are diffeomorphisms : $(\Gamma_t)^{-1}$ is $\mathcal{C}^1$ since it also describes an evolution between times t and 0 respectively. In particular, Γ_o is the identity map. Let us evaluate the variation of volumes near $t = 0$. For this purpose, select a small piece $\Pi \subset \mathbb{R}^3$ whose volume is measured by

$$|\Pi| = \int_\Pi dV.$$

Also put $\Pi_t = \Gamma_t(\Pi)$ (so that $\Pi_o = \Pi$ since $\Gamma_o = \mathrm{Id.}$). The volume of Π_t is measured by

$$|\Pi_t| = \int_{\Pi_t} dV = \int_{\Pi} \det(\Gamma_t')dV = \int_{\Pi} \det\partial_{\mathbf{r}}X(t,\mathbf{r})\ dV.$$

The rate of growth of these volumes is $(d/dt)|\Pi_t|$. We are interested in this *local* rate of growth for $t = 0$. This means that we have to compute the limit of (Leibniz' rule, cf.16.5)

$$\frac{1}{|\Pi|}\frac{d}{dt}\bigg|_{t=0} \int_{\Pi} \det\partial_{\mathbf{r}}X(t,\mathbf{r})\ dV = \frac{1}{|\Pi|}\int_{\Pi} \frac{\partial}{\partial t}\bigg|_{t=0} \det\partial_{\mathbf{r}}X(t,\mathbf{r})\ dV$$

when the piece Π shrinks to a point P (hence $|\Pi| \longrightarrow 0$). This limit is

$$\frac{\partial}{\partial t}\bigg|_{t=0} \det\partial_{\mathbf{r}}X(t,P)$$

(compare with 12.4). To simplify notations, introduce the matrices $U(t,\mathbf{r}) = \partial_{\mathbf{r}}X(t,\mathbf{r})$ (= Jac $\Gamma_t(\mathbf{r}) = \Gamma_t'(\mathbf{r})$...). By definition, we have

$$\frac{\partial}{\partial t}X(t,\mathbf{r}) = \mathbf{v}(X(t,\mathbf{r})).$$

We differentiate this expression with respect to $\mathbf{r}$ using the chain rule

$$\partial_{\mathbf{r}} \frac{\partial}{\partial t}X(t,\mathbf{r}) = \partial_{\mathbf{r}}\ \mathbf{v}(X(t,\mathbf{r})) = \mathrm{Jac}_{\mathbf{v}}(X(t,\mathbf{r}))\circ\partial_{\mathbf{r}}X(t,\mathbf{r}).$$

Permuting the derivations (assuming that X is $\mathcal{C}^2$, which will be the case if $\mathbf{v}$ is $\mathcal{C}^2$)

$$\frac{\partial}{\partial t}U(t,\mathbf{r}) = \mathrm{Jac}_{\mathbf{v}}(X(t,\mathbf{r}))\circ U(t,\mathbf{r}).$$

From this formula, we deduce the limited expansion

$$U(t,\mathbf{r}) = U(0,\mathbf{r}) + U'(0,\mathbf{r})\ t + o(t) =$$
$$= I + \mathrm{Jac}_{\mathbf{v}}(X(0,\mathbf{r}))\cdot t + o(t) = I + \mathrm{Jac}_{\mathbf{v}}(\mathbf{r})\ t + o(t),$$

and

$$\det U(t,\mathbf{r}) = \det(I + t\ \mathrm{Jac}_{\mathbf{v}}(\mathbf{r}) + o(t)) =$$
$$= 1 + t\ \mathrm{Tr}\ \mathrm{Jac}_{\mathbf{v}}(\mathbf{r}) + o(t)$$

(as in the proof of the lemma of 2.3). This gives

$$\frac{\partial}{\partial t}\bigg|_{t=0} \det U(t,\mathbf{r}) = \mathrm{Tr}\ \mathrm{Jac}_{\mathbf{v}}(\mathbf{r}) = \mathrm{div}\ \mathbf{v}(\mathbf{r}).$$

Thus the divergence of the vector field $\mathbf{v}$ at the point $\mathbf{r}$ appears as *local inflation rate for volumes in the stationary flow produced by the velocity field* $\mathbf{v}$.

10.4 SPECIAL FIELDS

When a vector field $\mathbf{u}$ does not derive from a scalar potential one can still try to look for a scalar function f such that f$\mathbf{u}$ derives from a scalar potential. Such a scalar function $f \neq 0$ is called an **integrating factor** for the field $\mathbf{u}$. If $\mathbf{u}$ admits the integrating factor f, then **curl**(f**u**) = 0. We have (cf.8.3)

$$\mathbf{curl}(f\mathbf{u}) = \nabla \wedge (f\mathbf{u}) = (\nabla f) \wedge \mathbf{u} + f\, \nabla \wedge \mathbf{u} = $$
$$= \mathbf{grad} f \wedge \mathbf{u} + f\mathbf{curl}\, \mathbf{u} \ .$$

Hence, the *existence of an integrating factor for the field* **u** implies an equation

$$f\, \mathbf{curl}\, \mathbf{u} = \mathbf{u} \wedge \mathbf{grad} f \quad \text{orthogonal to } \mathbf{u} \ .$$

Since $f \neq 0$, this implies that the field **u** itself is orthogonal to **curl u** (cf. the warning given in 8.4). This necessary condition is also *locally sufficient.*

Theorem. Let **u** be a $\mathcal{C}^1$ vector field in $\mathbb{R}^3$ with $\mathbf{u} \perp \mathbf{curl}\, \mathbf{u}$ (at all points of $\mathbb{R}^3$). Then there locally exist scalar functions f and g with f**u** = **grad**g. □

Proof. If we can find f and g as asserted, the implicit equations $g(x,y,z) = c$ will define a family of surfaces Σ_c having normal vectors parallel to **grad**g, hence parallel to **u**. In other words, the tangent planes to the surfaces Σ_c should be orthogonal to the vector field **u**. We have already studied this situation in 7.2 and the Frobenius theorem will give the *existence* of the family (Σ_c) under the assumption that **u** is orthogonal to **curl u**.

Let us call X, Y and Z the components of the field **u**, so that $X = X(x,y,z)$, ... The orthogonality assumption is

$$X(\partial_y Z - \partial_z Y) + Y(\partial_z X - \partial_x Z) + Z(\partial_x Y - \partial_y X) = 0 \ .$$

Hence

$$Z\partial_y X \ -X\partial_y Z - Y\partial_z X = Z\partial_x Y - Y\partial_x Z - X\partial_z Y$$

and

$$(Z\partial_y X \ -X\partial_y Z)/Z^2 - YZ^{-2}\partial_z X = (Z\partial_x Y - Y\partial_x Z)/Z^2 - XZ^{-2}\partial_z Y,$$

or

$$\partial_y(X/Z) - YZ^{-2}\partial_z X = \partial_x(Y/Z) - XZ^{-2}\partial_z Y \ .$$

Adding $XYZ^{-3}\partial_z Z$ to both sides, we obtain

$$\partial_y(X/Z) - (Y/Z)\partial_z(X/Z) = \partial_x(Y/Z) - (X/Z)\partial_z(Y/Z).$$

According to 7.2, this is a *necessary and sufficient condition* for the integrability of the system

$$\partial\zeta/\partial x = X(x,y,\zeta)/Z(x,y,\zeta),$$
$$\partial\zeta/\partial y = Y(x,y,\zeta)/Z(x,y,\zeta).$$

In other words, through each point of space, there will be a surface Σ of equation

$$z = \zeta(x,y) \ (= \zeta_c(x,y))$$

having tangent planes orthogonal to the field of components X/Z,

Y/Z and -1 (we are assuming that Z does not vanish in an open neighborhood of the point P under consideration). Equivalently, the surface Σ_c has tangent planes orthogonal to $\mathbf{u}$. Without loss of generality, we can assume that the "initial condition" for the surface Σ_c is given by $\zeta(a,b) = c$ for two fixed values a and b of x and resp. y. The function g having value c at all points of the surface Σ_c is a scalar field having the Σ_c for level surfaces. This function g is $\mathcal{C}^1$ by smooth dependence on initial conditions (a result quoted —but not proved— in the preceding section 10.3). The gradient of this field g must be parallel to the field $\mathbf{u}$. Hence we can find a scalar function f with

$$f\mathbf{u} = \mathbf{grad}g. \quad \blacksquare$$

Observe that the preceding condition makes it easy to quote a vector field $\mathbf{u}$ admitting no integrating factor. For example, check that the field $\mathbf{u} = {}^t(z,\ z^2-1,\ 0)$ admits no integrating factor (at no point is the $\mathbf{curl}\ \mathbf{u}$ orthogonal to $\mathbf{u}$).

Since current research (e.g. in plasma physics) makes a systematic use of the differential operators **grad**, div and **curl**, let us list a few useful computational observations.

If $\mathbf{u}$ is a vector field in $\mathbb{R}^3$ with components u_i in the canonical basis $(\mathbf{e}_i) \subset \mathbb{R}^3$, we can write

$$\mathbf{u} = (\mathbf{e}_3 \wedge \mathbf{u}) \wedge \mathbf{e}_3 + u_3\mathbf{e}_3.$$

Compute indeed the double cross product

$$\mathbf{e}_3 \wedge (\mathbf{e}_3 \wedge \mathbf{u}) = \mathbf{e}_3(\mathbf{e}_3 \cdot \mathbf{u}) - \mathbf{u}(\mathbf{e}_3 \cdot \mathbf{e}_3) = u_3\mathbf{e}_3 - \mathbf{u}.$$

On the other hand, if $\mathbf{a}$ is a constant vector and $\mathbf{u}$ is a $\mathcal{C}^1$ vector field,

$$\mathbf{curl}(\mathbf{a} \wedge \mathbf{u}) = \nabla \wedge (\mathbf{a} \wedge \mathbf{u}) = \mathbf{a}(\nabla\mathbf{u}) - (\mathbf{a}\nabla)\mathbf{u} =$$

$$= \mathbf{a} \cdot \mathrm{div}\mathbf{u} - \mathrm{Grad}\mathbf{u}(\mathbf{a}).$$

Taking $\mathbf{a} = \mathbf{e}_3$ in this formula, we get

$$\mathbf{curl}(\mathbf{e}_3 \wedge \mathbf{u}) = (\mathrm{div}\mathbf{u})\ \mathbf{e}_3 - \partial_z\mathbf{u}.$$

Proposition. Let $\mathbf{u}$ be a $\mathcal{C}^1$ vector field in $\mathbb{R}^3$ (or in any star-shaped subdomain) which is invariant under vertical z-translations and for which $\mathrm{div}\mathbf{u} = 0$. Then, there exists a scalar function φ with $\mathbf{e}_3 \wedge \mathbf{u} = \mathbf{grad}\varphi$ and so $\mathbf{u} = \mathrm{grad}\varphi \wedge \mathbf{e}_3 + u_3\mathbf{e}_3$. $\blacksquare$

PART TWO

INTEGRATION OF DIFFERENTIAL FORMS

CHAPTER 11
EXTERIOR DIFFERENTIAL FORMS

11.1 CHANGE OF VARIABLES IN DOUBLE INTEGRALS

We assume that the reader is familiar with multiple integrals and their basic properties, in particular with the change of variables formula. Let us recall explicitely the situation in the two variables case.

Let $D \subset \mathbb{R}^2$ be a compact domain (e.g. a square, a disk, etc.) and f a continuous scalar function on D. The integral

$$\iint_D f(x,y)\ dxdy$$

is defined in such a way as to give

$$\iint_D f(x,y)\ dxdy \geq 0 \quad \text{when} \quad f \geq 0.$$

If $\phi : D' \longrightarrow D$ is a change of variables, say

$$x = \varphi(\xi,\eta) \quad \text{and} \quad y = \psi(\xi,\eta),$$

it is known that

$$\iint_D f(x,y)\ dxdy = \iint_{D'} f(\phi(\xi,\eta)) \left| \frac{\partial(x,y)}{\partial(\xi,\eta)} \right| d\xi d\eta$$

$$= \iint_{D'} f(\phi(\xi,\eta))\ \left|\det\phi'(\xi,\eta)\right| d\xi d\eta\ .$$

How can we derive the preceding formula formally from the expressions

$$dx = \varphi'_\xi\ d\xi + \varphi'_\eta\ d\eta \quad \text{and} \quad dy = \psi'_\xi\ d\xi + \psi'_\eta\ d\eta\ ?$$

With any *distributive product* (*treating functions as scalars*) we shall have

$$dxdy = (\varphi'_\xi\ d\xi + \varphi'_\eta\ d\eta)(\psi'_\xi\ d\xi + \psi'_\eta\ d\eta) =$$

$$= \varphi'_\xi\psi'_\xi d\xi d\xi + \varphi'_\xi\psi'_\eta d\xi d\eta + \varphi'_\eta\psi'_\xi d\eta d\xi + \varphi'_\eta\psi'_\eta d\eta d\eta\ .$$

When $\det\phi' > 0$, we should find

$$dxdy = \frac{\partial(x,y)}{\partial(\xi,\eta)}\ d\xi d\eta = \varphi'_\xi\psi'_\eta d\xi d\eta - \varphi'_\eta\psi'_\xi d\xi d\eta\ .$$

Comparison shows that the correct formula would be obtained if we had

$$d\xi d\xi = 0,\ d\eta d\eta = 0 \text{ and } d\eta d\xi = -d\xi d\eta\ .$$

We shall introduce a product $\wedge$ of differential forms having these properties. The "cross" notation is there to recall the *anticommutative* property $d\eta\wedge d\xi = -d\xi\wedge d\eta$ also shared by the vector product in $\mathbb{R}^3$. But the similarity will end there, because the $\wedge$ product of differentials will be *associative*.

11.2 EXTERIOR DIFFERENTIALS IN $\mathbb{R}^n$

First order differentials have already been extensively considered in the first part of this book. Let us recall these notions and adapt them to the present context.

When $f : \mathbb{R}^n \longrightarrow \mathbb{R}$ is a $\mathcal{C}^1$ scalar field, the derivative or differential of f at a point $\mathbf{a} \in \mathbb{R}^n$ is the tangent linear map

$$f'(\mathbf{a}) : \mathbb{R}^n \longrightarrow \mathbb{R}$$

to f at **a**. This linear mapping has been denoted by $\mathbf{h} \longmapsto f'(\mathbf{a})\mathbf{h}$ so that the derivative itself is a mapping

$$f' : \mathbb{R}^n \longrightarrow \mathcal{L}(\mathbb{R}^n,\mathbb{R}) = \mathbb{R}_n.$$

The function of two variables $(\mathbf{a},\mathbf{h}) \longmapsto f'(\mathbf{a})\mathbf{h} = df(\mathbf{a},\mathbf{h})$ is the **differential of** f

$$df : \mathbb{R}^n \times \mathbb{R}^n \longrightarrow \mathbb{R}.$$

It is linear in the second variable.

The same conventions and notations are adopted for vector functions : if $F : \mathbb{R}^n \longrightarrow \mathbb{R}^m$, the derivative $F'(\mathbf{a}) : \mathbb{R}^n \longrightarrow \mathbb{R}^m$ is the tangent linear map and $dF(\mathbf{a},\mathbf{h}) = F'(\mathbf{a})\mathbf{h}$ is linear in its second variable

$$dF : \mathbb{R}^n \times \mathbb{R}^n \longrightarrow \mathbb{R}^m.$$

For example, let F denote the identity map in $\mathbb{R}^3$: $F(\mathbf{r}) = \mathbf{r}$ is a vector field with divergence 3. The derivative of F at any point **a** is the identity map $F'(\mathbf{a}) : \mathbf{h} \longmapsto F'(\mathbf{a})\mathbf{h} = \mathbf{h}$ having matrix

$$F'(\mathbf{a}) = \begin{pmatrix} \mathbf{grad}\ x \\ \mathbf{grad}\ y \\ \mathbf{grad}\ z \end{pmatrix}(\mathbf{a}) = \begin{pmatrix} 1 & 0 & 0 \\ 0 & 1 & 0 \\ 0 & 0 & 1 \end{pmatrix}.$$

In this case, the map F' is constant : the matrix $F'(\mathbf{a})$ is the identity matrix independently from the point **a**. Thus we can write

$$dF = \begin{pmatrix} dx \\ dy \\ dz \end{pmatrix} = \begin{pmatrix} 1 & 0 & 0 \\ 0 & 1 & 0 \\ 0 & 0 & 1 \end{pmatrix}$$

where e.g. dx is the constant field of linear forms (1,0,0)

$dx(\mathbf{a},\mathbf{h}) = h_1$ (first projection of **h** in the canonical basis).

With a slight abuse of notation, we can write

$$d\mathbf{r} = \begin{pmatrix} dx \\ dy \\ dz \end{pmatrix}.$$

Any field of linear forms in $\mathbb{R}^3$ can be expressed as a linear combination of the three basic fields (with variable coefficients)

$$(\mathbf{a},\mathbf{h}) \longmapsto u_1(\mathbf{a})dx + u_2(\mathbf{a})dy + u_3(\mathbf{a})dz.$$

(At the point **a**, dx, dy and dz constitute the dual basis of the canonical basis of $\mathbb{R}^3$.) The three coefficient functions may be grouped in a vector field

$$\mathbf{a} \longmapsto \mathbf{u}(\mathbf{a}) = \begin{pmatrix} u_1(\mathbf{a}) \\ u_2(\mathbf{a}) \\ u_3(\mathbf{a}) \end{pmatrix}$$

and the differential form is conveniently abbreviated

$$u_1dx + u_2dy + u_3dz = \mathbf{u}\cdot d\mathbf{r}.$$

Similarly, any field of linear forms in $\mathbb{R}^n$ can be expressed as a linear combination of the basic fields dx^i $(1 \le i \le n)$ with variable coefficients. In this way, any field of linear forms on $\mathbb{R}^n$ can be written

$$(\mathbf{a},\mathbf{h}) \longmapsto \omega(\mathbf{a},\mathbf{h}) = u_1(\mathbf{a})h^1 + \ldots + u_n(\mathbf{a})h^n = \sum u_i(\mathbf{a})\ dx^i(\mathbf{a},\mathbf{h}).$$

Fields of linear forms are called **differential forms of the first degree**. They are conveniently written

$$\omega = \sum u_i dx^i = \mathbf{u}\cdot d\mathbf{r}.$$

The set of differential forms of degree one is a vector space

$$\Omega^1(\mathbb{R}^n) \ni \omega = \mathbf{u}\cdot d\mathbf{r}$$

isomorphic to the vector space of vector fields **u** in $\mathbb{R}^n$. Let us stress that we are working with *smooth* coefficients : the differential forms that we consider have indefinitely differentiable coefficients.

Let us introduce *formally* the vector spaces $\Omega^p(\mathbb{R}^n)$ as follows.

An exterior differential form of degree $p = 2$ is a formal linear combination with $\mathcal{C}^\infty$ coefficients of the basic expressions

$$dx^i \wedge dx^j \quad (1 \le i < j \le n).$$

An exterior differential form of degree $p = 3$ is a formal linear combination with $\mathcal{C}^\infty$ coefficients of the basic expressions

$$dx^i \wedge dx^j \wedge dx^k \quad (1 \le i < j < k \le n), \text{ etc.}$$

For any choice of p-tuple of indices satisfying

$$1 \le i_1 < i_2 < \ldots < i_p \le n$$

we get a corresponding basic exterior differential form

$$dx^{i_1} \wedge dx^{i_2} \wedge \ldots \wedge dx^{i_p}.$$

We find in this way $\binom{n}{p}$ such basic elements of degree p in $\Omega^p(\mathbb{R}^n)$, since there are $\binom{n}{p}$ subsets with p elements in the set $\{1,\ldots,n\}$).

For example, for $p = n$, there is only one basic element $dx^1 \wedge dx^2 \wedge \dots \wedge dx^n$ and $\Omega^n(\mathbb{R}^n)$ consists of the multiples

$$f dx^1 \wedge dx^2 \wedge \dots \wedge dx^n \quad \text{where} \quad f \in \mathcal{C}^\infty.$$

for $p > n$, we have to admit that there are no basic elements of the required form and we put

$$\Omega^p(\mathbb{R}^n) = \{0\} \quad \text{if } p > n.$$

For completeness, let us take for $\Omega^o(\mathbb{R}^n)$ the space of $\mathcal{C}^\infty$ functions. In fact, $\Omega^o(\mathbb{R}^n)$ is a *ring* and $\Omega^p(\mathbb{R}^n)$ is a *module over this ring* $\Omega^o(\mathbb{R}^n)$. We are of course tempted to say that $\Omega^p(\mathbb{R}^n)$ is a vector space over $\Omega^o(\mathbb{R}^n)$ with basis given by the above described basic exterior differential forms of degree p, hence of dimension $\binom{n}{p}$ however, since $\Omega^o(\mathbb{R}^n)$ is not a *field* (in the technical meaning attributed to this word in algebra), the usual terminology is that of *free module of rank* $\binom{n}{p}$ *over* $\Omega^o(\mathbb{R}^n)$. But $\Omega^p(\mathbb{R}^n)$ is certainly a *vector space over* $\mathbb{R}$ *(or* $\mathbb{C}$*).* As such, its dimension is *infinite* when it is *not zero* (namely for $1 \le p \le n$).

Definition. The **space of exterior differentials** on $\mathbb{R}^n$ is

$$\Omega(\mathbb{R}^n) = \bigoplus_{p \ge 0} \Omega^p(\mathbb{R}^n).$$

In other words, an exterior differential form is a formal sum of homogeneous components $\omega = \omega_o + \omega_1 + \dots + \omega_n$ with $\omega_p \in \Omega^p(\mathbb{R}^n)$.

By construction, if we make a change of variables in the basic form

$$dx^1 \wedge \dots \wedge dx^n \in \Omega^n(\mathbb{R}^n)$$

of highest degree, say

$$x^1 = f^1(\xi^1,\dots,\xi^n),$$
$$\dots$$
$$x^n = f^n(\xi^1,\dots,\xi^n),$$

then, the expression of $dx^1 \wedge \dots \wedge dx^n$ in the new variables ξ^i is

$$\frac{\partial(f^1,\dots,f^n)}{\partial(\xi^1,\dots,\xi^n)} d\xi^1 \wedge \dots \wedge d\xi^n$$

(without absolute value for the Jacobian determinant !). To convince himself of this result, the reader should take the $n = 2$ case and make the explicit computation. This result will play an essential role for integration.

11.3 EXTERIOR PRODUCT IN $\Omega(\mathbb{R}^n)$

An algebraic structure can be introduced in the space $\Omega(\mathbb{R}^n)$ of exterior forms by defining —still in an ad hoc way— a product

$$\wedge : \Omega^p(\mathbb{R}^n) \times \Omega^q(\mathbb{R}^n) \longrightarrow \Omega^{p+q}(\mathbb{R}^n)$$

using the following rules. This product is required to be *associative and distributive*. For **monomials,** i.e. multiples of basic expressions, it is given by

$$(f dx^i \wedge dx^j \wedge \ldots) \wedge (g dx^k \wedge dx^\ell \wedge \ldots) = fg dx^i \wedge dx^j \wedge \ldots \wedge dx^k \wedge dx^\ell \wedge \ldots$$

Basic expressions can be reordered by using systematically the relations

$$dx^i \wedge dx^j = -dx^j \wedge dx^i$$

(note that these relations imply $dx^i \wedge dx^i = 0$). In other words, the product of monomials is simply obtained by multiplication of their coefficients and by multiplying in the sense of $\wedge$ their basic terms.

For example, in $\mathbb{R}^3$ the product of

$$dx \wedge dy \in \Omega^2(\mathbb{R}^3) \text{ with } dz \in \Omega^1(\mathbb{R}^3)$$

is $(dx \wedge dy) \wedge dz = dx \wedge dy \wedge dz$: this is the basic element of $\Omega^3(\mathbb{R}^3)$. Similarly, the exterior product of the monomials $f dx \wedge dy$ and $g dz$ would be $fg dx \wedge dy \wedge dz$. The exterior product of

$$f dy \text{ and } g dx \wedge dz$$

would be

$$fg\, dy \wedge (dx \wedge dz) = fg(dy \wedge dx) \wedge dz = -fg(dx \wedge dy) \wedge dz.$$

We hope that these few examples have clarified the definition. They show a *resemblance* of the exterior product $\wedge$ just introduced with the exterior product of vectors, e.g. in the rule

$$dx \wedge dy = -dy \wedge dx.$$

However, *associativity* is now postulated for the exterior product of differential forms. Note also that the exterior product $\wedge$ of exterior differential forms is not *anticommutative* because e.g.

$$(dx \wedge dy) \wedge dz = dx \wedge (dy \wedge dz) = -dx \wedge (dz \wedge dy) =$$
$$= -(dx \wedge dz) \wedge dy = +(dz \wedge dx) \wedge dy = +dz \wedge (dx \wedge dy)$$

($dx \wedge dy$ and dz *commute*: no minus sign appears when commuting them).

More generally, one can establish the following *skew commutation* rule

$$\omega_p \wedge \omega_q = (-1)^{pq}\, \omega_q \wedge \omega_p \quad \text{if} \quad \omega_p \in \Omega^p(\mathbb{R}^n),\ \omega_q \in \Omega^q(\mathbb{R}^n).$$

Also observe that the exterior product of two homogeneous forms $\omega_p \in \Omega^p(\mathbb{R}^n)$ and $\omega_q \in \Omega^q(\mathbb{R}^n)$ (of respective degrees p and q) must vanish if p+q > n since this product is an element of

$$\Omega^{p+q}(\mathbb{R}^n) = \{0\} \quad \text{when} \quad p+q > n.$$

11.4 EXTERIOR DIFFERENTIATION

The differential operator $d : f \longmapsto df$ associates a differential of degree one to a function. It is a linear operator

$$d : \mathcal{C}^\infty(\mathbb{R}^n) = \Omega^0(\mathbb{R}^n) \longrightarrow \Omega^1(\mathbb{R}^n).$$

We shall define linear operators

$$d = d_p : \Omega^p(\mathbb{R}^n) \longrightarrow \Omega^{p+1}(\mathbb{R}^n)$$

extending the preceding definition. The **exterior differentiation** operator d is then extended by additivity to the whole algebra of exterior differential forms $\Omega(\mathbb{R}^n)$. This means that we are defining

$$d : \Omega(\mathbb{R}^n) \longrightarrow \Omega(\mathbb{R}^n) , \sum \omega_p \longmapsto \sum d_p\omega_p \quad (\omega_p \in \Omega^p(\mathbb{R}^n)),$$

as soon as the homogeneous parts d_p of d are defined ($d_o = d$ is the usual differentiation operator).

Rule for exterior differentiation of a monomial $\omega_p \in \Omega^p(\mathbb{R}^n)$:

differentiate in the usual way the function coefficient and multiply $\wedge$ *by the basic element of which* ω_p *is a multiple.*

The result of this operation is another homogeneous differential form $d\omega_p \in \Omega^{p+1}(\mathbb{R}^n)$ (but this form will not be a monomial in general, as the next examples will immediately show).

Additivity is used to compute the exterior differential of a sum of monomials , namely an arbitrary element of $\Omega(\mathbb{R}^n)$.

The preceding rule is best illustrated in $\Omega(\mathbb{R}^3)$ and we shall determine completely the exterior differentiation in this case.

0) If f is a $\mathcal{C}^\infty$ function, $d_o = d$ is the usual differential operator and

$$df = \frac{\partial f}{\partial x}\, dx + \frac{\partial f}{\partial y}\, dy + \frac{\partial f}{\partial z}\, dz = \mathbf{grad}f \cdot d\mathbf{r}.$$

This is not new (and not typical to the three dimensional case !).

1) If $\omega \in \Omega^1(\mathbb{R}^3)$, we can write

$$\omega = u_1dx + u_2dy + u_3dz = \mathbf{u} \cdot d\mathbf{r}.$$

Using the rule for the exterior differential of the monomial u_1dx, we find

$$d(u_1dx) = du_1 \wedge dx = (\frac{\partial u}{\partial x}1dx + \frac{\partial u}{\partial y}1dy + \frac{\partial u}{\partial z}1dz) \wedge dx =$$

$$= \frac{\partial u}{\partial x}1dx \wedge dx + \frac{\partial u}{\partial y}1dy \wedge dx + \frac{\partial u}{\partial z}1dz \wedge dx =$$

$$= -\frac{\partial u_1}{\partial y} dx\wedge dy + \frac{\partial u_1}{\partial z} dz\wedge dx = -\partial_y u_1 \cdot dx\wedge dy + \partial_z u_1 dz\wedge dx.$$

The other two terms are treated similarly, and adding up all results, we find

$$d(\mathbf{u}\cdot d\mathbf{r}) = (\partial_y u_3 - \partial_z u_2) dy\wedge dz + (\partial_z u_1 - \partial_x u_3) dz\wedge dx + (\partial_x u_2 - \partial_y u_1) dx\wedge dy.$$

In this expression, we recognize the components of **curl u** and introducing a *formal vector differential of degree two*

$$d\sigma = \begin{pmatrix} dy\ dz \\ dz\wedge dx \\ dx\ dy \end{pmatrix},$$

we can write

$$d(\mathbf{u}\cdot d\mathbf{r}) = \mathbf{curl}\ \mathbf{u}\cdot d\sigma.$$

2) Finally, if $\omega \in \Omega^2(\mathbb{R}^3)$, we can write ω as the sum of multiples of the basic monomials

$$dy\wedge dz,\ dz\wedge dx \text{ and } dx\wedge dy$$

in this order (we write first the monomial where dx is missing, and the next monomials are obtained by cyclic permutation $x \mapsto y \mapsto z$). If we call these respective coefficients v_1, v_2 and v_3, we have

$$\omega = \mathbf{v}\cdot d\sigma \quad \text{with } \mathbf{v} = {}^t(v_1, v_2, v_3).$$

The exterior differential of the first monomial is

$$d(v_1 dy\wedge dz) = dv_1\wedge dy\wedge dz = (\partial_x v_1 \cdot dx + \ldots)\wedge dy\wedge dz =$$
$$= \partial_x v_1 \cdot dx\wedge dy\wedge dz.$$

The other two monomials give similar contributions and adding up the three results, we get

$$d(\mathbf{v}\cdot d\sigma) = (\partial_x v_1 + \partial_y v_2 + \partial_z v_3) dx\wedge dy\wedge dz.$$

We recognize

$$d(\mathbf{v}\cdot d\sigma) = \text{div}\ \mathbf{v}\ dV$$

where dV is an abbreviation for the third degree basic differential $dx\wedge dy\wedge dz$.

Let us summarize our results in a diagram.

$$\begin{array}{lcl}
\Omega^0(\mathbb{R}^3) & \ni & f \\
\downarrow & & \quad\searrow \\
\Omega^1(\mathbb{R}^3) & \ni & \mathbf{u}\cdot d\mathbf{r}\ ,\ \mathbf{grad} f\cdot d\mathbf{r} \\
\downarrow & & \quad\searrow \\
\Omega^2(\mathbb{R}^3) & \ni & \mathbf{v}\cdot d\sigma\ ,\ \mathbf{curl}\ \mathbf{u}\cdot d\sigma \\
\downarrow & & \quad\searrow \\
\Omega^3(\mathbb{R}^3) & \ni & g dV\ ,\ \text{div}\mathbf{v}\ dV\ .
\end{array}$$

As we see, a great unification among the differential operators **grad, curl** and div in $\mathbb{R}^3$ is achieved by using the exterior differential d. In particular, the basic identities

$$\mathbf{curl}\ \mathbf{grad} f = 0 \quad \text{and} \quad \text{div}\ \mathbf{curl}\ \mathbf{u} = 0$$

are equivalent to the single identity

$$d^2 = 0 \qquad (\text{or } d_{p+1} \circ d_p = 0).$$

This result is quite general.

Remark. Our definition of the exterior differential was based on the following properties

a) $d : \Omega \longrightarrow \Omega$ is *additive*,

b) for $f \in \mathcal{C}^\infty = \Omega^0$, $d(f\omega) = df \wedge \omega + f\, d\omega$,

c) if $\omega = df_1 \wedge df_2 \wedge \ldots \wedge df_p$, then $d\omega = 0$.

They *uniquely characterize* the operator d. In fact, the property c) shows that the differential of $dx^i \wedge dx^j \wedge \ldots$ vanishes, and b) now shows that

$$d(f\, dx^i \wedge dx^j \wedge \ldots) = (df) \wedge dx^i \wedge dx^j \wedge \ldots$$

in complete accordance with the **basic rule**. Alternatively, the exterior differential can also be *uniquely characterized* by the following properties

a) for $f \in \mathcal{C}^\infty = \Omega^0$, df is the usual 1-form : $df \in \Omega^1$,

b) for $\omega \in \Omega^p$, $d(\omega \wedge \eta) = d\omega \wedge \eta + (-1)^p \omega \wedge d\eta$,

c) $d^2 = 0$.

The interested reader will have to consult the references given in the bibliography for this point.

11.5 POINCARE THEOREM

Coming back to the general context of the dimension n, we can state the fundamental

Theorem (Lemma of Poincaré).

a) For any open set $D \subset \mathbb{R}^n$ and any $\omega \in \Omega^p(D)$

$$d^2\omega = d_{p+1} \circ d_p \omega = 0.$$

b) Conversely, if the open set $D \subset \mathbb{R}^n$ is *star shaped* with respect to a point (in particular if it is *convex*), the condition

$$d\eta = d_{p+1}\eta = 0 \text{ for some } \eta \in \Omega^{p+1}(\mathbb{R}^n)$$

implies that there exists a $\omega \in \Omega^p(\mathbb{R}^n)$ with $\eta = d\omega = d_p\omega$. □

It is convenient to introduce some terminology.

Definitions. Let $\omega \in \Omega^p(\mathbb{R}^n)$ be a differential form of degree p. Then we say that ω is **exact** if there exists an $\eta \in \Omega^{p-1}(\mathbb{R}^n)$ with $d\eta = \omega$. Such a differential form η is called **primitive of** ω. We say that ω is **closed** if $d\omega = 0$.

Alternatively, one could say that exact differentials are those which *derive from a potential* (*primitives* playing the role of *potentials*). At least, this way of speaking would extend the notion of potentials from $\mathbb{R}^3$ to $\mathbb{R}^n$. To recognize if a given differential form ω (of degree p) is exact, we compute its exterior differential : if it is exact, the result will be

$$d\omega = d(d\eta) = d^2\eta = 0$$

by the Poincaré theorem (part a). This shows that

$$\omega \text{ exact} \implies \omega \text{ closed},$$

and this necessary condition will also be sufficient in star-shaped domains (part b). For completeness, let us also record

$$\text{for any } \omega \ (\in \Omega^p(\mathbb{R}^n)),\ d\omega \ (\in \Omega^{p+1}(\mathbb{R}^n)) \text{ is closed.}$$

Proof of the Poincaré theorem.

Part a) has already been proved in the 3-dimensional case by the two identities **curl grad** = 0 and div **curl** = 0. We give now the proof in $\mathbb{R}^n$. It is enough to show that d^2 vanishes on monomials. Let $f dx_i \wedge dx_j \wedge \ldots = f(dx_i \wedge \ldots)$ be a multiple of a basic expression of a certain degree p. We have

$$d(f dx_i \wedge \ldots) = \sum D_k f \ dx_k \wedge dx_i \wedge \ldots$$

where it is enough to sum over indices k different from i, j,...

Taking the exterior differential once more

$$d^2(f dx_i \wedge \ldots) = \sum_{\ell,k} D_\ell D_k f \ dx_\ell \wedge dx_k \wedge dx_i \wedge \ldots.$$

Only indices $k \neq \ell$ occur in these expressions and terms cancel out in pairs since e.g. for $k \neq \ell$

$$D_\ell D_k f \ dx_\ell \wedge dx_k + D_k D_\ell f \ dx_k \wedge dx_\ell =$$

$$(D_\ell D_k f - D_k D_\ell f) \ dx_\ell \wedge dx_k = 0.$$

We have used once more the Schwartz theorem on the symmetry of second order derivatives (for $\mathcal{C}^2$ functions).

We turn to the proof of part b). Instead of writing a proof in the star shaped domains case, we shall only consider the case where D is a product of (open) intervals. This is not a serious restriction since the existence of primitives is really a local question (instead of making radial integrations, we shall simply consider variables separately). There is no loss in generality if we assume that $0 \in D$. This proof is made by induction on the dimension n of the space. The case $n = 1$ is obvious: if

$$\omega = f dx \in \Omega^1(D),$$

(the condition $d\omega = 0$ is automatically satisfied since $d\omega = \partial_x f dx \wedge dx = 0$) we can take for primitive of ω any $\eta = F \in \Omega^0(D)$ where F is a primitive of f, e.g.

$$F(x) = \int_0^x f(\xi) d\xi.$$

Let us assume now that the theorem is proved in $D_o \subset \mathbb{R}^{n-1}$ and introduce one more variable $x_n = t$ so that $D = D_o \times I \ni 0$, where $D_o = D \cap \mathbb{R}^{n-1}$. The first observation is the following:

If the differential form $\omega \in \Omega^p(D)$ is closed and does not contain dt, its coefficients are independent of t.

This is obvious since no compensation can occur between monomials obtained by taking partial derivatives with respect to t.

Now write the closed differential form $\omega \in \Omega^p(D)$ in the form

$$\omega = \omega_1 + \omega_o \wedge dt,$$

ω_1 based on monomials not containing dt, $\deg(\omega_o) = p-1$. Let us also denote by $\omega^{\sim} = \omega_1|_{t=0}$ the differential form obtained by putting $t=0$ and $dt=0$ in ω. In particular, $d(\omega^{\sim})$ contains all monomials of $d\omega$ which do not contain dt (with coefficients evaluated at $t=0$!). The assumption $d\omega = 0$ implies $d(\omega^{\sim}) = 0$. The induction assumption gives a primitive of $\omega^{\sim}$ in $D_o = \mathbb{R}^{n-1} \cap D$

$$\Omega^{\sim} \in \Omega^{p-1}(D_o), \ d(\Omega^{\sim}) = \omega^{\sim}.$$

Let us write more explicitely

$$\omega = \Sigma_{i_p < n} a_I dx^I + \Sigma_{i_p = n} a_I dx^I =$$

$$= \omega_1 + \Sigma_J a_{J,n} dx^J \wedge dt =$$

$$= \omega_1 + \omega_o \wedge dt.$$

The coefficients $a_I(x,t)$ of ω_1 correspond to basic monomials formed with coordinates of indices $I = (i_1 < \ldots < i_p)$ taken in $\{1,\ldots,n-1\}$ and correspondingly

$$\Omega^{\sim} = \Sigma\, a_J^{\sim} dx^J \quad (a_J^{\sim} \text{ independent of } t).$$

Let us define

$$A_J = a_J^{\sim} + (-1)^{p-1} \int_0^t a_{J,n}(x,\tau) d\tau,$$

$$\Omega = \sum_J A_J dx^J = \Omega^\sim + (-1)^{p-1}\sum_J \int_0^t a_{J,n}(x,\tau)d\tau \; dx^J \in \Omega^{p-1}(D)$$

(based on monomials without dt). The exterior differential of Ω is easily computed

$$d(\Omega - \Omega^\sim) =$$

$$= (-1)^{p-1}\sum_J \left(a_{J,n}(x,t)dt\wedge dx^J + \sum_i \int_0^t \partial_i a_{J,n}(x,\tau)d\tau \; dx^i\wedge dx^J\right) =$$

$$= \sum_J a_{J,n}(x,t)\; dx^J\wedge dt + \sum_{J,i} \int_0^t \partial_i a_{J,n}(x,\tau)d\tau \; dx^J\wedge dx^i =$$

$$= \omega_o\wedge dt + (\text{terms without dt and vanishing for } t{=}0).$$

In particular, we see that

$$(d\Omega - \omega)_{t=0} = d(\Omega - \Omega^\sim)_{t=0} + d(\Omega^\sim)_{t=0} - (\omega)_{t=0} =$$

$$= (\omega_o\wedge dt)_{t=0} + d(\Omega^\sim)_{t=0} - (\omega_1 + \omega_o\wedge dt)_{t=0} =$$

$$= d(\Omega^\sim) - \omega^\sim = 0.$$

But

$$d\Omega = \sum_J \partial_t A_J \; dt\wedge dx^J + \text{terms without dt} =$$

$$= \sum_J a_{J,n} \; dx^J\wedge dt + \text{terms without dt}$$

(since the $a_J^\sim$ are independent of t by construction). Thus,

1) $d\Omega - \omega = (d\Omega - \omega_o\wedge dt) - \omega_1$ does not contain dt,

2) $d(d\Omega - \omega) = d^2\Omega - d\omega = 0$ since by assumption ω is closed.

The first observation made in this proof shows that $d\Omega - \omega$ is independent from t. For t = 0, it vanishes and this proves

$$d\Omega - \omega = 0. \qquad ∎$$

Observe that the primitive Ω of ω that we have just constructed is based on momomials without dt (t is the last coordinate). In particular in $\mathbb{R}^3$, if $\omega = \mathbf{v}\cdot d\sigma$ is closed, i.e. $\operatorname{div}\mathbf{v} = 0$, there exists a potential $\mathbf{u}$ for $\mathbf{v}$: **curl** $\mathbf{u} = \mathbf{v}$ which is a *horizontal* vector field

$$\Omega = \mathbf{u}\cdot d\mathbf{r} = u_1 dx + u_2 dy \quad (u_3 = 0).$$

Here is a useful formula for finding such a vector potential in $\mathbb{R}^3$. If $\operatorname{div}\mathbf{v} = 0$, the following horizontal field $\mathbf{u}$

$$u_1(x,y,z) = \int_0^z v_2(x,y,\zeta)d\zeta - \int_0^y v_3(x,\eta,0)d\eta,$$

$$u_2(x,y,z) = -\int_0^z v_1(x,y,\zeta)d\zeta,$$

$$u_3(x,y,z) = 0$$

is a vector potential of **v** (it is characterized among all possible potentials of **v** by the two properties : **u** is horizontal and **u** vanishes at the origin). Let us check for example that that the third component of **curl u** is v_3. This third component is

$$\partial_x u_2 - \partial_y u_1 = -\int_0^z \left(\partial_x v_1 + \partial_y v_2\right)(x,y,\zeta)d\zeta + v_3(x,y,0)$$

but $\partial_x v_1 + \partial_y v_2 = -\partial_z v_3$ since div**v** = 0. We obtain

$$\partial_x u_2 - \partial_y u_1 = \int_0^z \partial_z v_3(x,y,\zeta)d\zeta + v_3(x,y,0) = v_3(x,y,z)$$

as expected !

CHAPTER 12
STOKES' FORMULA

12.1 ORIENTED PIECES IN $\mathbb{R}^n$ AND THEIR BOUNDARIES

Our purpose is not to give formal definitions of all notions used in this chapter. Rather, we intend to give enough intuitive support for a correct use of the Stokes formula (12.3). Readers rebutted by so many concepts can try to attack Stokes' formula first and come back to sections 12.1 and 12.2 later, when need arises...

Multiple integrals have to be evaluated over **pieces** of manifolds of suitable dimension. For example, triple integrals have to be computed on cubes, balls, etc. whereas double integrals have to be evaluated on parallelograms, spheres, etc. In each case, one must be able to parameterize the piece, or at least, cover it by images of parametrizations. Any parametrization gives a system of curvilinear coordinates in the piece.

More precisely, a p-**dimensional piece** $\Pi \subset \mathbb{R}^n$ is a compact subset which can be covered by images of parametrizations

$$\varphi : D \longrightarrow \Pi$$

having domains $D \subset \mathbb{R}^p$ of the form

$$\|\mathbf{x}\| < 1 \text{ and } x_1 \le 0, \ldots, x_k \le 0 \quad \text{(for some } 0 \le k \le p).$$

The following figure shows the possibilities for $p = 2 : D \subset \mathbb{R}^2$

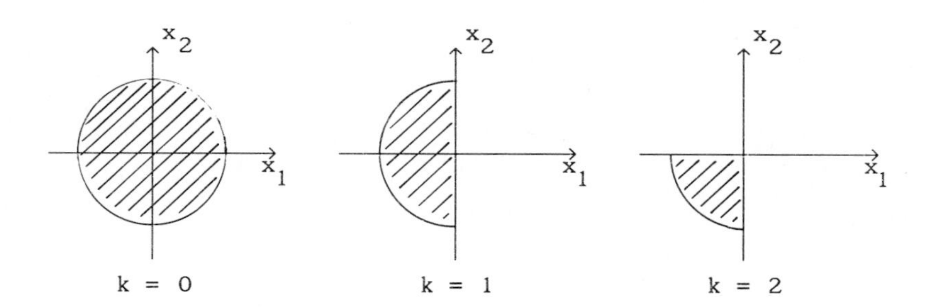

Fig.12.1

We shall only consider parametrizations φ which admit an extension $\varphi^{\sim}$ to an open neighborhood $D^{\sim}$ of D in $\mathbb{R}^p$ (with values in $\mathbb{R}^n$) with

- ▷ $\varphi^{\sim}$ is a homeomorphism of $D^{\sim}$ onto its image,
- ▷ $(\varphi^{\sim})'(\mathbf{a})$ is one to one (into) for all $\mathbf{a} \in D^{\sim}$.

These are typical parametrizations, but one could also consider domains D in the form of a square, a cube or a *simplex* if this is more convenient (without departing *really* from the previous typical cases). The second property required for parametrizations is usually called **immersion** property (of $\tilde{\varphi}$).

Two parametrizations

$$\varphi : D \longrightarrow \Pi \quad \text{and} \quad \phi : D' \longrightarrow \Pi$$

are said to induce the **same orientation**

(in the neighborhood of a point of Π) if the composite

$$\psi = \varphi^{-1} \circ \phi : D' \longrightarrow D$$

has a *positive Jacobian determinant*. This requirement introduces an equivalence relation between parametrizations, and a class of parametrizations with respect to this equivalence relation is called **orientation**. In each neighborhood of a point of Π, there are two possible choices of orientation.

Take a parametrization φ as before, and a point $\mathbf{a} \in D$ with image $\varphi(\mathbf{a}) \in \Pi$. Since parametrizations are *immersions*, the image of the canonical basis of $\mathbb{R}^p$ under the linear map

$$\varphi'(\mathbf{a}) \quad (\text{ or more precisely } \mathbf{h} \longmapsto \varphi(\mathbf{a}) + \varphi'(\mathbf{a})\mathbf{h}\ )$$

is a basis of the tangent space $T_{\varphi(\mathbf{a})}(\Pi)$ of the piece Π at the point $\varphi(\mathbf{a})$.

Lemma. Two parametrizations φ and ϕ will be in the same orientation class at the point $\varphi(\mathbf{a}) = \phi(\mathbf{b}) \in \Pi$ if the change of basis carrying

$$(\varphi'(\mathbf{a})(\mathbf{e}_i))_{1 \le i \le p} \quad \text{to} \quad (\phi'(\mathbf{b})(\mathbf{e}_i))_{1 \le i \le p}$$

has a positive determinant. □

Proof. Let us indicate the idea of proof with a diagram in which we use the simplifying notations

$$U = \varphi'(\mathbf{a}),\ \mathbf{t}_i = U(\mathbf{e}_i) \text{ (tangent vector to } \Pi \text{ at } \varphi(\mathbf{a})),$$

$$V = \phi'(\mathbf{b}),\ \tau_i = V(\mathbf{e}_i) \text{ (tangent vector to } \Pi \text{ at } \phi(\mathbf{b}) = \varphi(\mathbf{a})).$$

Here is the situation

$$\begin{array}{ccc} & \xrightarrow{U} & (\mathbf{t}_i) \\ (\mathbf{e}_i) & & \downarrow VU^{-1} \\ U^{-1}V \downarrow & \xrightarrow{V} & (\tau_i) \\ (\varepsilon_i) & \xleftarrow{U^{-1}} & \end{array}$$

In this diagram, we read the identity

$$U(U^{-1}V)U^{-1} = VU^{-1}.$$

Consequently $U^{-1}V$ and VU^{-1} have the *same* determinant (a fortiori, these determinants have the same *sign*!). ∎

Alternatively, we see that an orientation of the piece Π at the point $\varphi(\mathbf{a})$ is given by a class of bases of the tangent space $T_{\varphi(\mathbf{a})}(\Pi)$ with respect to the equivalence relation defined by positive determinant isomorphisms.

A **global orientation** of Π is given by a continuous choice of orientations at all of its points. Such a global orientation is given by a choice of a set of *coherent* parametrizations having union of images equal to the entire piece Π. By *coherence*, we mean the following property : if two parametrizations φ and ϕ have intersecting images, they induce the same orientations in the tangent spaces of the points common to both images. In other words, if

μ = restriction of φ^{-1} to this common part of the images,

ν = restriction of ϕ^{-1} to this common part of the images

(such *inverses* of parametrizations are usually called **chart maps** on Π), the functional determinant of $\mu\circ\nu^{-1}$ should remain positive

$$\det(\mu\circ\nu^{-1})' > 0 \text{ on } \mathrm{Image}(\varphi) \cap \mathrm{Image}(\phi) \subset \Pi.$$

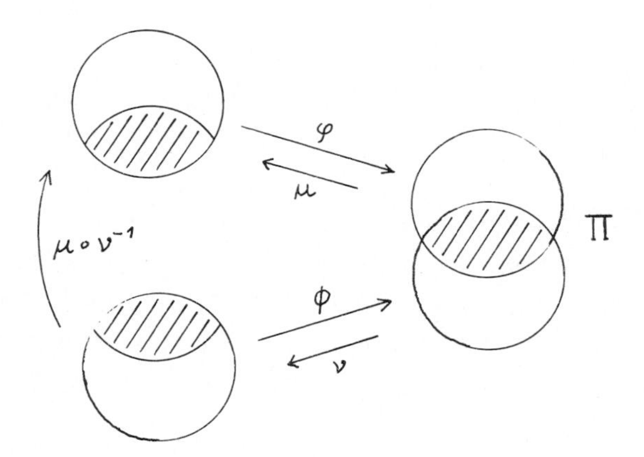

Fig.12.2

For example, a continuous field of local frames in the tangent spaces of Π determines an orientation of Π. A piece admitting a global orientation is said to be **orientable**. The following figure shows an example of an orientable piece and an example of a non orientable piece (both are 2-dimensional pieces in $\mathbb{R}^3$).

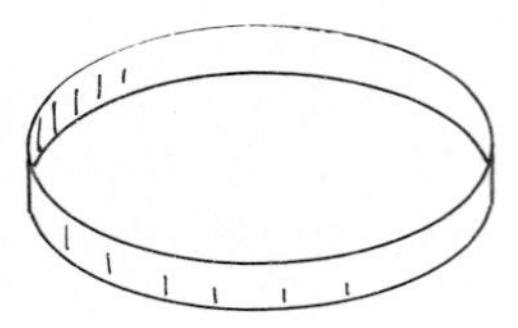

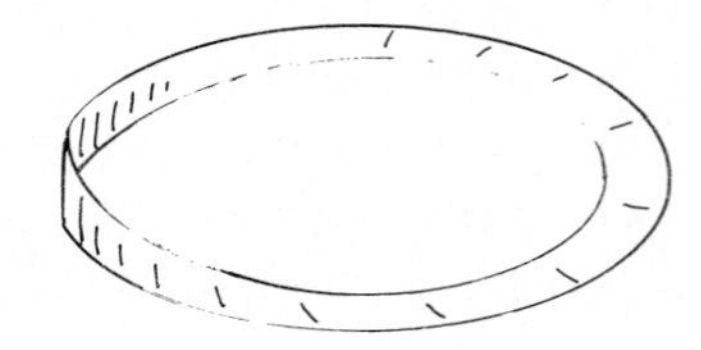

Usual band : orientable   Möbius band : non orientable

Fig.12.3

When a piece is *connected* and orientable, it is possible to select an orientation on it simply by prescribing *one basis of one tangent space to it.*

An **oriented piece** Π is simply a couple consisting of an (orientable) piece with a choice of orientation. Practically, an oriented piece is simply given by the couple : Π and one basis in one tangent space to each *component* of Π.

As we have defined them, pieces have **boundaries**. Indeed, if $k \geq 1$ in the domain D of a parametrization φ, the images of

$$D \cap \{\mathbf{x} \in \mathbb{R}^p : x_i = 0\} \qquad (1 \leq i \leq k)$$

are parts of **faces** of Π. If $\dim(\Pi) = p$, each face of Π is a piece of dimension p-1. The union of all the faces of Π constitutes its **boundary** ∂Π.

Let Π be an oriented piece of dimension p in $\mathbb{R}^n$. Let us explain how the boundary ∂Π inherits a canonical orientation from Π (in particular, ∂Π is always orientable). Take any smooth point $\mathbf{m} \in \partial\Pi$ and any basis $(\mathbf{t}_i)_{1\leq i\leq p-1}$ of the tangent space $T_{\mathbf{m}}(\partial\Pi)$. The tangent space at **m** *of* Π contains $T_{\mathbf{m}}(\partial\Pi)$ with codimension 1 and has a basis consisting of

$$(\nu, \mathbf{t}_1, \ \ldots \ , \mathbf{t}_{p-1})$$

where ν represents any vector pointing outside Π. For example, one can take for ν the tangent vector of the coordinate curve

$$t = x_i \longmapsto \varphi(x_1, \ldots, t{=}x_i, \ldots, x_p)$$

at the origin (we are assuming that the point $\mathbf{m} = \varphi(\mathbf{a})$ lies on the face containing the image of $D \cap \{\mathbf{x} \in \mathbb{R}^p : x_i = 0\}$). The convention is now the following *the* **induced orientation** *of the face of* $\partial\Pi$ *containing the point* $\mathbf{m}$ *is given by the basis* $(\mathbf{t}_1, \dots, \mathbf{t}_{p-1})$ *exactly when the orientation of* Π *at* $\mathbf{m}$ *is given by* $(\nu, \mathbf{t}_1, \dots, \mathbf{t}_{p-1})$.

Observe that the exterior normal has to be written *first*. The only justification for this convention is Stokes' formula : it holds with this choice of signs ! It is important to determine correctly the induced orientation on *each face* of $\partial\Pi$ since *one* error may completely alter a sum containing orientation dependent terms (a *physical intuition* may help in determining *one global sign* —e.g. energy is *positive* for a freely moving particle— but will usually not help for individual terms !).

For example, let $\Pi \subset \mathbb{R}^2$ be a two-dimensional piece, oriented by the choice of the canonical basis $\mathbf{e}_1$, $\mathbf{e}_2$ of $\mathbb{R}^2$. What is the induced orientation of the boundary $\partial\Pi$? As the following figure shows, this boundary $\partial\Pi$ is generally a union of one dimensional pieces (curves) γ_i. Select one smooth point P_i on each γ_i. Then

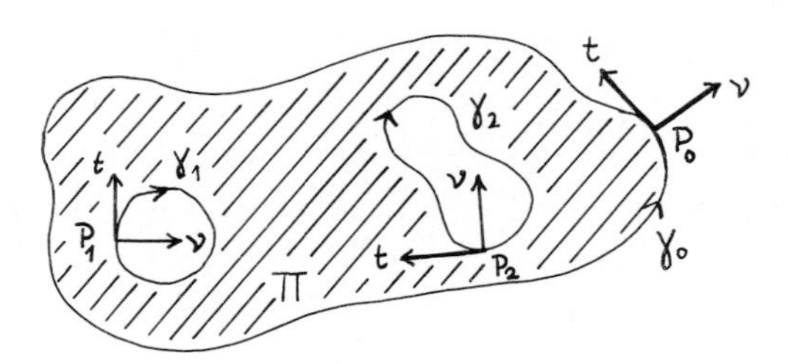

Fig.12.4

the choice of tangent vector $\mathbf{t}_i$ to γ_i at P_i gives the correct induced orientation on this component of the boundary, provided $(\nu, \mathbf{t}_i)$ has the same orientation as $(\mathbf{e}_1, \mathbf{e}_2)$. In other words, the determinant formed with the components of ν and $\mathbf{t}_i$ should be positive. In this case —$\Pi \subset \mathbb{R}^2$ with the canonical orientation of $\mathbb{R}^2$— the induced orientation of the boundary $\partial\Pi$ is obtained by parametrizations *which leave* Π *on the left !* This *rule of thumb* is easy to apply in $\mathbb{R}^2$. The analogue for 2-dimensional pieces of $\mathbb{R}^3$ is the *corkscrew law*. A piece of surface $\Pi \subset \mathbb{R}^3$ is conveniently oriented by a choice of normal $\mathbf{n}$ to it : a basis $(\mathbf{t}_i)$ of a tangent space $T_{\mathbf{m}}$ of a point $\mathbf{m} \in \Pi$ is in the orientation defined by $\mathbf{n}$ if $(\mathbf{n}, \mathbf{t}_1, \mathbf{t}_2)$ is in the canonical orientation of $\mathbb{R}^3$ (namely, if the determinant formed with the components of $\mathbf{n}, \mathbf{t}_1, \mathbf{t}_2$ is positive). If now $\mathbf{m}$ is a smooth point of the boundary $\partial\Pi$, and ν is the exterior normal in $T_{\mathbf{m}}$ of $\partial\Pi$, the tangent **t** to $\partial\Pi$ at **m** will give the induced orientation of the boundary, provided $(\nu,\mathbf{t})$ is in the orientation of Π. This means that the determinant of $\mathbf{n}$, ν, $\mathbf{t}$ (in this order!) should be positive.

12.2 INTEGRATION OF FORMS ON ORIENTED PIECES

When $\omega \in \Omega^p(\mathbb{R}^n)$ is a differential form of degree p, its integral over a p-dimensional piece Π is defined as follows. For simplicity, we assume that Π is covered by one parametrization (in general, we would subdivide Π in smaller pieces) :

$$\varphi : D \longrightarrow \Pi .$$

Using $\varphi = {}^t(\varphi^1,\ldots,\varphi^n)$ and $dx^i = d\varphi^i = \sum (\partial\varphi^i/\partial\xi^j)d\xi^j$, the form ω will have an expression $\omega_\xi \in \Omega^p(\mathbb{R}^p)$ in the variables ξ of the form

$$\omega_\xi = g(\xi^1,\ldots,\xi^p)\ d\xi^1\wedge\ldots\wedge d\xi^p.$$

We put

$$\int_\Pi \omega = \int\!\!\ldots\!\!\int_D g(\xi^1,\ldots,\xi^p)\ d\xi^1\ldots d\xi^p$$

if the parametrization φ belongs to the given orientation of Π, or

$$\int_\Pi \omega = -\int\!\!\ldots\!\!\int_D g(\xi^1,\ldots,\xi^p)\ d\xi^1\ldots d\xi^p$$

if the parametrization φ is not in the orientation of Π.

In other words, $\int_\Pi \omega = \int_D \omega_\xi$ is computed by means of the rule

$$d\xi^1\ldots d\xi^p = \begin{cases} d\xi^1\wedge\ldots\wedge d\xi^p & \textit{if the } \xi^j \textit{ give the orientation of } \Pi \\ -\,d\xi^1\wedge\ldots\wedge d\xi^p & \textit{if not.}\end{cases}$$

This rule will conveniently be abbreviated by

$$d\xi^1\ldots d\xi^p = |d\xi^1\wedge\ldots\wedge d\xi^p| = \pm d\xi^1\wedge\ldots\wedge d\xi^p$$

with an orientation dependent sign.

The preceding definition is meaningful since it is independent of the choice of parametrization. Let indeed

$$\psi : (\eta^1,\ldots,\eta^p) \longmapsto \psi(\eta^1,\ldots,\eta^p) : D'\longrightarrow \Pi$$

be another parametrization of Π. With $\phi = \varphi^{-1}\circ\psi$ we can consider that

$$\psi = \varphi\circ\phi : D' \xrightarrow{\phi} D \xrightarrow{\varphi} \Pi.$$

Hence

$$\omega_\eta = \det(\phi'(\eta))\ g(\xi^1(\eta),\ldots,\xi^p(\eta))\ d\eta^1\wedge\ldots\wedge d\eta^p ,$$

more briefly $\omega_\eta = \det(\phi')\omega_\xi$.

Now, the formula for the change of variables in multiple integrals gives

$$\int_{\Pi} \omega = \int_{D} \omega_{\xi} = \int \ldots \int_{D} g(\xi)\, d\xi^1 \ldots d\xi^p =$$

$$= \int \ldots \int_{D'} g(\phi(\eta))\,|\det(\phi')|\, d\eta^1 \ldots d\eta^p = \int_{D'} \omega_{\eta}$$

provided $\det(\phi') > 0$, namely φ and ψ are in the same orientation class. In the contrary case, two compensating signs are to be introduced in the preceding chain of equalities, thus preserving the extreme terms.

12.3 STOKES' FORMULA, EXAMPLES

Let Π be an oriented piece in $\mathbb{R}^n$ and $\partial\Pi$ its oriented boundary. If the dimension of Π is p, we can integrate p-forms ω on Π and **Stokes'theorem** asserts that provided ω admits a primitive η (of degree p-1)

$$\int_{\Pi} \omega = \int_{\partial\Pi} \eta \qquad (\text{where } d\eta = \omega).$$

Several **comments** should be made. In order to be valid, this formula must be applied to forms which are continuous over the pieces in question : ω must be at least continuous in a neighborhood of the piece Π and η must consequently be at least $\mathcal{C}^1$ in a neighborhood of Π.

The simplest case of Stokes' formula is the case $n = 1$, $\Pi = I$: compact interval $[a,b]$ so that $\partial\Pi = \partial I = \{b\} - \{a\}$ and

$$\int_{I} f dx = [F]_{\partial I} = F(b) - F(a) \qquad (\text{if } F' = f).$$

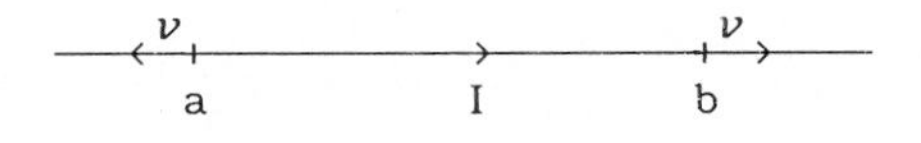

Fig.12.5

(We interpret a 0-dimensional integration as an evaluation at the end points.) This is the usual —most useful— integration formula, valid with the same regularity restrictions. For example, consider the following fallacious computation. Since $F = -1/x$ has derivative $f = 1/x^2$, we might be tempted to write

$$\int_{-1}^{+1} dx/x^2 = [-1/x]_{-1}^{+1} = -1 -(+1) = -2.$$

This is simply silly since the integrand $1/x^2$ is positive over the whole integration interval $[-1,1]$! The truth is that the integral of $1/x^2$ over $[-1,1]$ *does not even exist !*

Stokes' formula also shows that if $\omega \in \Omega^p(\mathbb{R}^n)$ admits a primitive η, the integral of ω on any piece Π only depends on the values of a primitive on the boundary $\partial\Pi$ of Π.

In the case of dimension 2, a typical case of the Stokes theorem is concerned with 2-forms ω admitting a primitive

$$\eta = udx + vdy \in \Omega^1(\mathbb{R}^2).$$

Namely, $\omega = d\eta = (\partial v/\partial x - \partial u/\partial y)\, dx\wedge dy$ and then

$$\int_{\Pi} (\partial v/\partial x - \partial u/\partial y)\, dx\wedge dy = \int_{\mathcal{C}=\partial\Pi} udx + vdy.$$

This formula is **Riemann's formula**, usually written

$$\iint_{\Pi} (\partial v/\partial x - \partial u/\partial y)\, dxdy = \oint_{\mathcal{C}=\partial\Pi} udx + vdy$$

where the correct orientation of the boundary curve(s) $\mathcal{C} = \partial\Pi$ has to be chosen.

Let us give the typical cases of Stokes' formula in $\mathbb{R}^3$.

1) When the 1-form $\omega = \mathbf{u}\cdot\mathbf{dr} \in \Omega^1(\mathbb{R}^3)$ admits a primitive (i.e. when the vector field $\mathbf{u}$ derives from a scalar potential f) $\eta = f$, say $d\eta = df = \mathbf{grad} f\cdot\mathbf{dr} = \mathbf{u}\cdot\mathbf{dr}$, we can write

$$\int_{\mathcal{C}} \mathbf{u}\cdot\mathbf{dr} = [f]_{\partial\mathcal{C}} = f(Q) - f(P)$$

if $\mathcal{C}$ is a curve linking P and Q (origin at P and extremity at Q).

2) Take now a 2-form $\omega = \mathbf{v}\cdot\mathbf{d\sigma} \in \Omega^2(\mathbb{R}^3)$ admitting a primitive $\eta = \mathbf{u}\cdot\mathbf{dr} \in \Omega^1(\mathbb{R}^3)$, namely $\mathbf{v} = \mathbf{curl\ u}$ derives from the vector potential $\mathbf{u}$. Then we have

$$\int_{\Sigma} \mathbf{v}\cdot\mathbf{d\sigma} = \int_{\Sigma} \mathbf{curl\ u}\cdot\mathbf{d\sigma} = \int_{\mathcal{C}=\partial\Sigma} \mathbf{u}\cdot\mathbf{dr}$$

for any piece of surface Σ admitting $\mathcal{C}$ as its boundary : the flux of $\mathbf{v}$ through Σ is equal to the circulation of the vector potential $\mathbf{u}$ on the boundary $\mathcal{C}$ of Σ. Observe that in general, the boundary of Σ will be made up of several components (recall the discussion in 12.1 for pieces $\Sigma \subset \mathbb{R}^2$).

3) A 3-form $\omega = gdV \in \Omega^3(\mathbb{R}^3)$ having primitive $\eta \in \Omega^2(\mathbb{R}^3)$ must be of the form $\omega = \mathrm{div}\ \mathbf{v}\cdot dV$ (i.e. $\eta = \mathbf{v}\cdot\mathbf{d\sigma}$) and in this case, Stokes' formula leads to

$$\int_V gdV = \int_V \mathrm{div}\ \mathbf{v}\ dV = \int_{\Sigma=\partial V} \mathbf{v}\cdot\mathbf{d\sigma}.$$

All these formulas represent multiple integrals : as soon as a correct parametrization of the pieces is chosen, they become usual multiple integrals and Stokes' theorem (when applicable) shows how to *economize one integration* by use of the boundary.

Finally, let us rewrite the Stokes formula in the following form

$$\int_{\Pi} d\eta = \int_{\partial\Pi} \eta$$

for any form η of degree p-1 and any piece Π of dimension p. Formally, the exterior differential d can be shifted to the piece as a boundary operator ∂. With the other notation

$$\int_{\Pi} \omega = \langle \Pi,\omega \rangle,$$

the Stokes' formula reads

$$\langle \Pi, d\eta \rangle = \langle \partial\Pi, \eta \rangle.$$

This way of writing the formula exhibits a *duality* between d and ∂. More precisely, the piece Π can be identified to the linear form on $\Omega^p(\mathbb{R}^n)$ that it produces by integration :

$$\Pi \in (\Omega^p(\mathbb{R}^n))^* \quad \text{is the linear form} \quad \omega \longmapsto \int_{\Pi} \omega .$$

Then Stokes' theorem states that the *transposed* of the operator

$$d = d_{p-1} : \Omega^{p-1}(\mathbb{R}^n) \longrightarrow \Omega^p(\mathbb{R}^n) \text{ is}$$

is simply the boundary operator $^t d = \partial : (\Omega^p(\mathbb{R}^n))^* \longrightarrow (\Omega^{p-1}(\mathbb{R}^n))^*$ in this model.

The preceding interpretation also explains the terminology : a piece Π is **closed** when its boundary $\partial\Pi$ is empty (corresponds to the zero linear form). A form ω is correspondingly called *closed* when $d\omega = 0$.

12.4 BACK TO THE DEL OPERATOR

A couple of extra applications of the Stokes formula are still interesting. For example,

$$d(f dy\wedge dz) = df\wedge dy\wedge dz = \frac{\partial f}{\partial x} dx\wedge dy\wedge dz = f'_x dV$$

leads to the vector relation

$$d(f d\sigma) = \begin{pmatrix} d(f dy\wedge dz) \\ d(f dz\wedge dx) \\ d(f dx\wedge dy) \end{pmatrix} = \begin{pmatrix} \partial f/\partial x \ dV \\ \partial f/\partial y \ dV \\ \partial f/\partial z \ dV \end{pmatrix} = \mathbf{grad} f \ dV.$$

Hence the Stokes formula gives

$$(1) \qquad \iiint_V \mathbf{grad} f \ dV = \iint_{\Sigma=\partial V} f d\sigma.$$

Recall that we have also proved

$$(2) \qquad \iiint_V \operatorname{div} \mathbf{v} \, dV = \iint_{\Sigma=\partial V} \mathbf{v} \cdot d\sigma.$$

Finally, it is easy to check (in components)

$$d(d\sigma \wedge \mathbf{u}) = \mathbf{curl}\ \mathbf{u} \, dV,$$

giving rise to the vector relation

$$(3) \qquad \iiint_V \mathbf{curl}\ \mathbf{u} \, dV = \iint_{\Sigma=\partial V} d\sigma \wedge \mathbf{u}.$$

These three formulas lead to integral interpretations of the differential operators **grad,** div and **curl**.

Fix a point P in space, and take a small element of volume V centered at P (a small ball of radius ε, or a cube of side ε would do). Let $|V| = \iiint_V dV$ be the measure of the volume V. For $\varepsilon \longrightarrow 0$, the quotients $\frac{1}{|V|}\iiint_V f dV$ have a limit, namely $f(P)$: compute indeed the difference

$$\frac{1}{|V|}\iiint_V f dV - f(P) = \frac{1}{|V|}\iiint_V f dV - f(P)\frac{1}{|V|}\iiint_V dV =$$

$$= \frac{1}{|V|}\iiint_V [f - f(P)] dV$$

is arbitrarily small when ε is small if f is continuous at P

$$\frac{1}{|V|} \left|\iiint_V f dV - f(P)\right| \leq \operatorname*{Max}_{Q \in V} |f(Q) - f(P)|.$$

Our contention applies to each component of the vector formulas (1) and (3) provided that the functions that we integrate are continuous at P. Let us denote by $V \to P$ the situation of small elements of volume centered at P with diameter $\varepsilon \longrightarrow 0$. We have found

$$(4) \qquad \mathbf{grad} f(P) = \lim_{V \to P} \frac{1}{|V|}\iint_{\Sigma=\partial V} f d\sigma.$$

Similarly,

$$(5) \qquad \operatorname{div} \mathbf{v}(P) = \lim_{V \to P} \frac{1}{|V|}\iint_{\Sigma=\partial V} \mathbf{v} \cdot d\sigma,$$

$$(6) \qquad \mathbf{curl}\ \mathbf{u}\ (P) = \lim_{V \to P} \frac{1}{|V|}\iint_{\Sigma=\partial V} d\sigma \wedge \mathbf{u}.$$

These formulas express the differential operators as limits of quotients. As such, they correspond to the usual definition of the derivative by means of the limit of differential quotients.

Formula (5) is particularly intuitive. It gives the divergence of a vector field **v** at a point P as the limit of flux of **v** through a small sphere around P (suitably divided by the measure of the volume of the ball). This gives a precise meaning to the interpretation of div**v** as the local productivity of **v** at P.

Also note the mnemonics

$$\nabla f(P) = \lim_{V\to P} \frac{1}{|V|}\iint_{\Sigma=\partial V} d\sigma\ f, \tag{7}$$

$$\nabla \mathbf{v}(P) = \lim_{V\to P} \frac{1}{|V|}\iint_{\Sigma=\partial V} d\sigma\cdot\mathbf{v}, \tag{8}$$

$$\nabla\wedge\mathbf{u}(P) = \lim_{V\to P} \frac{1}{|V|}\iint_{\Sigma=\partial V} d\sigma\wedge\mathbf{u}, \tag{9}$$

allowing one to write symbolically

$$\nabla\ \top\ .\ (P) = \lim_{V\to P} \frac{1}{|V|}\iint_{\Sigma=\partial V} d\sigma\ \top\ .$$

for $\top$ equal to the usual, scalar or vector product.

12.5 AREA ELEMENT

We have introduced the *vector 2-form* $d\sigma = \begin{pmatrix} dy\wedge dz \\ dz\wedge dx \\ dx\wedge dy \end{pmatrix}$ in a rather formal way. Our purpose now is to show its connection with a *surface element*.

For this, we first have to recall how areas are computed.

Let S be a portion of surface described by an equation

$$z = f(x,y),\ f \in \mathcal{C}^1 \text{ in a region } S_{xy} = pr_z(S) \subset \mathbb{R}^2.$$

The coordinates of a point $P \in S$ are the components of the vector

$$\mathbf{r} = \overrightarrow{OP} = \begin{pmatrix} x \\ y \\ f(x,y) \end{pmatrix}.$$

To compute the area of S we introduce a mesh of squares in the projection S_{xy} of S and compute Riemann sums of areas of parallelograms tangent to the surface (projecting on the squares of S_{xy}). If the sides of the bottom squares are Δx and Δy, the *vector* sides of the parallelograms on top are

$$\frac{\partial\mathbf{r}}{\partial x}\Delta x \quad\text{and}\quad \frac{\partial\mathbf{r}}{\partial y}\Delta y \quad\text{respectively.}$$

Since

$$\frac{\partial\mathbf{r}}{\partial x} = \begin{pmatrix} 1 \\ 0 \\ f'_x \end{pmatrix} \quad\text{and}\quad \frac{\partial\mathbf{r}}{\partial y} = \begin{pmatrix} 1 \\ 0 \\ f'_y \end{pmatrix},$$

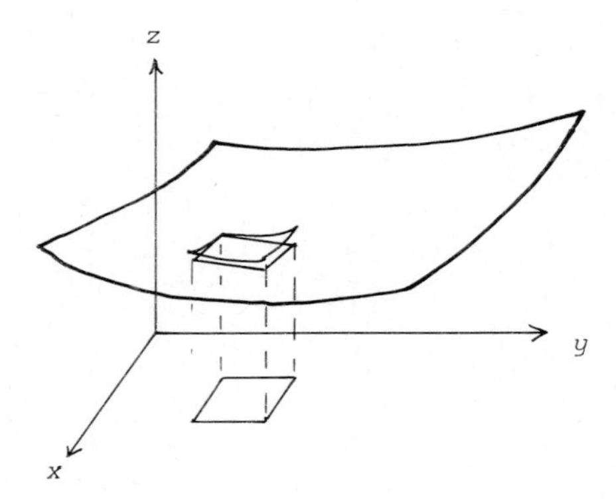

Fig.12.6

the area of such a parallelogram is the norm of

$$\frac{\partial \mathbf{r}}{\partial x}\Delta x \wedge \frac{\partial \mathbf{r}}{\partial y}\Delta y = \begin{vmatrix} \mathbf{e}_1 & \mathbf{e}_2 & \mathbf{e}_3 \\ 1 & 0 & f'_x \\ 0 & 1 & f'_y \end{vmatrix} = \begin{pmatrix} -f'_x \\ -f'_y \\ 1 \end{pmatrix} \Delta x \Delta y.$$

The area of S is by definition the limit of the sums of such elementary areas

$$\text{Area}(S) = |S| = \lim_{\Delta x, \Delta y \to 0} \sum \| \frac{\partial \mathbf{r}}{\partial x}\Delta x \wedge \frac{\partial \mathbf{r}}{\partial y}\Delta y \| =$$

$$= \lim \sum \left\| \frac{\partial \mathbf{r}}{\partial x} \wedge \frac{\partial \mathbf{r}}{\partial y} \right\| \Delta x \Delta y = \lim \sum \|{}^t(-f'_x, -f'_y, 1)\| \Delta x \Delta y =$$

$$= \iint_{S_{xy}} \|{}^t(-f'_x, -f'_y, 1)\| \, dxdy .$$

We have found

$$(1) \qquad |S| = \iint_{S_{xy}} \sqrt{(1 + (f'_x)^2 + (f'_y)^2)} dxdy.$$

As we have donė in (12.2), we write $dxdy = |dx \wedge dy|$ and we compute the *vector differential form* ω for which

$$\|\omega\| = \left\| \begin{pmatrix} -f'_x \\ -f'_y \\ 1 \end{pmatrix} \right\| dxdy = \left\| \begin{pmatrix} -f'_x \\ -f'_y \\ 1 \end{pmatrix} \right\| |dx \wedge dy| = \left\| \begin{pmatrix} -f'_x \\ -f'_y \\ 1 \end{pmatrix} dx \wedge dy \right\|$$

namely ω is *defined by*

$$\omega = \begin{pmatrix} -f'_x dx \wedge dy \\ -f'_y dx \wedge dy \\ dx \wedge dy \end{pmatrix}.$$

The first component of ω is

$$\omega_1 = -f'_x \, dx \wedge dy = -df \wedge dy = dy \wedge df = dy \wedge dz$$

since $dz = df$ on S. This shows that ω is the vector 2-form $d\sigma$ and we have shown

$$(2) \qquad |S| = \iint_S \|d\sigma\|.$$

Remark. The reader will have noted the analogy between the computation of the area of the surface S (in $\mathbb{R}^3$) and the computation of the length of a curve (in $\mathbb{R}^2$ or $\mathbb{R}^3$). However, if it is possible to make Riemann sums of lengths of secants in the case of curves, four points of S projecting on the vertices of a square will not usually lie in the same plane. This is why we had to look at the tangent parallelograms. In the case of curves, this would correspond to approximating the length of the curve by the addition of small lengths of pieces of tangents to the curve (cf. Fig.12.7). Anyway, the Riemann sums converge to the same limit, namely the integral

$$|\mathcal{C}| = \int_I \sqrt{(1 + (f'_x)^2)} \, dx.$$

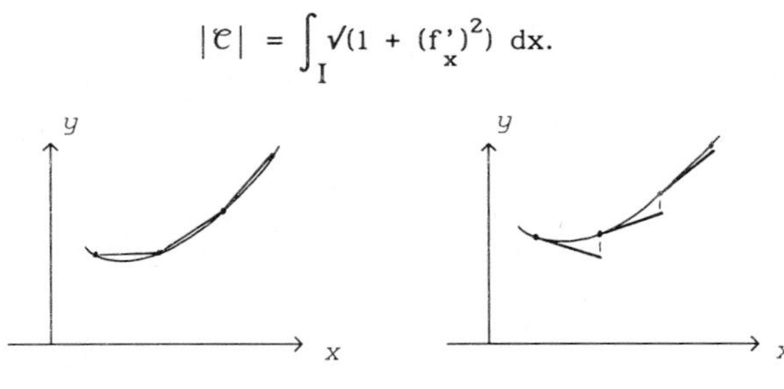

Fig.12.7

The square root under the integral sign is the norm of the vector $(-f'_x,1) = \mathbf{grad}(y - f(x))$. The curve $\mathcal{C}$ is the level 0 curve of the scalar function $y - f(x)$. Hence the vector $(-f'_x,1)$ is orthogonal to $\mathcal{C}$. Similarly, in the surface case, the vector $(-f'_x,-f'_y,1)$ is the gradient of the scalar function $z - f(x,y)$ and hence is orthogonal to the surface S

$$\mathbf{n}(x,y) = \mathbf{grad}(z-f(x,y)) = (-f'_x,-f'_y,1) \perp S.$$

Let us come to the case of a piece of surface described by an implicit equation $F(x,y,z) = 0$ ($F \in \mathcal{C}^1$). To be able to link this description to the preceding one, we assume that F'_z does not vanish (in the neighborhood of the point that we are considering), say $F'_z > 0$. As follows from the implicit function theorem, this allows one to solve the implicit relation $F = 0$ locally with respect to z. Namely, there exists a $\mathcal{C}^1$ function f of two variables such that $z = f(x,y)$ parameterizes locally the surface in question. This means that $F(x,y,f(x,y)) = 0$ (identically). Differentiating this identity, we get

$$F'_x dx + F'_y dy + F'_z(f'_x dx + f'_y dy) = 0$$

and

$$(F'_x + F'_z f'_x)dx + (F'_y + F'_z f'_y)dy = 0.$$

This implies that the coefficients of dx and dy vanish identically

$$f'_x = -F'_x/F'_z \quad \text{and} \quad f'_y = -F'_y/F'_z.$$

This gives the desired connection with the preceding description of a surface. The area of the portion S of the surface is the double integral of the absolute value of

$$\sqrt{(1 + (f'_x)^2 + (f'_y)^2)}\, dx\wedge dy =$$

$$= \frac{(1 + (f'_x)^2 + (f'_y)^2) dx\wedge dy}{\sqrt{(1 + (f'_x)^2 + (f'_y)^2)}} = \frac{F'_z(1 + (f'_x)^2 + (f'_y)^2) dx\wedge dy}{\sqrt{((F'_z)^2 + (F'_x)^2 + (F'_y)^2)}}.$$

But $F'_z(f'_x)^2 dx\wedge dy = F'_z(-F'_x/F'_z)f'_x dx\wedge dy = -F'_x df\wedge dy = F'_x dy\wedge dz$ and similarly for the other terms. We find thus a symmetric formula for the area

$$(3) \qquad |S| = \iint_S \frac{|F'_x dy\wedge dz + F'_y dz\wedge dx + F'_z dx\wedge dy|}{\sqrt{((F'_x)^2 + (F'_y)^2 + (F'_z)^2)}}.$$

Observing as before that

$$\mathbf{grad}\ F = (F'_x, F'_y, F'_z) = \mathbf{n}(x,y,z)$$

is orthogonal to S, we can rewrite (3) as

$$(4) \qquad |S| = \iint_S \frac{|\mathbf{n}\cdot d\sigma|}{\|\mathbf{n}\|} = \iint_S |\nu_S \cdot d\sigma|$$

where $\nu_S = \mathbf{n}/\|\mathbf{n}\|$ denotes a vector normal to S of unit length. We have obtained the following differential expressions for the area element

$$dS = \|d\sigma\| = |\nu_S \cdot d\sigma|.$$

More precisely, instead of $d\sigma$, we should denote by $d\sigma|_S$ the restriction of the vector two form $d\sigma$ to the surface S.

Formula (4) is also valid —and especially useful— when the surface S is given in parametric form

$$x = \xi(u,v),\ y = \eta(u,v),\ z = \zeta(u,v).$$

In particular, it would be valid for a parametrized branch of a singular point, where the three partial derivatives of F vanish.

Let us also introduce a scalar differential form of the second degree

$$dS_{or} = \nu_S \cdot d\sigma$$

where the orientation of the unit normal ν_S to S is chosen in such a way that followed by the orientation of S_{or} it gives the canonical orientation of $\mathbb{R}^3$. It is then possible to define an oriented area (or *algebraic area*) of S by

$$\mathrm{Area}(S_{or}) = \iint_{S_{or}} dS_{or} = \iint_{S_{or}} \nu_S \cdot d\sigma = \pm|S|.$$

Conversely, we can write

$$d\sigma = \nu_S \cdot dS_{or}.$$

It is even possible to introduce a **vector area** by

$$\mathbf{Area}(S_{or}) = \iint_{S_{or}} d\sigma \, .$$

The following discussion shows that the three components of this vector area are the *oriented areas of the three projections* of the piece of surface. It is obtained by integrating the corresponding result for parallelograms.

Lemma. Let **h** and **k** be two vectors in $\mathbb{R}^3$, $\Pi_{\mathbf{h},\mathbf{k}}$ the parallelogram that they define. We have

$$\mathbf{Area}(\Pi_{\mathbf{h},\mathbf{k}}) = \mathbf{h} \wedge \mathbf{k} = \begin{vmatrix} \mathbf{e}_1 & \mathbf{e}_2 & \mathbf{e}_3 \\ h_1 & h_2 & h_3 \\ k_1 & k_2 & k_3 \end{vmatrix} = \begin{vmatrix} h_2 & h_3 \\ k_2 & k_3 \end{vmatrix} \mathbf{e}_1 + \dots \qquad \square$$

Proof. The first component of the vector integral of $d\sigma$ over $\Pi_{\mathbf{h},\mathbf{k}}$ is

$$\int_{\Pi_{\mathbf{h},\mathbf{k}}} dy \wedge dz.$$

To compute this integral, we use the parametrization (in the correct orientation)

$$\Pi_{\mathbf{h},\mathbf{k}} \ni \mathbf{r} = u\mathbf{h} + v\mathbf{k} \qquad (0 \le u,v \le 1).$$

Hence

$$y = h_2 u + k_2 v \, , \qquad dy = h_2 du + k_2 dv \, ,$$
$$z = h_3 u + k_3 v \, , \qquad dz = h_3 du + k_3 dv \, ,$$

and

$$dy \wedge dz = (h_2 k_3 - h_3 k_2) du \wedge dv.$$

The first component of the proposed integral is thus

$$\int_{\Pi_{\mathbf{h},\mathbf{k}}} dy \wedge dz = (h_2 k_3 - h_3 k_2) \int_0^1 \int_0^1 du dv = h_2 k_3 - h_3 k_2.$$

This is the first component of the vector product. The other components are treated similarly. ■

In particular, we can write

$$\mathbf{Area}(\Pi_{\mathbf{h},\mathbf{k}}) = \pm|\Pi_{yz}|\mathbf{e}_1 \pm |\Pi_{zx}|\mathbf{e}_2 \pm |\Pi_{xy}|\mathbf{e}_3$$

where $\Pi_{yz} = pr_x(\Pi_{\mathbf{h},\mathbf{k}})$ is the projection in the (y,z)-plane of $\Pi_{\mathbf{h},\mathbf{k}}$ (parallel to the Ox-axis). Of course

$$\mathbf{Area}(\Pi_{yz}) = \mathrm{Area}(\Pi_{yz})\mathbf{e}_1 = (h_2k_3 - h_3k_2)\mathbf{e}_1.$$

The coefficient of this expression is a bilinear form in $\mathbf{h}$ and $\mathbf{k}$ which can be described in terms of the constant fields of linear forms dx, dy and dz :

$$\mathrm{Area}(\Pi_{yz}) = h_2k_3 - h_3k_2 = dy(\mathbf{h})dz(\mathbf{k}) - dz(\mathbf{h})dy(\mathbf{k}).$$

But now, if f and g are linear forms over any vector space V, we can define two bilinear forms $f \otimes g$ and $f \wedge g = f \otimes g - g \otimes f$ by the formulas

$$f \otimes g\ (\mathbf{v},\mathbf{w}) = f(\mathbf{v})g(\mathbf{w}),$$

$$f \wedge g\ (\mathbf{v},\mathbf{w}) = f(\mathbf{v})g(\mathbf{w}) - f(\mathbf{w})g(\mathbf{v}).$$

With these definitions,

$$\mathrm{Area}(\Pi_{yz}) = dy \wedge dz = dy \otimes dz - dz \otimes dy$$

evaluated on the couple $(\mathbf{h},\mathbf{k})$. One can thus write

$$\mathbf{h} \wedge \mathbf{k} = \mathbf{Area}(\Pi_{\mathbf{h},\mathbf{k}}) = d\boldsymbol{\sigma}(\mathbf{h},\mathbf{k}).$$

This gives an interpretation of the vector 2-form $d\boldsymbol{\sigma}$ as a constant field of (vector) bilinear forms as indicated.

For later reference, let us give the expression of the vector 2-form $d\boldsymbol{\sigma}$ in spherical coordinates (notations of 9.5).

The three parameters r, ϑ, φ in this order, give the canonical orientation of $\mathbb{R}^3$ (minus the Oz-axis). We denote by F the map

$$\begin{pmatrix} r \\ \vartheta \\ \varphi \end{pmatrix} \longmapsto \begin{pmatrix} x = r\sin\vartheta\cos\varphi \\ y = r\sin\vartheta\sin\varphi \\ z = r\cos\vartheta \end{pmatrix}.$$

Its differential is given by

$$d\mathbf{r} = \begin{pmatrix} dx \\ dy \\ dz \end{pmatrix} = \mathrm{Jac}(F) \begin{pmatrix} dr \\ d\vartheta \\ d\varphi \end{pmatrix},$$

with

$$\begin{cases} dx = \sin\vartheta\cos\varphi\, dr + r\cos\vartheta\cos\varphi\, d\vartheta - r\sin\vartheta\sin\varphi\, d\varphi, \\ dy = \sin\vartheta\sin\varphi\, dr + r\cos\vartheta\sin\varphi\, d\vartheta + r\sin\vartheta\cos\varphi\, d\varphi, \\ dz = \cos\vartheta\, dr - r\sin\vartheta\, d\vartheta\ . \end{cases}$$

One can then compute the basis of second degree differential forms

$$\begin{cases} dy\wedge dz = r^2\sin^2\vartheta\cos\varphi\, d\vartheta\wedge d\varphi + r\sin\vartheta\cos\vartheta\cos\varphi\, d\varphi\wedge dr - r\sin\varphi\, dr\wedge d\vartheta\ , \\ dz\wedge dx = r^2\sin^2\vartheta\sin\varphi\, d\vartheta\wedge d\varphi + r\sin\vartheta\cos\vartheta\sin\varphi\, d\varphi\wedge dr + r\cos\varphi\, dr\wedge d\vartheta\ , \\ dx\wedge dy = r^2\sin\vartheta\cos\vartheta\, d\vartheta\wedge d\varphi - r\sin^2\vartheta\, d\varphi\wedge dr. \end{cases}$$

Observe that

$$d\sigma = \begin{pmatrix} dy\wedge dz \\ dz\wedge dx \\ dx\wedge dy \end{pmatrix} = M \begin{pmatrix} d\vartheta\wedge d\varphi \\ d\varphi\wedge dr \\ dr\wedge d\vartheta \end{pmatrix}$$

with the matrix M of *2 by 2 minors* (suitable signs) of Jac(F). From these expressions it follows —in particular— that on a sphere of radius r centered at the origin

$$d\sigma|_S = r^2 \begin{pmatrix} \sin^2\vartheta\ \cos\varphi \\ \sin^2\vartheta\ \sin\varphi \\ \sin\vartheta\ \cos\vartheta \end{pmatrix} d\vartheta\wedge d\varphi \qquad (dr = 0).$$

On this same sphere,

$$\mathbf{r}\cdot d\sigma|_S = r^3 \begin{pmatrix} \sin\vartheta\ \cos\varphi \\ \sin\vartheta\ \sin\varphi \\ \cos\ \vartheta \end{pmatrix} \cdot \begin{pmatrix} \sin^2\vartheta\ \cos\varphi \\ \sin^2\vartheta\ \sin\varphi \\ \sin\vartheta\ \cos\vartheta \end{pmatrix} d\vartheta\wedge d\varphi =$$

$$= r^3(\sin^3\vartheta + \sin\vartheta\ \cos^2\vartheta)\ d\vartheta\wedge d\varphi =$$

$$= r^3\sin\vartheta\ d\vartheta\wedge d\varphi\ .$$

On these formulas, one can check

$$\|d\sigma|_S\| = r^2(\sin^4\vartheta + \sin^2\vartheta\ \cos^2\vartheta)^{1/2}\ |d\vartheta\wedge d\varphi| =$$

$$= r^2|\sin\vartheta|d\vartheta d\varphi = \left|\frac{\mathbf{r}}{r}\cdot d\sigma|_S\right| = |dS_{or}|.$$

In fact $|\sin\vartheta| = \sin\vartheta > 0$ since $0 < \vartheta < \pi$.

12.6 APPLICATION TO ARCHIMEDES' PRESSURE

The Stokes theorem is certainly one of the profound results of mathematics. Since it is based on the exterior differentiation operator for which an intuitive interpretation is not immediately available, it may appear as very mysterious ! However, anybody (since Archimedes!) sitting in his bath experiences a corollary of this integral transformation...

Let us compute the total force acting on a solid immersed in a homogeneous fluid, by integrating all local forces on its surface, produced by the pressure field in the fluid. For this, we introduce a vertical Oz-coordinate axis and express the pressure as a scalar function $p = f(z)$.

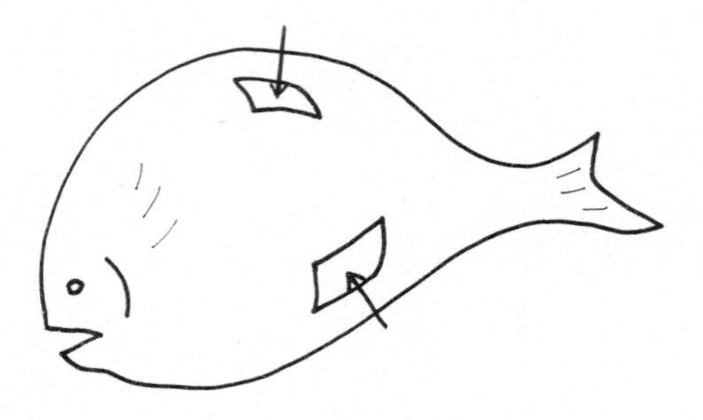

Fig.12.8

Here are the physical data relevant to the situation

$p = f(z) = -\rho gz$ where ρ is the density of the fluid and g the "acceleration of gravity" constant,

$\mathbf{df} = -p d\boldsymbol{\sigma}$: elementary force acting on the surface element $d\boldsymbol{\sigma}$,

$\mathbf{f} = \int_{\Sigma=\partial V} \mathbf{df}$: total force acting on the immersed volume V due to the field of pressure p.

Using one form of the Stokes theorem (cf.formula (1) of 12.4), we get

$$\mathbf{f} = \int_{\Sigma=\partial V} \mathbf{df} = \int_{\Sigma=\partial V} -p d\boldsymbol{\sigma} = \int_V -\mathbf{grad}p \, dV.$$

Since $p = -\rho gz$, $\mathbf{grad}p = -\rho g\mathbf{e}_3$ and

$$\mathbf{f} = \int_V \rho g\mathbf{e}_3 dV = \rho|V|g\mathbf{e}_3 = mg\mathbf{e}_3$$

provided the density ρ is constant. Here $m = \rho|V|$ is the mass of fluid that would occupy the portion of volume V (if it could) and mg is the weight of *displaced fluid.* The Archimedes force is a vertical force having for magnitude the weight of displaced fluid.

It is possible to go one step further and to determine the application point of the preceding force. For this purpose, let us fix an origin and compute the total moment of force with respect to this origin produced by the elementary forces $\mathbf{df}$. The elementary moment of $\mathbf{df}$ is

$$d\mu = \mathbf{r}\wedge\mathbf{df} \quad \text{(vector product } \wedge \text{)}$$

and the total moment of force is

$$\mu = \int_{\Sigma=\partial V} \mathbf{r}\wedge\mathbf{df} = \int_{\Sigma=\partial V} -\mathbf{df}\wedge\mathbf{r} = \int_{\Sigma=\partial V} d\boldsymbol{\sigma}\wedge p\mathbf{r}.$$

Using the Stokes formula as given in (3) of 12.4,

$$\mu = \int_V \mathbf{curl}(p\mathbf{r})dV.$$

We have seen (cf.8.3)

$$\mathbf{curl}(p\mathbf{r}) = \mathbf{grad}p \wedge \mathbf{r} + p\ \mathbf{curl}\ \mathbf{r}$$

and since

$$\mathbf{curl}\ \mathbf{r} = 0,\ \mathbf{grad}p = -\rho g\mathbf{e}_3,$$

we obtain

$$\mu = \int_V -\rho g\mathbf{e}_3 \wedge \mathbf{r}\ dV = \rho g\int_V \mathbf{r}dV \wedge \mathbf{e}_3.$$

By definition, the integral

$$\int_V \mathbf{r}dV = \mathbf{r}_G|V|$$

gives the vector position $\mathbf{r}_G$ of the center of gravity of the volume V and finally

$$\mu = \rho g|V|\ \mathbf{r}_G \wedge \mathbf{e}_3.$$

This global moment of force will vanish when $\mathbf{r}_G\ //\ \mathbf{e}_3$, namely when the origin is chosen above the center of gravity G of the volume V (this center of gravity being computed in a homogeneous V, is of course independent of the particular nature of the possibly inhomogeneous solid being immersed). In this way, we see that the application point of the Archimedes force is situated on the vertical line going through the gravity point of V.

12.7 AN EXAMPLE

We shall compute the flow of the vector field $\mathbf{u}(\mathbf{r}) = \mathbf{r}$ through the unit sphere S $(r = 1)$ in several ways. This unit sphere is oriented by the choice of a normal *exterior* to the unit ball B $(r \leq 1)$ and inherits the induced orientation $S = \partial B$ if the ball B has the canonical orientation of $\mathbb{R}^3$. This flow is

$$\Phi = \int_{S_{or}=\partial B_{or}} \mathbf{u}d\sigma = \int_{S_{or}=\partial B_{or}} \mathbf{r}d\sigma\ .$$

Here are several ways of computing it.

First method. Considering that $S_{or} = \partial B_{or}$ (with orientation) Stokes formula can be used in the form of the divergence theorem. We have

$$d(\mathbf{r}d\sigma) = \mathrm{div}\mathbf{r}\ dV = 3\ dV,$$

and hence

$$\int_{\partial B_{or}} \mathbf{r}d\sigma = \int_{B_{or}} d(\mathbf{r}d\sigma) = 3\int_{B_{or}} dV.$$

Since the orientation of the unit ball is given by the coordinates x, y, z, this integral is

$$3\iiint_B dxdydz = 3\ \frac{4}{3}\pi = 4\pi.$$

(We shall have to use the *measure of the volume* of the unit ball $4\pi/3$ repeatedly!)

Second method. Since the vector fields $\mathbf{u}_n(\mathbf{r}) = r^n\mathbf{r}$ take the same values as $\mathbf{u}_o$ (former field simply denoted by $\mathbf{u}$ in the preceding computation) on $S = \partial B$, they must produce the same flow through S. For $n \geq 0$, Stokes'formula in the form of the divergence formula is valid as before and we need the following result.

Lemma. For any integer n, one has $\mathrm{div}(r^n\mathbf{r}) = (n+3)r^n$ (in $\mathbb{R}^3$). □

Proof. a) Starting with $r^2 = x^2 + y^2 + z^2$ we find

$$2r(\partial r/\partial x) = 2x \quad \text{hence} \quad \partial r/\partial x = x/r$$

and similar expressions for the other partial derivatives of r. Hence

$$\begin{aligned}\mathrm{div}(r^n\mathbf{r}) &= \partial_x(r^n x) + \partial_y(r^n y) + \partial_z(r^n z) = \\ &= nr^{n-1}(x/r)x + r^n + \ldots = \\ &= nr^{n-2}(x^2 + y^2 + z^2) + 3r^n = nr^n + 3r^n.\end{aligned}$$

b) (Variant) Using the ∇ operator,

$$\begin{aligned}\mathrm{div}(r^n\mathbf{r}) = \nabla(r^n\mathbf{r}) &= \nabla(\overset{\uparrow}{r^n}\mathbf{r}) + \nabla(r^n\overset{\uparrow}{\mathbf{r}}) = \mathbf{grad}r^n\cdot\mathbf{r} + r^n\mathrm{div}\mathbf{r} = \\ &= nr^{n-1}\,\frac{\mathbf{r}}{r}\cdot\mathbf{r} + r^n\,3 = (n+3)r^n.\end{aligned}$$ ■

Applying the formula just found in the Stokes theorem, we get

$$\begin{aligned}\int_{S_{or}} r^n\mathbf{r}d\sigma &= \int_{B_{or}} \mathrm{div}(r^n\mathbf{r})dV = (n+3)\int_{B_{or}} r^n dV = \\ &= (n+3)\iiint_B r^n dxdydz.\end{aligned}$$

This integral is best computed using spherical coordinates (cf.9.5)

$$\iiint_B r^n dxdydz = 4\pi\int_0^1 r^n\, r^2 dr = 4\pi\left[\frac{r^{n+3}}{n+3}\right]_0^1 = 4\pi/(n+3).$$

Let us observe that the use of Stokes' formula —although illegal for $n < 0$— would lead to a correct result for all integers $n \neq -3$ (we shall come back to this crucial case below).

Third method. Without using Stokes' formula, the computation could be made as follows. The unit sphere S is the union of two pieces

S^+ : Northern hemisphere ($z \geq 0$),

S^- : Southern hemisphere ($z \leq 0$).

Each of these is the image of a parametrization. Consider e.g. the Northern hemisphere with the parametrization

$$\begin{pmatrix} x \\ y \end{pmatrix} \longmapsto \begin{pmatrix} x \\ y \\ z = \sqrt{(1-x^2-y^2)} \end{pmatrix}$$

defined in the domain

$$D = D_{xy} : x^2 + y^2 \le 1$$

of the $\mathbb{R}^2$ plane. This parametrization is in the correct orientation and we could similarly parameterize the Southern hemisphere by using y and x (in this order to be in the correct orientation)

$$S = S^+ \cup S^- \quad \text{and} \quad S_{or} = S^+_{xy} + S^-_{yx} .$$

It will be enough to evaluate the integral over the Northern hemisphere (the other one being treated similarly). The parametrization gives

$$dz = \frac{1}{2}(1-x^2-y^2)^{-1/2}(-2xdx - 2ydy) = -\frac{xdx + ydy}{\sqrt{(1-x^2-y^2)}} .$$

In this way,

$$d\sigma\Big|_S = \begin{pmatrix} dy\wedge dz \\ dz\wedge dx \\ dx\wedge dy \end{pmatrix}_S = \begin{pmatrix} xdx\wedge dy \\ ydx\wedge dy \\ zdx\wedge dy \end{pmatrix} \frac{1}{z} = \begin{pmatrix} x \\ y \\ z \end{pmatrix} \frac{dx\wedge dy}{z}$$

where $z = \sqrt{(1-x^2-y^2)}$. We find

$$\mathbf{r}d\sigma\Big|_S = \frac{dx\wedge dy}{z}[x^2 + y^2 + (1-x^2-y^2)] = \frac{dx\wedge dy}{\sqrt{(1-x^2-y^2)}} .$$

Coming back to the integral over S^+

$$\iint_{x^2+y^2\le 1} \frac{dxdy}{\sqrt{(1-x^2-y^2)}} = \int_{-1}^{+1} dx \int_{-\rho}^{\rho = \sqrt{(1-x^2)}} \frac{dy}{\sqrt{(\rho^2-y^2)}} .$$

After an easy computation, one gets

$$\int_{S^+} \mathbf{r}d\sigma = 2\pi = \int_{S^-} \mathbf{r}d\sigma$$

whence the same final result with this method.

Fourth method. In spherical coordinates, the restriction of $\mathbf{r}d\sigma$ to the unit sphere is simply given in the (ϑ,φ)-coordinates by the expression $\sin\vartheta \; d\vartheta\wedge d\varphi$. Thus we find

$$\int_{S_{or}} \mathbf{r}d\sigma = \int_{D_{or}} \sin\vartheta \; d\vartheta\wedge d\varphi = \iint_D \sin\vartheta \; |d\vartheta\wedge d\varphi| =$$

$$= \iint_D \sin\vartheta \; d\vartheta d\varphi = \int_0^{2\pi} d\varphi \int_0^{\pi} \sin\vartheta \; d\vartheta =$$

$$= 2\pi \int_0^{\pi} \sin\vartheta \; d\vartheta = 4\pi \int_0^{\pi/2} \sin\vartheta \; d\vartheta = 4\pi.$$

Fifth method. The lemma proved in the second method shows that the differential form $r^{-3}\,\mathbf{r}d\sigma$ is *closed*

$$d(r^{-3}\,\mathbf{r}d\sigma) = \mathrm{div}(r^{-3}\mathbf{r})dV = 0.$$

But this differential form is *not exact* in the whole space (or in $\mathbb{R}^3-\{0\}$). However, it must be exact in any star-shaped or convex region. In particular, it must be exact in the half space $z > 0$ (this is a convex region)

$$\omega = r^{-3}\mathbf{r}d\sigma = d\eta \quad \text{for some} \quad \eta \in \Omega^1(z>0).$$

Equivalently, the field $\mathbf{u} = \mathbf{r}/r^3$ must derive from a vector potential $\mathbf{v}$ in the upper half space :

$$\mathbf{r}/r^3 = \mathbf{curl}\ \mathbf{v} \quad \text{if} \quad r^{-3}\mathbf{r}d\sigma = d\eta \quad \text{with} \quad \eta = \mathbf{v}\cdot d\mathbf{r}.$$

Here is one method for determining such a primitive. Let us work in spherical coordinates. A few computations show that

$$\begin{aligned} \mathbf{r}d\sigma &= r^3(\sin^3\vartheta + \sin\vartheta\cos^2\vartheta)d\vartheta\wedge d\varphi + \\ &\quad + r^2(\sin^2\vartheta\cos\vartheta - \sin^2\vartheta\cos\vartheta)d\varphi\wedge dr + \\ &\quad + r^2(-\sin\vartheta\sin\varphi\cos\varphi + \sin\vartheta\sin\varphi\cos\varphi)dr\wedge d\vartheta = \\ &= r^3\sin\vartheta\ d\vartheta\wedge d\varphi. \end{aligned}$$

Hence

$$\omega = r^{-3}\mathbf{r}d\sigma = \sin\vartheta\ d\vartheta\wedge d\varphi = d(c - \cos\vartheta)\wedge d\varphi$$

for any value of the constant c. At least, $\eta = (c - \cos\vartheta)d\varphi$ *looks like a primitive of* ω ! However, recall that the angle φ is *not defined* for $\vartheta = 0$ and the differential $d\varphi$ is —a priori— not defined on the Oz-axis. Taking $c = 1$ leads to a coefficient that vanishes for $\vartheta = 0$ (positive Oz-axis) and the 1-form

$$\eta^+ = (1 - \cos\vartheta)d\varphi$$

is well defined for $z > 0$ (or even in a neighborhood $0 < \vartheta < 3\pi/4$ of S^+ : this is star-shaped with respect to any point on the positive Oz-axis). We can now apply the Stokes formula to S^+ and its boundary

$$\partial S^+_{xy} = \mathcal{C}_\varphi \ : \text{unit circle oriented by the parameter } \varphi.$$

Hence we find

$$\int_{S^+_{xy}} r^{-3}\mathbf{r}d\sigma = \int_{S^+_{xy}} d\eta^+ = \int_{\partial S^+_{xy}} \eta^+ = \int_{\mathcal{C}_\varphi} (1 - \cos\vartheta)d\varphi = 2\pi$$

(since $\vartheta = \pi/2$ on $\mathcal{C}$). The contribution from the southern hemisphere is the same (another good exercise for practising orientations and Stokes' formula!...)

Remark. We have seen that $\mathbf{r}/r^3$ derives from a vector potential. It is interesting to note that $\mathbf{curl}\ \mathbf{r}/r^3 = 0$, so that $\mathbf{r}/r^3$ also derives from a scalar potential in suitable regions. More precisely,

$$\mathbf{r}/r^3 = \mathbf{grad}(-1/r) \quad \text{in } \mathbb{R}^3 - \{0\}.$$

CHAPTER 13
INTRODUCTION TO FUNCTIONS OF ONE COMPLEX VARIABLE

13.1 COMPLEX NUMBERS AND FUNCTIONS

Complex numbers have been created to make −1 a *square* : a complex unit i with $i^2 = -1$ is introduced *formally* and complex numbers are formal sums $z = x+iy$ with x and y real numbers. The set of complex numbers $\mathbb{C}$ is identified to the real plane $\mathbb{R}^2$ and we obtain a natural identification

$$\mathbb{R}^2 \longrightarrow \mathbb{C} : \binom{x}{y} \longmapsto x+iy = z.$$

We assume that the reader is already familiar with the operations on complex numbers : addition and multiplication are defined in $\mathbb{C}$ and this set is a *commutative field*. Nevertheless, we review a few definitions and properties to explain our notations.

When a complex number z is written $z = x+iy$ with two real numbers x and y, x is called **real part** of z and denoted by Re(z). Similarly, y is the **imaginary part** of z and denoted by Im(z) (observe that this imaginary part is still a *real* number!).

The **conjugate** $\bar{z}$ of the complex number $z = x+iy$ is defined by $\bar{z} = x-iy$. By definition

$$z = \mathrm{Re}(z) + i\ \mathrm{Im}(z),$$
$$\bar{z} = \mathrm{Re}(z) - i\ \mathrm{Im}(z)$$

and hence

$$\mathrm{Re}(z) = (z + \bar{z})/2,$$
$$\mathrm{Im}(z) = (z - \bar{z})/(2i).$$

The **absolute value** of the complex number $z = x+iy$ is the norm of the corresponding vector in $\mathbb{R}^2$

$$|z| = \left\|\binom{x}{y}\right\| = \sqrt{(x^2 + y^2)},$$

and $|z|^2 = z\bar{z} = x^2 + y^2$. In particular

$$|z| = 0 \Longleftrightarrow z = 0$$

and if $z \neq 0$, $1/z = \bar{z}/|z|^2$. Let us also recall that if the complex number z has absolute value $|z| = r$, then it can be written

$$z = re^{i\varphi} \quad \text{with} \quad \varphi \in \mathbb{R}$$

The real number φ is the **argument** of z : it is well determined mod 2π. For example, one can decide to choose $0 \leq \varphi < 2\pi$ (but it may be more convenient to choose the principal determination for which $-\pi < \varphi \le \pi$). Finally, let us also recall the **Euler formula**

$$e^{i\varphi} = \cos\varphi + i\sin\varphi.$$

Let $z = a+ib \in \mathbb{C}$ be a complex number. Multiplication by z in $\mathbb{C} \cong \mathbb{R}^2$ produces a linear transformation whose matrix in the canonical basis $\mathbf{e}_1 = 1$, $\mathbf{e}_2 = i$ is given by

$$M_z = \begin{pmatrix} a & -b \\ b & a \end{pmatrix}$$

and in particular, multiplication by i (a quarter turn in the positive direction) is given by the matrix

$$J = \begin{pmatrix} 0 & -1 \\ 1 & 0 \end{pmatrix}.$$

Complex functions $F : \mathbb{C} \longrightarrow \mathbb{C}$ can be viewed as maps $\mathbb{R}^2 \longrightarrow \mathbb{R}^2$ according to the preceding identifications. Namely, if $F = u + iv$ with two real functions $u, v : \mathbb{C} \longrightarrow \mathbb{R}$ we can identify F with the vector field of components u and v in $\mathbb{C} \cong \mathbb{R}^2$

$$F : \mathbb{C} \longrightarrow \mathbb{C} \quad \text{or} \quad \mathbb{R}^2 \longrightarrow \mathbb{R}^2 .$$
$$z \longmapsto F(z) \qquad \begin{pmatrix} x \\ y \end{pmatrix} \longmapsto \begin{pmatrix} u(x,y) \\ v(x,y) \end{pmatrix}$$

As before,

$$u = \mathrm{Re}(F) = (F + \overline{F})/2, \ v = \mathrm{Im}(F) = (F - \overline{F})/2i.$$

In principle at least, the study of complex functions could be replaced by the study of couples of real functions in $\mathbb{R}^2$ (or vector fields in $\mathbb{R}^2$). However, the possibility of multiplying couples (*and dividing* by non zero couples) brings up new phenomena which are best understood in the complex domain.

13.2 COMPLEX DIFFERENTIABILITY

When F is a complex function (defined in some open domain of the complex field), we can examine the notion of complex differentiability. As usual, we assume that F is defined in the whole of $\mathbb{C}$ and fix a point $z \in \mathbb{C}$. If F is differentiable in the real sense at this point, we shall have a limited expansion of the form

$$F(z + h) = F(z) + \alpha h_1 + \beta h_2 + o(|h|)$$

with some complex numbers α and β ($h = h_1 + ih_2$). In fact,

$$\alpha = D_1F(z), \ \beta = D_2F(z)$$

(compare with 9.4).

Writing

$$h_1 = \mathrm{Re}(h) = (h + \overline{h})/2 \text{ and } h_2 = (h - \overline{h})/2i,$$

the linear part of the limited expansion can be rewritten

$$\alpha h_1 + \beta h_2 = (\frac{\alpha}{2} + \frac{\beta}{2i})h + (\frac{\alpha}{2} - \frac{\beta}{2i})\bar{h} = \frac{1}{2}(\alpha - i\beta)h + \frac{1}{2}(\alpha + i\beta)\bar{h}.$$

Let us introduce the *notations*

$$\partial_z F = \frac{1}{2}(D_1F - iD_2F) , \partial_{\bar{z}} F = \frac{1}{2}(D_1F + iD_2F).$$

Then we shall have

$$F(z+h) = F(z) + \partial_z F(z)h + \partial_{\bar{z}} F(z)\bar{h} + o(|h|).$$

Definition. We say that F is **$\mathbb{C}$-differentiable** at the point z if F has a limited expansion

$$F(z+h) = F(z) + L(h) + o(|h|)$$

with a $\mathbb{C}$-linear part $h \longmapsto L(h)$. □

Since any $\mathbb{C}$-linear map $\mathbb{C} \longrightarrow \mathbb{C}$ is given by multiplication by a complex number L (corresponding to the 1×1 matrix of this map) we shall write more simply $L(h) = Lh$, $L = F'(z)$. This last notation is supported by the fact that if F is $\mathbb{C}$-differentiable at z, then the differential *quotients* formed with complex increments have limit L

$$\frac{F(z+h) - F(z)}{h} - L = \frac{1}{h} o(|h|) \longrightarrow 0 \quad \text{when} \quad h \longrightarrow 0.$$

When it exists, this limit will be denoted by F'(z)

$$\frac{F(z+h) - F(z)}{h} \longrightarrow F'(z).$$

One way of computing it consists in taking real increments :

$$F'(z) = \frac{\partial F}{\partial x}(z).$$

Alternatively, we could take purely imaginary increments $h = ik$ $(k \in \mathbb{R})$:

$$\frac{F(z+ik) - F(z)}{ik} \longrightarrow \frac{1}{i}\frac{\partial F}{\partial y}(z) = L = F'(z).$$

Existence of a limit for the complex differential quotients immediately leads to the **Cauchy-Riemann relations**

(CR1) $$\frac{\partial F}{\partial x}(z) = \frac{1}{i}\frac{\partial F}{\partial y}(z).$$

Separating real and imaginary parts in this complex relation, we get

(CR2) $$\frac{\partial u}{\partial x}(z) = \frac{\partial v}{\partial y}(z) \quad \text{and} \quad \frac{\partial v}{\partial x}(z) = -\frac{\partial u}{\partial y}(z).$$

With the notation just introduced for $\partial_{\bar{z}}F(z)$, the relation (CR1) is equivalent to

(CR3) $$\partial_{\bar{z}} F(z) = 0.$$

The equalities (CR2) show that the Jacobian matrix of F' (viewed as real vector map) has a special form

$$\begin{pmatrix} a & -b \\ b & a \end{pmatrix} \quad \text{with } a = \frac{\partial u}{\partial x}(z) \text{ and } b = \frac{\partial v}{\partial x}(z).$$

This precisely means that this Jacobian matrix is the matrix of the multiplication by the complex number

$$L = a+ib = \frac{\partial u}{\partial x}(z) + i\frac{\partial v}{\partial x}(z) = \frac{\partial F}{\partial x}(z).$$

Reading the coefficients a and b in the second column of the Jacobian matrix, we would get

$$L = a+ib = \frac{\partial v}{\partial y}(z) - i\frac{\partial u}{\partial y}(z) = -i\left(\frac{\partial u}{\partial y}(z) + i\frac{\partial v}{\partial y}(z)\right) = \frac{1}{i}\frac{\partial F}{\partial y}(z).$$

One can check without any difficulty that the matrices of the form $\begin{pmatrix} a & -b \\ b & a \end{pmatrix}$ commute with $J = \begin{pmatrix} 0 & -1 \\ 1 & 0 \end{pmatrix}$. Conversely, if a matrix

$$M = \begin{pmatrix} a & c \\ b & d \end{pmatrix}$$

commutes with the special matrix J, i.e. $M \circ J = J \circ M$, we also get the two relations $c = -b$ and $d = a$. This corresponds to the fact that M must commute with the quarter turn (or multiplication by i) and thus corresponds to a $\mathbb{C}$-linear map : $M = M_{a+ib}$ is the matrix of a multiplication operator in $\mathbb{C}$.

Altogether, we have proved the following result.

Theorem. For a complex function $F = u + iv : \mathbb{C} \longrightarrow \mathbb{C}$ (with real u and v), and a point $z \in \mathbb{C}$, the following conditions are equivalent

i) F is $\mathbb{C}$-differentiable at the point z,

ii) F is differentiable (in the real sense) and

(CR1) $$\frac{\partial F}{\partial x}(z) = \frac{1}{i}\frac{\partial F}{\partial y}(z),$$

iii) $F = u + iv$ is differentiable (in the real sense) and

(CR2) $$\frac{\partial u}{\partial x}(z) = \frac{\partial v}{\partial y}(z) \quad \& \quad \frac{\partial v}{\partial x}(z) = -\frac{\partial u}{\partial y}(z),$$

iv) F is differentiable (in the real sense) and the Jacobian matrix of F at the point z has the form

$$\begin{pmatrix} a & -b \\ b & a \end{pmatrix},$$

v) F is differentiable (in the real sense) and the Jacobian matrix of F at the point z commutes with $\begin{pmatrix} 0 & -1 \\ 1 & 0 \end{pmatrix}$. ∎

For example, this theorem shows that the function $F(z) = \bar{z}$ is not $\mathbb{C}$-differentiable although it is differentiable in the real sense: $F(z) = F(x+iy) = x-iy$ has partial derivatives

$$\partial_x F(z) = 1 \quad \text{and} \quad \partial_y F(z) = -i$$

not satisfying the Cauchy-Riemann relations. More directly, differential quotients can be evaluated

$$\frac{F(z+h) - F(z)}{h} = [\overline{(z+h)} - \bar{z}]/h = [\bar{z} + \bar{h} - \bar{z}]/h = \bar{h}/h.$$

But if we write $h = re^{i\varphi}$ in polar form, the complex numbers $\bar{h}/h = e^{-2i\varphi}$ have no limit for $|h| \to 0$ (observe that $\bar{h}/h$ has constant modulus 1, with an argument -2φ varying in an arbitrary way). But this function F is differentiable in the real sense with a

Jacobian matrix $\begin{pmatrix} 1 & 0 \\ 0 & -1 \end{pmatrix}$ (a matrix which is not of the required form for $\mathbb{C}$-differentiability).

In order to be able to give interesting **examples**, let us note the following obvious properties of $\mathbb{C}$-differentiability.

Proposition. If F and G : $\mathbb{C} \longrightarrow \mathbb{C}$ are two functions which are $\mathbb{C}$-differentiable at a point z, then

a) F + G is $\mathbb{C}$-differentiable at z with

$$(F+G)'(z) = F'(z) + G'(z),$$

b) FG is $\mathbb{C}$-differentiable at z with

$$(FG)'(z) = F'(z)G(z) + F(z)G'(z),$$

c) if $G(z) \neq 0$, F/G is $\mathbb{C}$-differentiable at z with

$$(F/G)'(z) = [F'(z)G(z) - F(z)G'(z)]/G(z)^2,$$

d) the composite $F \circ G$ is also $\mathbb{C}$-differentiable with

$$(F \circ G)'(z) = F'(G(z)) \cdot G'(z). \qquad \blacksquare$$

These rules are immediately verified (using differential quotients) as in the one real variable case.

Since the identity function $F(z) = z$ is $\mathbb{C}$-linear, it is $\mathbb{C}$-differentiable with derivative given by $h \longmapsto h = 1 \cdot h$. In other words, the derivative of $F(z) = z$ is $F'(z) = 1$ (at all points z). Using part b) of the above proposition, we conclude that z^2 and by induction that z^n ($n \in \mathbb{N}$) are also $\mathbb{C}$-differentiable with

$$(z^n)' = nz^{n-1}.$$

Consequently, all polynomials are $\mathbb{C}$-differentiable at all points. We also conclude from part c) of the above proposition that all rational functions $F(z) = P(z)/Q(z)$ (P and Q polynomials) are $\mathbb{C}$-differentiable at all points where the denominator Q does not vanish.

The **exponential function** $e^z = e^{x+iy} = e^x \cdot e^{iy}$ is also $\mathbb{C}$-differentiable at all points z since the Cauchy-Riemann relations are easily verified:

$$\partial_x e^z = \partial_x (e^x \cdot e^{iy}) = e^x \cdot e^{iy} = e^z \quad \text{and} \quad \partial_y (e^z) = ie^z.$$

Alternatively, one could check the Cauchy-Riemann relations for

$$u(x,y) = \mathrm{Re}(e^z) = e^x \cos y \quad \text{and} \quad v(x,y) = \mathrm{Im}(e^z) = e^x \sin y.$$

Moreover, we see that

$$(e^z)' = e^z$$

just as in the real case.

Since $\cos z = (e^{iz} + e^{-iz})/2$, it follows from the preceding example that $\cos z$ is $\mathbb{C}$-differentiable at all points with

$$(\cos z)' = -\sin z.$$

Similar formulas hold for the trigonometric functions

$$\sin z = (e^{iz} - e^{-iz})/2i \implies (\sin z)' = \cos z,$$

$$\operatorname{tg} z = \sin z/\cos z \implies (\operatorname{tg} z)' = 1/\cos^2 z = 1 + \operatorname{tg}^2 z,$$

and for the hyperbolic functions

$$\operatorname{Ch} z = (e^z + e^{-z})/2, \quad \operatorname{Sh} z = (e^z - e^{-z})/2,$$

which are also $\mathbb{C}$-differentiable with

$$(\operatorname{Ch} z)' = \operatorname{Sh} z \ , \quad (\operatorname{Sh} z)' = \operatorname{Ch} z \ .$$

Finally, the part d) of the proposition shows how to construct more sophisticated examples of $\mathbb{C}$-differentiable functions. For example, the function $F(z) = e^{-1/z}$ is $\mathbb{C}$-differentiable at all points $z \neq 0$ with $(e^{-1/z})' = z^{-2}e^{-1/z}$.

Observation. If F is $\mathbb{C}$-differentiable at a point $z \in \mathbb{C}$ with, say $F'(z) = a+ib = \partial_x F(z) = -i\partial_y F(z)$ then the Jacobian determinant in the real sense is

$$\Delta_F(z) = \begin{vmatrix} a & -b \\ b & a \end{vmatrix} = a^2 + b^2 = |a+ib|^2 = |F'(z)|^2 \geq 0.$$

13.3 HOLOMORPHIC FUNCTIONS

Definition. When D is an open subset of $\mathbb{C}$, a complex function $F : D \longrightarrow \mathbb{C}$ is called **holomorphic** when F is $\mathbb{C}$-differentiable at all points $z \in D$ and F' is continuous in D. ■

It can be shown that if a function F is $\mathbb{C}$-differentiable at all points of an open subset $D \subset \mathbb{C}$, then F' is automatically continuous in D. However —for the sake of simplicity— we take the continuity assumption as part of the definition of holomorphy.

As we have seen in 13.2, if F and G are holomorphic in an open subset D of $\mathbb{C}$, so is their sum and their product. For the quotient, we have to remove from D the zeros of the denominator...

Proposition. Let D be an open subset of $\mathbb{C}$ and $F : D \longrightarrow \mathbb{C}$ be $\mathcal{C}^1$ in the real sense. Then F is holomorphic in D precisely when the differential form $\omega = Fdz = Fdx + iFdy$ is closed. □

Proof. The exterior differential of $\omega = Fdz = Fdx + iFdy$ is

$$d\omega = dF\wedge dx + idF\wedge dy = \partial_y F\cdot dy\wedge dx + i\partial_x F\cdot dx\wedge dy =$$
$$= (i\partial_x F - \partial_y F)dx\wedge dy = i(\partial_x F + i\partial_y F)dx\wedge dy.$$

In particular, we see that $d\omega = 0$ is equivalent to $\partial_x F + i\partial_y F = 0$. Consequently, ω is closed precisely when the Cauchy-Riemann relations are satisfied. ■

For the sake of completeness, let us note that if $F = u + iv$ is the decomposition of F in real and imaginary parts, then the differential form $\omega = Fdz = Fdx + iFdy$ is equal to

$$\omega = Fdz = (u+iv)dx + i(u+iv)dy = udx - vdy + i(vdx + udy).$$

This means that the differential form $\omega = Fdz$ could be identified to the couple of *real* differential forms

$$Re(\omega) = udx - vdy \quad \text{and} \quad Im(\omega) = vdx + udy.$$

The importance of holomorphic functions is obvious from the above proposition since Stokes' theorem gives

$$\int_{\partial\Pi} Fdz = \int_{\Pi} d\omega = 0$$

for any holomorphic function F in the open set $D \subset \mathbb{C}$ and any piece $\Pi \subset D$ (recall that pieces are compact sets so that $\Pi \subset D$ implies $\partial\Pi \subset D$). In the following picture, $\partial\Pi = \gamma - \gamma_1 - \gamma_2$ so that

$$\int_{\gamma - \gamma_1 - \gamma_2} Fdz = 0$$

and this means that

$$\int_{\gamma} Fdz = \int_{\gamma_1} Fdz + \int_{\gamma_2} Fdz.$$

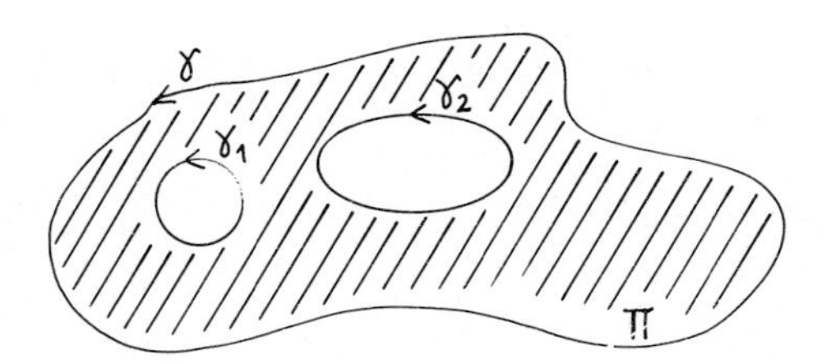

Fig.13.1

Let us make an explicit computation which will play a crucial theoretical role. Denote by $B = B_{\leq r}(a)$ the closed disc defined by $|z-a| \leq r$. Its boundary ∂B is the circle $|z-a| = r$ (positively oriented since B has the canonical orientation). Let us evaluate the integrals

$$\oint_{\partial B} (z-a)^n dz = \oint_{|z-a|=r} (z-a)^n dz.$$

If $n \in \mathbb{N}$ is a positive integer, the function $(z-a)^n$ is holomorphic everywhere so that Stokes' formula for the disc B gives

$$\oint_{\partial B} (z-a)^n dz = 0.$$

If n is a negative integer, Stokes' formula can still be applied in an annulus of the type $r_1 \le |z-a| \le r_2$ where $(z-a)^n$ is holomorphic and the conclusion is

$$\oint_{|z-a|=r_2} (z-a)^n dz = \oint_{|z-a|=r_1} (z-a)^n dz.$$

In other words, the integral

$$\oint_{|z-a|=r} (z-a)^n dz$$

is still independent of $r > 0$. This does not tell us its value... But we can easily compute the explicit value by the parametrization $t \longmapsto z(t) = a + re^{it}$ $(0 \le t \le 2\pi)$ of the positively oriented circle of radius r and center a. We find $dz(t) = ire^{it}dt$ and then

$$\oint_{|z-a|=r} (z-a)^n dz = \int_0^{2\pi} (re^{it})^n ire^{it} dt = ir^{n+1}\int_0^{2\pi} e^{it(n+1)} dt.$$

Remember that the result should be *independent from* r ! Fortunately, the integral of the exponential can be evaluated by means of a primitive. More precisely, there are two cases

a) if $n+1 = 0$, $\int_0^{2\pi} e^{it(n+1)} dt = \int_0^{2\pi} dt = 2\pi$,

b) if $n+1 \ne 0$, $\int_0^{2\pi} e^{it(n+1)} dt = \frac{1}{i(n+1)} \Big[e^{it(n+1)} \Big]_{t=0}^{t=2\pi} = 0.$

We have found

$$\oint_{|z-a|=r} (z-a)^n dz = \begin{cases} 2\pi i & \text{if } n = -1 \\ 0 & \text{if } n \ne -1 \end{cases}$$

which is indeed independent of r in all cases (and coïncides with the expected value for positive n !).

Theorem (Cauchy). Let $F : D \longrightarrow \mathbb{C}$ be a holomorphic function in an open subset $D \subset \mathbb{C}$. If the closed disc $B_{\le r}(a)$ defined by $|z-a| \le r$ is contained in D, then

$$F(t) = \frac{1}{2\pi i} \oint_{|z-a|=r} \frac{F(z)\, dz}{z - t} \quad \text{as soon as} \quad |t-a| < r. \qquad \square$$

Proof. Let us write

$$\frac{F(z)}{z-t} = \frac{F(z) - F(t)}{z - t} + \frac{F(t)}{z-t}$$

$$= G(z) + F(t)\,\frac{1}{z-t} .$$

The function G is first defined for $z \ne t$, but can be extended continuously by the convention $G(t) = F'(t)$ since by assumption, the differential quotients tend to this limit for $z \to t$. The integral

$$\frac{1}{2\pi i}\oint_{|z-a|=r}\frac{F(z)\,dz}{z-t}$$

can then be evaluated as sum of two terms. Consider first

$$F(t)\,\frac{1}{2\pi i}\oint_{|z-a|=r}\frac{dz}{z-t}\,.$$

The Stokes theorem applied to a region

$$\Pi_\varepsilon\ :\ |z-a|\le r \quad \text{and} \quad |z-t|\ge\varepsilon$$

for small enough $\varepsilon > 0$ (a picture shows that any $\varepsilon < r-|t-a|$ is adequate) furnishes

$$F(t)\,\frac{1}{2\pi i}\oint_{|z-a|=r}\frac{dz}{z-t} = F(t)\,\frac{1}{2\pi i}\oint_{|z-t|=\varepsilon}\frac{dz}{z-t}\,.$$

But we have already computed such an integral just before the statement of the theorem (crucial case $n = -1$). We find the value $F(t)$ for contribution of this term. The proof will be concluded as soon as we have shown that the other contribution, given by

$$\frac{1}{2\pi i}\oint_{|z-a|=r}G(z)\,dz$$

vanishes. As before, Stokes' theorem shows that

$$\oint_{|z-a|=r}G(z)\,dz = \oint_{|z-t|=\varepsilon}G(z)\,dz$$

for small enough $\varepsilon > 0$. But the continuity of the function G implies that it remains bounded, say $|G| \le M$ in $B_{\le r}(a)$. Consequently,

$$\left|\oint_{|z-a|=r}G(z)\,dz\right| = \left|\oint_{|z-t|=\varepsilon}G(z)\,dz\right| \le M\,2\pi\varepsilon.$$

Since this is arbitrarily small, the integral $\oint_{|z-a|=r}G(z)\,dz$ can only vanish. ∎

The applications of Cauchy's theorem are both important in theory and in practice.

13.4 THEORETICAL APPLICATIONS OF CAUCHY'S THEOREM

Without going into all details, let us indicate how many properties of holomorphic functions can be derived from the Cauchy formula just proved in the preceding section.

The formula

$$F(\underset{\uparrow}{t}) = \frac{1}{2\pi i}\oint_{|z-a|=r}\frac{F(z)\,dz}{z-\underset{\uparrow}{t}} \quad \text{as soon as} \quad |t-a| < r$$

exhibits the dependence in t of the function F through a parametric integral. Moreover, once the values of F on the circle $|z-a| = r$ are known, the integral can be computed and hence the values of F inside the same circle are completely determined.

Here is a **list of properties** of holomorphic functions that follow from Cauchy's integral formula once the classical results for parametric integrals have been proved (cf.16.5) :

a) F(a) is the average of the values $F(a+\varepsilon e^{it})$ of F on small circles of radius ε centered at a (simply parameterize these circles and use Cauchy's formula);

b) F is *smooth*, i.e. infinitely differentiable : the derivatives of F are obtained by derivation under the integral sign

$$F^{(k)}(t) = \frac{k!}{2\pi i} \oint_{|z-a|=r} \frac{F(z)\,dz}{(z-t)^{k+1}} \quad \text{for every } k \in \mathbb{N};$$

c) F is *analytic*, i.e. can be expanded in a convergent power series around any point of its domain of definition : expand $(z-t)^{-1}$ as uniformly convergent Taylor series around the point a using

$$(z-t)^{-1} = (z-a - t+a)^{-1} = (z-a)^{-1}\left(1 - \frac{t-a}{z-a}\right)^{-1}$$
$$= (z-a)^{-1} \sum_{n\geq 0} \left(\frac{t-a}{z-a}\right)^n \quad \text{(geometric series)},$$

and integrate term by term;

d) F is *harmonic* : since F is continuously differentiable as many times as we like (by b)), the following computations based on the Cauchy-Riemann relations are legitimate for the real part u = Re(F) (and similarly for the imaginary part v = Im(F))

$$\partial^2 u/\partial x^2 = \partial_x(\partial u/\partial x) = \partial_x(\partial v/\partial y) = \partial_x \partial_y v =$$
$$= \partial_y \partial_x v = \partial_y(-\partial_y u) = -\partial^2 u/\partial y^2$$

(notice that we have used the Schwarz theorem 7.1 for the permutation of partial derivatives).

13.5 PRACTICAL APPLICATIONS OF CAUCHY'S THEOREM

Let us show how the methods of 13.3 permit one to determine the value of certain definite integrals even when no primitive of the integrand is known.

The principle is the following. If F is holomorphic in $\mathbb{C}$ except at a finite number of points, say F holomorphic in $\mathbb{C} - A$ where A is finite (or at least consists in isolated points), we define the **residue of** F **at the point** $a \in A$ by

$$\text{Res}_a(F) = \frac{1}{2\pi i} \oint_{|z-a|=\varepsilon} F(z)\, dz$$

where $\varepsilon > 0$ is small enough (smaller than the distance of a to the other points of A). If Π is a piece not containing any point of A on its boundary and $\gamma = \partial\Pi$ (disjoint from A), the integral

$$\oint_\gamma F(z)\, dz$$

is easily evaluated in function of the residues of F at the points $a \in A \cap \Pi$. Choose indeed small open discs B(a) centered at the points of $A \cap \Pi$ (these should be disjoint with closure in Π) and apply Stokes' theorem to the piece $\Pi - \bigcup_a B(a)$ over which F is holomorphic. As we have seen in 13.3, we shall get

$$\oint_{\partial\Pi} F(z)\, dz = \sum_{a\in A\cap\Pi} \oint_{\partial B(a)} F(z)\, dz = 2\pi i \sum_{a\in A\cap\Pi} \text{Res}_a(F).$$

It is usually quite easy to determine the residues at the singular points a of a function F simply by writing a limited expansion of F in the neighborhood of a. This will be the case if F is a quotient of two holomorphic functions (with denominator vanishing at the points of A) : limited expansions of numerator and denominator at a point $a \in A$ will lead to an expansion of F near $a \in A$

$$F(z) = c_{-n}(z-a)^{-n} + \ldots + c_{-1}/(z-a) + c_o + c_1(z-a) + \ldots$$

which we write in the form

$$F(z) = c_{-n}(z-a)^{-n} + \ldots + c_{-1}/(z-a) + G(z)$$

with a bounded continuous function G (defined e.g. by its series expansion valid in a neighborhood of the point a). The residue of F at a is simply the coefficient c_{-1} in this expansion. Indeed, by definition, this residue is given by

$$2\pi i\ \text{Res}_a(F) = \oint_{|z-a|=\varepsilon} F(z)\, dz$$

and this integral can be evaluated as a sum of integrals of G(z) and $c_j(z-a)^j$ $(-n \le j \le -1)$. As we have already seen in the proof of the Cauchy theorem, $\oint_{|z-a|=\varepsilon} G(z)\, dz$ is arbitrarily small when ε is small. But it is independent of $\varepsilon > 0$ (ε being small) since it is equal to $\oint_{|z-a|=\varepsilon} \left[F - \Sigma_{j<0}\ c_j(z-a)^j\right] dz$. Consequently, this integral of G can only vanish. There remain the other terms

$$\oint_{|z-a|=\varepsilon} c_j(z-a)^j dz.$$

They all vanish except for the crucial exponent $j = -1$ where

$$\oint_{|z-a|=\varepsilon} c_j(z-a)^j dz = 2\pi i\ c_{-1} \text{ if } j = -1.$$

We have thus found

$$\mathrm{Res}_a(F) = c_{-1} = c_{-1}(a).$$

For example, if $F(z) = \mathrm{ctg}z = \cos z/\sin z$, we have near the origin

$$F(z) = (1 - z^2/2! + \dots) / (z - z^3/3! + \dots) =$$
$$= 1/z + c_1 z + c_3 z^3 + \dots = 1/z + G(z).$$

Consequently, $\oint_{|z|=r} \mathrm{ctg}z\, dz = 2\pi i\ \mathrm{Res}_o(\mathrm{ctg}z) = 2\pi i$

provided $r > 0$ is small enough ($r < \pi$ will be enough in order that the disc of radius r would not contain other zeros of the denominator sinz of ctgz).

Let us give only one simple example of the method of residues for the computation of a definite integral. Let us explain this method on

$$\int_{-\infty}^{\infty} \frac{dx}{1 + x^2}.$$

For this integral, we consider the function $F(z) = 1/(1 + z^2)$ and the pieces Π_R illustrated in the fig.

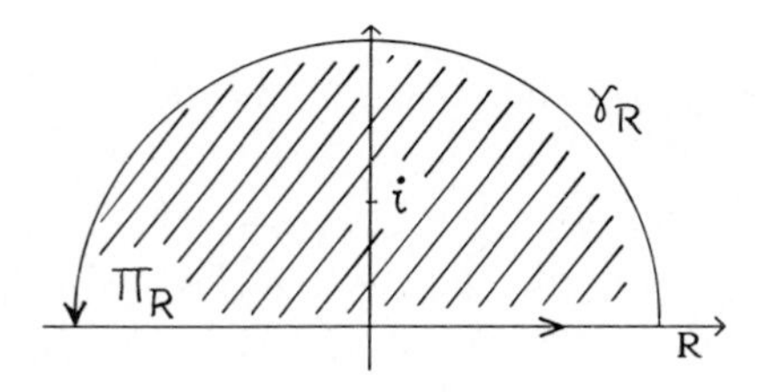

Fig.13.2

Let us put $\gamma_R = \partial\Pi_R$. Since the function F has two singularities, namely at the points $1 + z^2 = 0$, $z = \pm i$, we shall assume that the number R is greater than 1 (with eventually $R \longrightarrow \infty$). The residue formula gives

$$\int_{\gamma_R} \frac{dz}{1 + z^2} = 2\pi i\ \mathrm{Res}_i\left(\frac{1}{1 + z^2}\right).$$

The residue in question is easily determined by the limited expansion near $z = i$ (one more occasion to realize the importance of limited expansions!)

$$\frac{1}{1 + z^2} = (z-i)^{-1}(z+i)^{-1} = (z-i)^{-1}(a_o + a_1(z-i) + \dots).$$

The coefficient c_{-1} of $(z-i)^{-1}$ is precisely a_o, and this constant term is the value $(z+i)^{-1}_{z=i} = 1/2i$. Having determined this residue, we have obtained

$$\int_{\gamma_R} \frac{dz}{1 + z^2} = 2\pi i/(2i) = \pi.$$

The relevance of the preceding computation is due to the following circumstance. The result is independent of R and hence also valid in the limit

$$\pi = \lim_{R\to\infty} \int_{\gamma_R} \frac{dz}{1 + z^2} = \lim_{R\to\infty} \left(\int_{-R}^{R} + \int_{\gamma'_R} \right).$$

Here, the curvilinear integral over γ_R is decomposed into its part over the real axis and its part over the semi-circle in the upper half of $\mathbb{C}$. It is easy to show that

$$\int_{\gamma'_R} \ldots \longrightarrow 0 \quad \text{when} \quad R \longrightarrow \infty .$$

The standard majoration $|1 + z^2|^{-1} \le M/|z|^2$ (valid for $|z| \longrightarrow \infty$) indeed gives

$$\left| \int_{\gamma'_R} \ldots \right| \le (M/R^2)\ \pi R = \pi M/R \longrightarrow 0$$

since $|z| = R$ on γ'_R .

13.6 THE COMPLEX LOGARITHM : A PARADOX

Here is a funny *deduction*. Consider the chain of implications

$$-1 = -1 \Longrightarrow -1/1 = 1/-1 \Longrightarrow \sqrt{-1/1} = \sqrt{1/-1} \Longrightarrow$$
$$\Longrightarrow \sqrt{-1}/\sqrt{1} = \sqrt{1}/\sqrt{-1} \Longrightarrow (\sqrt{-1})^2 = (\sqrt{1})^2 \Longrightarrow -1 = 1$$

(i.e. we run into trouble, and normal people do not find it *funny* after all...). The problem is to define a *good* square root function in the complex domain with e.g. $\sqrt{(a/b)} = \sqrt{a}/\sqrt{b}$. In fact, the preceding fallacious deductions shows that *such a function cannot be defined in the whole of* $\mathbb{C}$. To define $\sqrt{z}$, we have to consider the solutions of $t^2 = z$. With $z = re^{i\varphi}$, we can select $\sqrt{r} \ge 0$ and take

$$t = \pm\sqrt{r}\ e^{i\varphi/2} \quad (= \sqrt{r}\ e^{i\varphi/2} \quad \text{or} \quad \sqrt{r}\ e^{i(\varphi/2 + \pi)}).$$

A *function* —being a *map*— must consist in choices : which of the two possible values $f(z) = \pm\sqrt{r}\ e^{i\varphi/2}$ is selected? (When $|z| = r \ne 0$ there are *two distinct* possible values.) Such choices *cannot* be made *continuously* around the circle $|z| = 1$! It is easy to understand why : if we start with the choice $\sqrt{1} = 1$ and go around the unit circle in the positive trigonometric direction, i.e. take $z = e^{i\varphi}$ with $0 \leq \varphi < 2\pi$, continuity forces us to choose

$$\sqrt{z} = e^{i\varphi/2} \qquad (\text{e.g. } \sqrt{-1} = i)$$

but when $\varphi \longrightarrow 2\pi$, $e^{i\varphi/2} \longrightarrow e^{i\pi} = -1$ and the values of this choice present a jump around $z = 1$.

Similar problems arise with cubic roots, etc.

In general, we would like to be able to define fractional, or even real powers of a complex number by the formula

$$z^a = e^{a\log z} \quad \text{if} \quad a \in \mathbb{R}.$$

This shows that our definitions will be meaningful as soon as we have defined the single function logz. The requirement is

$$z = e^{\log z} = e^w$$

and if we try to solve this basic relation for $w = a + ib$, we come across the identity

$$e^w = e^{a + ib} = e^a e^{ib} = z = |z| e^{i\arg(z)} .$$

This requires

$$e^a = |z|, \qquad \text{namely} \qquad a = \log|z|,$$

and

$$b = \arg(z) = \varphi + k \cdot 2\pi \qquad \text{with any } k \in \mathbb{Z}.$$

In other words, *any logarithm of* z has the form

$$\log z = \log|z| + i\arg(z) = \log|z| + i\varphi + k \cdot 2i\pi \qquad (k \in \mathbb{Z}).$$

Whenever a continuous choice of argument is possible, this choice will determine a *continuous* choice of logarithm and consequently a $\mathcal{C}^1$ and holomorphic choice of log (Cor. of the local inversion theorem in 6.5). All real powers will also be defined by this choice.

For example, the choice $-\pi < \varphi < \pi$ in $\mathbb{C} -]-\infty,0]$ defines what is commonly called **principal determination of the argument** and denoted by Arg(z). Notice that this Argument is not defined for negative numbers. The corresponding logarithm is the **principal determination** Log of the logarithm. This special choice also defines **principal determinations of all real powers** (but there is no canonical notation for them).

Other choices are possible —or even preferable— in certain circumstances. For example, it is useful to consider the determination of the argument for which $0 < \varphi < 2\pi$. This determination is not defined for positive real numbers... It would be possible to choose the argument according to the rule

$$0 \leq \varphi < 2\pi \qquad \text{for all} \qquad z \in \mathbb{C} - \{0\},$$

but observe that this choice presents a *jump* along the positive real axis...

Let us come back to the paradox given at the beginning of this section. Since we would like to be able to speak of $\sqrt{-1}$, it is necessary to choose a domain (in which the argument can be continuously chosen) containing -1. As a possible domain, let us first delete the negative imaginary axis and choose correspondingly

$$-\pi/2 < \varphi' < 3\pi/2 \quad \text{for} \quad \log' \quad \text{and resp.} \ \sqrt{}'.$$

Deleting similarly the positive imaginary axis, we could define

$$-3\pi/2 < \varphi'' < \pi/2 \quad \text{for} \ \log'' \quad \text{and resp.} \ \sqrt{}''.$$

With these choices, we have

$$\sqrt{}'1 = 1 = \sqrt{}''1.$$

In particular, we have for both determinations

$$\sqrt{(-1/1)} = \sqrt{-1} = (\sqrt{-1})/1 = \sqrt{-1}/\sqrt{1}$$

(I hope that the reader enjoys my prudence). But

$$\varphi'(-1) = \pi \quad \Longrightarrow \quad \sqrt{}'-1 = e^{i\pi/2} = i,$$

whereas

$$\varphi''(-1) = -\pi \quad \Longrightarrow \quad \sqrt{}''-1 = e^{i(-\pi)/2} = 1/i = -i.$$

In particular, we see that

$$\sqrt{}'(1/-1) = \sqrt{}'-1 = i = -1/i = -\sqrt{}'1/\sqrt{}'-1,$$

or

$$\sqrt{}'(1/-1) = \sqrt{}'-1 = i = 1/-i = \sqrt{}''1/\sqrt{}''-1.$$

Similarly, $\sqrt{}''(1/-1) \neq \sqrt{}''1/\sqrt{}''-1$ so that we cannot make the deductions of the beginning of this section (this is fortunate !).

The moral of these considerations is that *extreme care has to be taken when logarithms, roots,... are dealt with in the complex domain!*

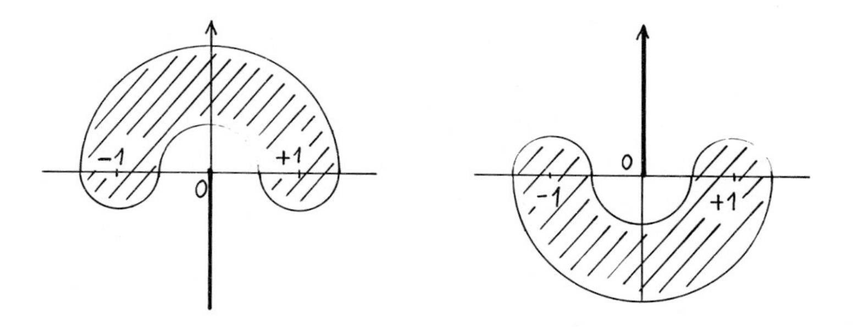

Fig.13.3

CHAPTER 14
APPLICATIONS OF INTEGRATION

14.1 TWO EASY LEMMAS

Several applications of integration are based on two easy results that we establish now.

Let us observe first that there is a real $\mathcal{C}^1$ function φ on the real line with the following properties

$$\varphi(x) = \begin{cases} 1 & \text{for } x \leq 0 \\ \text{monotonously decreasing} & \\ 0 & \text{for } x \geq 1 \end{cases}$$

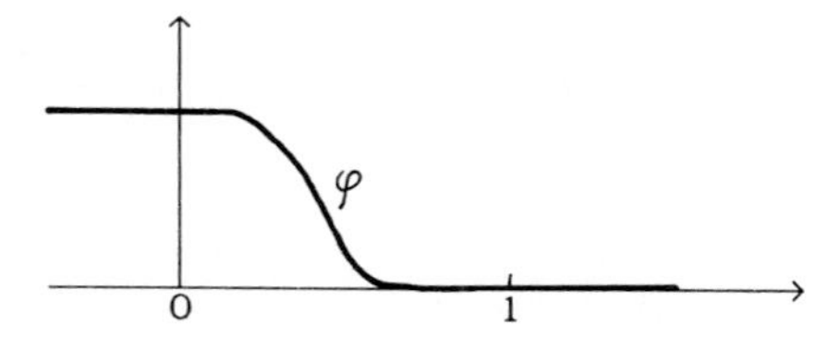

Fig.14.1

and so $\varphi \geq 0$, $\psi = -\varphi' \geq 0$, and ψ vanishes outside $[0,1]$. (It would not be difficult to construct $\mathcal{C}^\infty$ functions φ, ψ with the same properties.) In the space $\mathbb{R}^n$, it is useful to know that there exist $\mathcal{C}^1$ (resp.$\mathcal{C}^\infty$) functions for which

$$\begin{cases} \Phi \geq 0 \text{ everywhere} \\ \Phi(\mathbf{a}) > 0 \\ \Phi(\mathbf{x}) = 0 \text{ for } \|\mathbf{x} - \mathbf{a}\| > \varepsilon \end{cases}$$

whenever $\mathbf{a} \in \mathbb{R}^n$ and $\varepsilon > 0$ are given. For this purpose, simply consider the preceding function φ and put

$$\Phi(\mathbf{x}) = \varphi(\|\mathbf{x} - \mathbf{a}\|/\varepsilon).$$

The integral of a function Φ of the preceding type is > 0. In fact we have the following general result.

Lemma 1. If $f : \mathbb{R}^n \longrightarrow \mathbb{R}$ is continuous and positive with vanishing integral over some n-dimensional piece $\Pi \subset \mathbb{R}^n$, i.e. $\int_\Pi f \, dV = 0$, then f vanishes identically in the interior of Π. □

Proof. If a continuous function f is such that $f(\mathbf{a}) > 0$ at some interior point $\mathbf{a} \in \Pi$, there will be a small ball B (or a cube) in Π, centered at the point $\mathbf{a}$ with $f(\mathbf{x}) \geq \varepsilon = f(\mathbf{a})/2$ (for all $\mathbf{x} \in B$). Consequently $\int_\Pi f \, dV \geq \int_B f \, dV \geq \int_B \varepsilon \, dV \geq \varepsilon \mathrm{Vol}(B) > 0$. ■

Lemma 2. Let $f : \mathbb{R}^n \longrightarrow \mathbb{R}$ (or $\mathbb{C}$) be a continuous function and Π an n-dimensional piece such that

$$\int_\Pi f(x)\varphi(x)dV = 0$$

for all $\mathcal{C}^1$ (or $\mathcal{C}^\infty$) functions φ which vanish outside some ball $B \subset \Pi$. Then f vanishes identically in the interior of Π. □

Proof. We show that if the continuous function f does not vanish at an interior point $\mathbf{a}$ of Π, then there exists a $\mathcal{C}^1$ (resp. $\mathcal{C}^\infty$) φ as in the statement for which $\int_\Pi f\varphi dV = \int_B f\varphi dV \neq 0$. Separating real and imaginary parts, we can always assume that f is real valued. Changing signs if necessary we can also assume that $f(\mathbf{a}) > 0$. In this case, f will be positive in a ball B centered at $\mathbf{a}$. Taking a smooth positive function φ vanishing outside B and with $\varphi(\mathbf{a}) > 0$, we obtain a positive function $f\varphi \geq 0$ (vanishing outside B) with

$$f\varphi(\mathbf{a}) > 0 \quad \text{hence} \quad \int_\Pi f\varphi dV = \int_B f\varphi dV \neq 0$$

by lemma 1. ■

14.2 DIFFERENTIAL OPERATORS IN CURVILINEAR COORDINATES

In this section, we are going to present the computation of the divergence and Laplacian operators in curvilinear coordinates. (The gradient has already been treated in 9.2.) We recall the notations for the introduction of curvilinear coordinates in some open region $D \subset \mathbb{R}^n$ (cf. 9.4). Let $F : U \longrightarrow D$ be a diffeomorphism

$$\mathbb{R}^n \supset U \ni \xi \longmapsto \mathbf{x} = F(\xi) \in D.$$

We shall call (ε_i) the field of local frames associated to these curvilinear coordinates : if $(\mathbf{e}_i)$ is the canonical basis of $\mathbb{R}^n$, then $\varepsilon_j = F'(\xi)\mathbf{e}_j$ is tangent to the jth coordinate curve (the jth column of the Jacobian matrix $F'(\xi)$). If $\mathbf{u}$ is a vector field defined in D, it is natural to introduce the vector field $\mathbf{v}$ in U by the relation

$$F'(\xi)\mathbf{v}(\xi) = \mathbf{u}(\mathbf{x}(\xi)).$$

If $\mathbf{u}$ has the components u^i, namely if $\mathbf{u} = u^i\mathbf{e}_i$ (summation convention) and similarly $\mathbf{v}$ the components v^i, $\mathbf{v} = v^i\mathbf{e}_i$, we shall have

$$u^i\mathbf{e}_i = \mathbf{u} = F'\cdot\mathbf{v} = F'\cdot v^i\mathbf{e}_i = v^iF'\mathbf{e}_i = v^i\varepsilon_i.$$

This shows that the v^i are also the components of the field $\mathbf{u}$ in the *local frame* (ε_i) attached to the given curvilinear coordinates.

Proposition. With the preceding notations, the divergence of the $\mathcal{C}^1$ vector field **u** can be computed according to the formula

$$\operatorname{div} \mathbf{u} = \partial u^i/\partial x^i = g^{-1/2}(\partial/\partial\xi^j)(g^{1/2}v^j) \text{ (summation convention).} \ \square$$

Proof. It is enough to establish the given equality at all points of D (or U). Thus, the problem is a local one. For each smooth function φ vanishing outside a small ball B, we can compute the integral (over D or the small ball B)

$$\int_D \varphi \operatorname{div}\mathbf{u} \, dV = \sum_i \int_B \varphi \, \partial u^i/\partial x^i \, dV.$$

Here, $dV = dV(x) = dx^1 \ldots dx^n$ and each of these integrals can be evaluated by parts (in the x^i variable). Since φ vanishes on the boundary of B, the integrated terms will vanish and we find

$$\int \varphi \operatorname{div}\mathbf{u} \, dV = \sum -\int \partial\varphi/\partial x^i \; u^i \, dV.$$

Let us now use the change of variables given by F. We should give a new name to the new functions $\varphi \circ F$ and $\mathbf{u} \circ F$ on $U = F^{-1}(D)$, but let us make the usual abuse of using the same letters for them. For example, the chain rule for differentiation gives

$$\partial\varphi/\partial x^i = \sum_j \partial\varphi/\partial\xi^j \cdot \partial\xi^j/\partial x^i.$$

But now $(\partial\xi^j/\partial x^i)$ is the inverse of the Jacobian matrix $(\partial x^i/\partial\xi^j)$ of F'. This implies —by contravariance— that

$$\sum_i \partial\xi^j/\partial x^i \cdot u^i = v^j.$$

We have proved

$$\int_D \varphi \operatorname{div}\mathbf{u} \, dV = \sum_j -\int_U \partial\varphi/\partial\xi^j \; v^j \sqrt{g} \, dV(\xi).$$

In fact, the integral can again be evaluated on a small ball of U (since φ and consequently $\partial\varphi/\partial\xi^j$ vanish outside such a small ball) and we can again integrate by parts with respect to the jth variable ξ^j

$$\int_D \varphi \operatorname{div}\mathbf{u} \, dV = \sum_j \int_U \varphi \, (\partial/\partial\xi^j) \, (v^j \sqrt{g}) \, dV(\xi).$$

$$= \int_U \varphi \sum_j \sqrt{g}^{-1}(\partial/\partial\xi^j) \, (v^j \sqrt{g}) \sqrt{g} \, dV(\xi).$$

In this expression, we recognize $\sqrt{g}\, dV(\xi) = dV(x)$ and we can come back to the original variables

$$\int_D \varphi \operatorname{div}\mathbf{u} \, dV = \int_D \varphi \sum_j \sqrt{g}^{-1}(\partial/\partial\xi^j) \, (v^j \sqrt{g}) \, dV.$$

Since this equality holds for all smooth functions φ vanishing outside a small ball, the lemma 2 of 14.1 implies that

$$\operatorname{div}\mathbf{u} = \sum_j \sqrt{g}^{-1}(\partial/\partial\xi^j) \, (v^j \sqrt{g}). \qquad \blacksquare$$

Corollary. Let f be a $\mathcal{C}^2$ scalar field in D. Then the Laplacian Δf of f can be computed in curvilinear coordinates according to the formula

$$\Delta f = \sum_{ij} g^{-1/2}(\partial/\partial\xi^j)(g^{ij}g^{1/2}\partial f/\partial\xi^i). \qquad \square$$

Proof. Since $\Delta f = \mathrm{div}\,\mathbf{grad}f$, we simply have to determine the components v^j of the vector field $\mathbf{u} = \mathbf{grad}f$. But we have already done this in 9.2

$$v^j = \sum_i g^{ij}\partial f/\partial\xi^i.$$

The announced result follows (cf.9.6 for abuse of notation !) ■

14.3 THE GREEN-RIEMANN FORMULA

The Stokes formula has several consequences of fundamental importance for harmonic functions. Let us give a few typical examples in $\mathbb{R}^3$ (the interested reader can try to obtain similar results in $\mathbb{R}^2$... or $\mathbb{R}^n$!).

Let f and g be two $\mathcal{C}^2$ scalar fields in (some open subset of) $\mathbb{R}^3$. The identity

$$\nabla(f\nabla g) = \nabla f\cdot\nabla g + f\Delta g,$$

namely

$$\mathrm{div}(f\,\mathbf{grad}g) = \mathbf{grad}f\cdot\mathbf{grad}g + f\Delta g$$

leads to the integral transformation formula

$$\int_{\Sigma=\partial V} f\nabla g\cdot d\sigma = \int_V (\mathbf{grad}f\cdot\mathbf{grad}g + f\Delta g)dV$$

for 3-dimensional pieces V. Exchanging f and g in this formula and subtracting the two results, we obtain the **Green-Riemann** formula

$$\int_{\Sigma=\partial V} (f\nabla g - g\nabla f)\cdot d\sigma = \int_V (f\Delta g - g\Delta f)dV.$$

If $\mathbf{n}$ denotes the outside normal of V (on Σ), $d\sigma = \mathbf{n}|d\sigma|$ and $\nabla f\cdot\mathbf{n} = \partial f/\partial n$ is the normal derivative of f (on Σ). Thus we can also write

$$\int_V (f\Delta g - g\Delta f)dV = \int_{\Sigma=\partial V} (f\ \partial g/\partial n - g\ \partial f/\partial n)|d\sigma|.$$

Taking $g \equiv 1$ in this formula, we see that in particular

$$\int_V \Delta f\ dV = \int_{\Sigma=\partial V} \partial f/\partial n\ |d\sigma|$$

and if f is harmonic, i.e. $\Delta f = 0$,

$$\int_{\Sigma=\partial V} \partial f/\partial n\ |d\sigma| = 0.$$

We have proved the first part of the following statement.

Theorem. Let f be a harmonic function in a domain containing a piece $V \subset \mathbb{R}^3$. Then

$$\int_{\Sigma=\partial V} \partial f/\partial n\ |d\sigma| = 0.$$

Moreover, if g is any harmonic function with the same normal derivative as f over Σ, then f - g is constant in each component of V. □

Proof. In the identity

$$\int_{\Sigma=\partial V} f\nabla g \cdot d\sigma = \int_V (\mathbf{grad} f \cdot \mathbf{grad} g + f\Delta g) dV$$

take f = g harmonic. We obtain

$$\int_V \|\mathbf{grad} f\|^2 dV = \int_{\Sigma=\partial V} f\nabla f \cdot d\sigma = \int_\Sigma f\ \partial f/\partial n |d\sigma| .$$

This identity shows that

$$\partial f/\partial n = 0 \text{ on } \Sigma \implies \int_V \|\mathbf{grad} f\|^2 dV = 0.$$

By lemma 1 of 14.1, this shows that **gradf** $\equiv 0$ in V hence f is constant in each component of V. These observations prove the second part of the theorem and conclude the proof. ■

Let us reformulate the result of the theorem in a slightly different form. Assume that we are looking for a harmonic function f in a domain containing a piece V, with a prescribed normal derivative $\partial f/\partial n$ on $\Sigma = \partial V$. In other words, assume that we are trying to solve a problem of the type

$$\Delta f = 0 \text{ in } V \quad \text{and} \quad \partial f/\partial n = g \text{ on } \Sigma = \partial V$$

where g is a given function on Σ. Then this problem will have a solution *only if the condition* $\int_\Sigma g d\sigma = 0$ *is satisfied*. Moreover, two solutions can only differ by a constant (in each component of V). It can be proved that the given condition is indeed *sufficient* for the existence of a solution (at least when V and g are sufficiently regular...)

At this point, it is possible to give a few sharper results on uniqueness of potentials (cf.8.7).

Theorem. Let **u** and **v** be two $\mathcal{C}^1$ vector fields defined in some open domain of $\mathbb{R}^3$ with

$$\mathbf{curl}\ \mathbf{u} \equiv \mathbf{curl}\ \mathbf{v}\ ,\ \operatorname{div} \mathbf{u} \equiv \operatorname{div} \mathbf{v}.$$

Let $V \subset \mathbb{R}^3$ be a *convex* 3-dimensional piece and **n** the exterior normal of V defined on $\Sigma = \partial V$. Then either of the following two conditions will ensure **u** = **v** in V

a) $\mathbf{u}\cdot\mathbf{n} = \mathbf{v}\cdot\mathbf{n}$ on Σ,

b) $\mathbf{u} \wedge \mathbf{n} = \mathbf{v} \wedge \mathbf{n}$ on Σ. □

Proof. Replacing **u** by **u** - **v**, we are reduced to the case **v** = 0. Then **curl u** = 0 implies that **u** = **gradf** (since V is convex). Now,

$\Delta f = \text{div}\,\mathbf{grad}f = \text{div}\,\mathbf{u} = 0$ so that f is a harmonic function. Under the assumption a), $\partial f/\partial n = \mathbf{grad}f\cdot\mathbf{n} = \mathbf{u}\cdot\mathbf{n} = 0$ on Σ. We have seen that this implies that f is constant in V (being convex, V has only one component!). Finally, $\mathbf{u} = \mathbf{grad}f = 0$ in V as expected in the case $\mathbf{v} = 0$. Under the assumption b), still with $\mathbf{v} = 0$, we shall have $\mathbf{grad}f = \mathbf{u}$ *parallel to* $\mathbf{n}$ on Σ. This shows that Σ is a level surface of the harmonic function f. Using the *maximum principle for harmonic functions*, we see that f is necessarily constant in V and thus $\mathbf{u} = 0$ in V. ■

The preceding proof is incomplete in the sense that we have used the maximum principle for harmonic functions to deduce b). The reader interested in the topic of harmonic functions will have to consult a book on partial differential equations, e.g. the widely available :
Courant R., Hilbert D. "Methoden der Mathematischen Physik" vol.1 (Springer-Verlag Taschenbücher Band 30, 1968).

14.4 CALCULUS OF VARIATIONS : THE SIMPLEST CASE

We are going to consider a problem of extrema for certain numerical functions defined over function spaces. Here is the simplest typical case to be treated.

Let $f : \mathbb{R}^3 \longrightarrow \mathbb{R}$ be a given smooth function and $I = [a,b]$ a compact interval. For $\mathcal{C}^1$ differentiable functions $\varphi : I \longrightarrow \mathbb{R}$, we can consider the integrals

$$F(\varphi) = \int_I f(x,\varphi(x),\varphi'(x))dx$$

as defining a mapping on the set of $\mathcal{C}^1$ functions $\mathcal{F}$

$$F : \mathcal{F} \longrightarrow \mathbb{R}\ ,\ \varphi \longmapsto F(\varphi).$$

The **question** is : how can we determine the extrema of F (when they exist) ? More precisely, we shall try to determine the extrema of the restriction of F on a subvariety $\mathcal{F}_{\alpha\beta} \subset \mathcal{F}$ defined by a fixed boundary condition

$$\varphi \in \mathcal{F} \quad \text{satisfies} \quad \varphi(a) = \alpha,\ \varphi(b) = \beta.$$

Assume that $\varphi \in \mathcal{F}$ is an extrema and take an increment $h \in \mathcal{F}$ with $h(a) = 0$, $h(b) = 0$. Then all $\varphi + th \in \mathcal{F}_{\alpha\beta}$ $(t \in \mathbb{R})$ and

$$t \longmapsto F(\varphi+th) \quad \text{has an extremum for } t = 0.$$

We infer that the corresponding directional derivative must vanish

$$(d/dt)_{t=0}F(\varphi + th) = 0.$$

Let us compute this directional derivative

$$(d/dt)_{t=0}\int_I f(x,\varphi(x)+th(x),\varphi'(x)+th'(x))dx.$$

Leibniz' rule for the derivation of a parametric integral (to be established in 16.5) requires a computation of the corresponding partial derivative of the integrand

$$(\partial/\partial t)_{t=0}\ f(x,\varphi(x)+th(x),\varphi'(x)+th'(x)) =$$
$$= D_2f(\ldots)h(x) + D_3f(\ldots)h'(x).$$

Thus we must have

$$\int_I [D_2f\cdot h + D_3f\cdot h']dx = 0$$

for all differentiable functions h which vanish at the extremities of I.

Lemma. Let $I = [a,b]$ be a compact interval, A and $B : I \longrightarrow \mathbb{R}$ two continuous functions such that

$$\int_I [A(x)\cdot h(x) + B(x)\cdot h'(x)]dx = 0$$

for all $\mathcal{C}^1$ functions h vanishing at the extremities of I. Then

$$B \text{ is } \mathcal{C}^1 \quad \text{and} \quad B' = A. \qquad \square$$

Proof. Observe that if $A = B'$,

$$\int_I [A(x)\cdot h(x) + B(x)\cdot h'(x)]dx = \int_I [B'\cdot h + B\cdot h']dx =$$
$$= \int_I [B\cdot h]'dx = [B\cdot h]_{\partial I} = 0.$$

The lemma asserts that the *sufficient condition* $A = B'$ is also *necessary* for the vanishing of the integral. For the proof, let us introduce the primitive C vanishing at a of the function A. Then

$$\int_I A(x)\cdot h(x)dx = [C(x)\cdot h(x)]_a^b - \int_I C(x)\cdot h'(x)dx = -\int_I C(x)\cdot h'(x)dx$$

since h vanishes at the extremities of I. Our assumption is now

$$\int_I [B - C]\cdot h'dx = 0$$

for all $\mathcal{C}^1$ functions h vanishing at the extremities of I. But we can write

$$B - C = m + g$$

with $m = (b-a)^{-1}\int_I[B-C]dx$ (average of B–C on I) and $\int_I g dx = 0$. In particular, the $\mathcal{C}^1$ function $h(x) = \int_a^x g(t)dt$ vanishes at both ends of the interval I and $B - C = m + h'$. Thus we must have

$$0 = \int_I [m + h']h'dx = m[h]_{\partial I} + \int_I (h')^2dx = \int_I (h')^2dx.$$

By lemma 1 of 14.1, this implies $g = h' = 0$, and consequently

$$B - C = m \text{ (constant).}$$

Since $C' = A$ by construction, we also conclude $B \in \mathcal{C}^1$ and $B' = A$.■

The consequence of this lemma in our context is

$$D_2 f = (d/dx) D_3 f$$

or more explicitely

$$E_f(\varphi) = D_2 f(x,\varphi(x),\varphi'(x)) - (d/dx) D_3 f(x,\varphi(x),\varphi'(x)) = 0.$$

This is a (non linear) differential equation satisfied by the extrema φ of our problem : it is the **Euler equation** for φ.

Since we now know that $D_3 f(\ldots)$ is differentiable, it is legitimate to perform an integration by parts in

$$\int_I [D_2 f \cdot h + D_3 f \cdot h'] dx = 0$$

thus obtaining the equivalent conditions

$$\int_I [D_2 f - (D_3 f)'] \cdot h dx = 0.$$

The Euler equation would now be obtained by an application of lemma 2 of 14.1. It is quite remarkable that the vanishing of the single *directional derivatives* are sufficient to imply the Euler equation for the function φ realizing the extremum.

Let us summarize the result obtained.

Theorem. Let $f : \mathbb{R}^3 \longrightarrow \mathbb{R}$ be a $\mathcal{C}^1$ function and $I = [a,b]$ a compact interval. Call $F : \mathcal{F} \longrightarrow \mathbb{R}$ the *functional*

$$\varphi \longmapsto F(\varphi) = \int_I f(x,\varphi(x),\varphi'(x)) dx.$$

If φ is an extremum of the restriction of F to the subvariety $\mathcal{F}_{\alpha\beta}$ consisting of $\mathcal{C}^1$ functions φ with $\varphi(a) = \alpha$, $\varphi(b) = \beta$, then φ is a solution of the Euler equation $E_f(\varphi) = 0$. *In extenso*

$$(d/dx) D_3 f(x,\varphi(x),\varphi'(x)) - D_2 f(x,\varphi(x),\varphi'(x)) = 0. \qquad ■$$

14.5 CALCULUS OF VARIATIONS : THE GENERAL CASE

The extrema problems that arise in many applications have the following type. Let $E = \mathbb{R}^n$, $f : \mathbb{R} \times E \times E \longrightarrow \mathbb{R}$ (or $\mathbb{C}$) be a $\mathcal{C}^1$ function and $I = [a,b]$ a compact interval. What are the extrema of the *functional*

$$F(\varphi) = \int_I f(t,\varphi(t),\varphi'(t)) dt$$

when restricted to the set of $\mathcal{C}^1$ parameterized curves $\varphi : I \longrightarrow E$ linking two given points **a** and **b** of E ? Again, the conditions

$$\varphi(a) = \mathbf{a}, \ \varphi(b) = \mathbf{b}$$

define a subvariety $\mathcal{F}_{\mathbf{ab}}$ of the space $\mathcal{F}$ of $\mathcal{C}^1$ curves $I \longrightarrow E$ and the same method as in the previous section will show that the solutions φ will satisfy the general Euler equations

$$(d/dt)\mathbf{grad}_{\mathbf{y}}f - \mathbf{grad}_{\mathbf{x}}f = 0.$$

Here, we imagine that the variables of f are written in the following form

$$f = f(t,\mathbf{x},\mathbf{y}) = f(t,x_1,\ldots,x_n,y_1,\ldots,y_n)$$

so that e.g. $\mathbf{grad}_{\mathbf{x}}f$ represents the line vector with components

$$\partial f/\partial x_1 \ , \ \ldots \ , \ \partial f/\partial x_n.$$

In particular, the Euler equations give a system of n coupled differential equations for the components of φ

$$[(d/dt)\partial f/\partial y_i - \partial f/\partial x_i](t,\varphi(t),\varphi'(t)) = 0 \quad (1 \le i \le n).$$

In some cases, an extra condition in the form

$$G(\varphi) = \int_I g(t,\varphi(t),\varphi'(t))dt = 0 \text{ (constant)}$$

is required for all functions φ to be considered. In such a case, the method of the Lagrange parameter is applicable (cf.5.5). Its justification in the present context can be given as follows. Take two increments h and k : $I \longrightarrow E$ (vanishing at the extremities) and consider the functions $\varphi + \tau h + \sigma k$ (in a neighborhood of the extremal φ). Thus

$$F(\varphi + \tau h + \sigma k) \text{ must have an extrema for } \sigma = \tau = 0$$

when σ and τ are restricted to vary in accordance with the condition $G(\varphi + \tau h + \sigma k) = 0$. In this form, we recognize a classical problem to which the Lagrange method is ideally suited : we have to look for *unconditional extrema* of

$$(\tau,\sigma) \longmapsto [F - \lambda G](\varphi + \tau h + \sigma k).$$

The annulation of the derivatives $(\partial/\partial\tau)_{\tau=0}$, $(\partial/\partial\sigma)_{\sigma=0}$ will lead to the equations

▷ $\int_I [E_f(\varphi) - \lambda E_g(\varphi)]h\,dt = 0,$

▷ $\int_I [E_f(\varphi) - \lambda E_g(\varphi)]k\,dt = 0,$

▷ $G(\varphi) = 0.$

The first (resp. second) relation shows that λ is independent from k (resp.h) and we find that φ is a solution of the Euler equation $E_{f-\lambda g}(\varphi) = 0$ relative to $f - \lambda g$.

14.6 A TYPICAL EXAMPLE

Let us look for a curve in the plane $\mathbb{R}^2$ going through two fixed points and maximizing the area from the line linking these two points. Assume for example that I = [0,1] and $\varphi \in \mathcal{C}^1$ satisfies $\varphi(0) = \varphi(1) = 0$. How can we maximize the area below the graph of φ keeping the length of this graph constant ? In formulas

$$\int_0^1 \varphi(t)dt \quad \text{to maximize with respect to}$$

$$\int_0^1 \sqrt{(1 + \varphi'(t)^2)}\ dt = \ell \quad \text{kept constant .}$$

Here, we have to take

$$f(t,x,y) = x \quad \text{and} \quad g(t,x,y) = \sqrt{(1 + y^2)} - \ell$$

so that

$$f(t,\varphi(t),\varphi'(t)) = \varphi(t)\ \&\ g(t,\varphi(t),\varphi'(t)) = \sqrt{(1 + \varphi'(t)^2)} - \ell.$$

The Euler equation relative to $H = f - \lambda g$ is obtained by computing

$$\left(\frac{d}{dt}\frac{\partial}{\partial y} - \frac{\partial}{\partial x}\right)(f - \lambda g).$$

Let us introduce the following notational convention. When F is a function of t, x and y, and φ is a given $\mathcal{C}^1$ function (of t) we denote by F_φ the function $t \longmapsto F(t,\varphi(t),\varphi'(t))$. In particular, even if F is independent of t, F_φ may depend on t and, if $\varphi \in \mathcal{C}^2$, we shall have

$$\frac{d}{dt}F_\varphi = \left(\frac{\partial F}{\partial x}\right)_\varphi \varphi' + \left(\frac{\partial F}{\partial y}\right)_\varphi \varphi''.$$

Coming back to our problem with $H = f - \lambda g$ independent of t, we observe that

$$\frac{d}{dt}\left(H - y\frac{\partial H}{\partial y}\right)_\varphi = \left(\frac{\partial H}{\partial x}\right)_\varphi \varphi' + \left(\frac{\partial H}{\partial y}\right)_\varphi \varphi'' - \varphi''\left(\frac{\partial H}{\partial y}\right)_\varphi - \varphi'\frac{d}{dt}\left(\frac{\partial H}{\partial y}\right)_\varphi =$$

$$= \left(\frac{\partial H}{\partial x}\right)_\varphi \varphi' - \varphi'\left(\frac{\partial H}{\partial x}\right)_\varphi = 0$$

by the Euler equation. This proves that

$$\left(H - y\frac{\partial H}{\partial y}\right)(t,\varphi(t),\varphi'(t)) = c \text{ (a constant).}$$

With

$$H = x - \lambda\sqrt{(1 + y^2)} - \lambda\ell$$

the preceding equation is

$$\varphi - \lambda\sqrt{(1 + \varphi'^2)} - \lambda\ell - \varphi'(-\lambda)\varphi'/\sqrt{\ldots} = c$$

and a line of computation leads to

$$(\varphi - c - \lambda\ell)d\varphi/\sqrt{(\lambda^2 - (\varphi - c - \lambda\ell)^2)} = dt.$$

From here follows

$$(\varphi - c - \lambda\ell)^2 + (t - d)^2 = \lambda^2 .$$

This is the equation of a circle of radius $|\lambda|$. To respect the conditions $\varphi(0) = \varphi(1) = 0$, we have to take $d = 1/2$ and the radius will be computable in terms of the imposed arc length ℓ. (Once more we see that the value of the Lagrange parameter has a nice interpretation for the extremum.)

The preceding solution only makes sense if the length ℓ satisfies $1 \leq \ell \leq \pi/2$. If $\ell < 1$, the problem has no solution. If $\ell > \pi/2$, the analytic solution found is a circle which is no more a functional graph (two values of $\varphi(t)$ occur when $t < 0$ or $t > 1$). A better and more symmetrical way of treating this extremum problem is to look for a closed parameterized curve

$$t \longmapsto \Phi(t) = (\varphi(t), \psi(t)) \qquad (\Phi(0) = \Phi(1) = 0)$$

of given length and maximizing the inner area. The Euler equations for φ and ψ will lead to circles.

Anyway, the preceding method only gives candidates for extrema, without proving that they indeed give minima. The **isoperimetric inequality** states more precisely that the length ℓ of a simple closed rectifiable curve and the enclosed area S satisfy

$$\ell^2 \geq 4\pi S \ ,$$

with equality only for circles. We shall come back to this question with a more satisfactory answer based on the theory of Fourier series (cf.24.5).

PART THREE

FUNCTION SPACES

CHAPTER 15
UNIFORM CONVERGENCE: THEORY

15.1 SIMPLE CONVERGENCE

Let us start with a sequence of numerical functions

$$f_n : I \longrightarrow \mathbb{R} \text{ (or } \mathbb{C})$$

where I is any set (in practice, I will often be an interval in $\mathbb{R}$, or a compact i.e. *closed* and *bounded*, subset of $\mathbb{R}^n$). When there exists a function $f : I \longrightarrow \mathbb{R}$ (resp. $\mathbb{C}$) such that

$$\text{for each } x \in I, \ f_n(x) \longrightarrow f(x) \ (\text{for } n \longrightarrow \infty),$$

we say that (f_n) **converges simply** or **pointwise** to f. The meaning of this situation is now reviewed in a few examples.

First example. Let $f_n(x) = x^n$ on $I = [0,1]$. Then,

$$f_n(x) = x^n \longrightarrow 0 \text{ for each } x \in [0,1[$$

whereas

$$f_n(1) = 1 \longrightarrow 1.$$

Here, we have to take for limit function f the discontinuous function jumping from 0 to 1 at the point 1

$$f(x) = \begin{cases} 0 \text{ for } 0 \leq x < 1 \\ 1 \text{ for } x = 1 \end{cases}$$

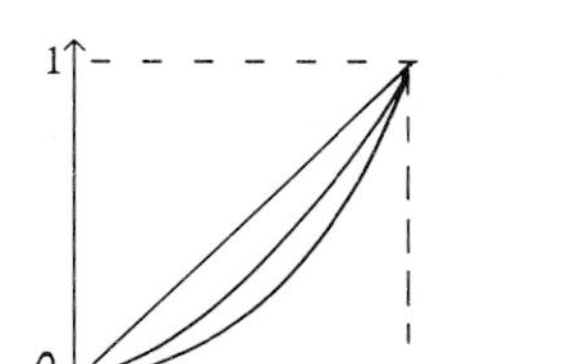

Fig.15.1

Second example. Let $(a_n)_{n\geq 1}$ be any sequence of positive numbers, and let f_n be the (continuous) function on $x \geq 0$ defined by

$$f_n(x) = \begin{cases} na_n x \text{ for } 0 \leq x \leq 1/n \\ a_n(2-nx) \text{ for } 1/n \leq x \leq 2/n \\ 0 \text{ for } x \geq 2/n \ . \end{cases}$$

For each $x > 0$, we shall have

$$2/n < x \quad \text{for large } n\text{'s}$$

hence $f_n(x) = 0$ for these n's. This proves that $f_n(x) \longrightarrow 0$ (as $n \to \infty$) *first for* $x > 0$, but obviously *also for* $x = 0$ since $f_n(0) = 0$ (for all n) ! The limit function f is the zero function *independently from the values* a_n (in particular, even if $a_n \to \infty$). Here, each f_n is continuous and

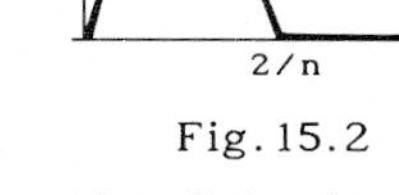

Fig.15.2

so is f. But

$$\int_0^1 f_n(x)dx = a_n/n \quad \text{whereas} \quad \int_0^1 f(x)dx = 0.$$

Taking $a_n = n$ for all n, we see that

$$\lim_{n\to\infty} \int_0^1 f_n(x)dx = 1 \quad \text{is different from} \quad \int_0^1 f(x)dx = \int_0^1 \lim_{n\to\infty} f_n(x)dx$$

(taking $a_n = n^2$, the integrals $\int_0^1 f_n(x)dx = n$ diverge !).

These two examples show that the notion of simple convergence (although quite natural) has pathological consequences...

15.2 UNIFORM CONVERGENCE

Let us introduce a stronger notion of convergence.

If two functions f and g : I $\longrightarrow$ $\mathbb{R}$ (or $\mathbb{C}$) have a bounded difference, we define their *distance* to be the number

$$d(f,g) = \operatorname*{Sup}_{x\in I} |f(x) - g(x)|.$$

Definition. We say that the sequence of functions (f_n) **converges uniformly** (on I) to f when $d(f_n,f) \longrightarrow 0$.

Uniform convergence of a sequence (f_n) to f thus means that for any given $\varepsilon > 0$, we have

$$|f_n(x) - f(x)| \le \varepsilon \quad \text{for all } x \in I$$

as soon as n *is large enough*, namely for all $n \ge N$ for some integer N depending on the proximity requirement ε (and the f_n's !) More geometrically, we can say that the graph of f_n should lie in a small *tubular neighborhood* of the graph of f (cf. Fig.15.3 where the real case is sketched). In the two examples above, uniformity failed since $d(f_n,f) = 1$ for all n in the first case and $d(f_n,f) = a_n$ in the second case.

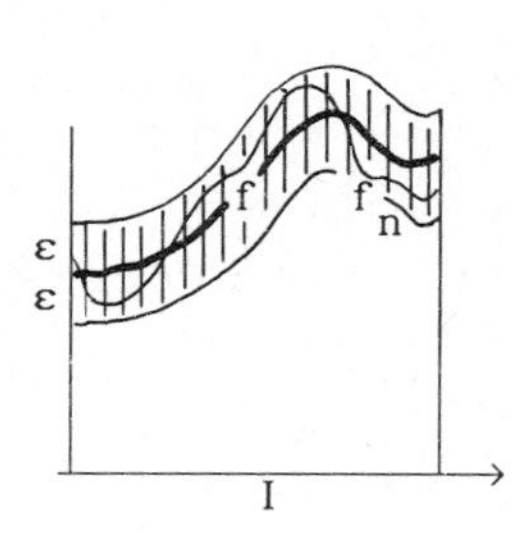

Fig.15.3

From now on, we shall only consider bounded functions and let us denote by $\mathcal{C}_b(I)$ the vector space of *bounded continuous* functions over I. We shall use the **norm of uniform convergence** over this vector space $\mathcal{C}_b(I)$: it is defined by

$$\mathcal{C}_b(I) \longrightarrow \mathbb{R} : f \longmapsto \|f\| = \|f\|_I = \operatorname*{Sup}_{x \in I} |f(x)| \ .$$

This *norm* has the characteristic properties

1. $\|f\| > 0$ for all $f \neq 0$,
2. $\|\lambda f\| = |\lambda| \, \|f\|$ for $\lambda \in \mathbb{R}$ (or $\mathbb{C}$),
3. $\|f + g\| \leq \|f\| + \|g\|$

(for f and $g \in \mathcal{C}_b(I)$). Observe that the second property implies that $\|0\| = 0$ (take $\lambda = 2$!) and thus we could equally well replace 1. by 1'.

1'. $\|f\| \geq 0$ and $(\|f\| = 0 \Longleftrightarrow f = 0)$.

We shall still use the **uniform distance** over $\mathcal{C}_b(I)$ defined by

$$d(f,g) = \|f - g\| = \operatorname*{Sup}_{x \in I} |f(x) - g(x)|.$$

This *distance* has the characteristic properties

1. $d(f,g) \geq 0$ and $\{ d(f,g) = 0 \Longleftrightarrow f = g \}$,
2. $d(f,g) = d(g,f)$,
3. $d(f,g) \leq d(f,h) + d(h,g)$

(for f, g and $h \in \mathcal{C}_b(I)$). Conversely, the norm can be recovered from the distance via the formula $\|f\| = d(f,0)$.

With these definitions, $f_n \longrightarrow f$ uniformly on I exactly when

$$d(f_n,f) = \|f_n - f\| \longrightarrow 0.$$

(Although the notations rarely exhibit it, both norm and distance *depend on the choice of the set* I *on which the supremum is estimated.*)

Basic theorem. Let I be a subset of $\mathbb{R}^n$ (or $\mathbb{C}^n$) and $(f_n) \subset \mathcal{C}_b(I)$ a sequence which converges uniformly to a certain function f on I. Then

a) f is continuous and bounded, i.e. $f \in \mathcal{C}_b(I)$,

b) $\int_\Pi f_n(x)dx \longrightarrow \int_\Pi f(x)dx$ over any compact piece $\Pi \subset I$.

□

Proof. By uniform convergence assumption

$$|f_n(x) - f(x)| \leq 1 \quad \text{for all } x \in I$$

as soon as n *is large enough*, e.g. for all $n \geq N_1$, and this shows

$$|f(x)| \leq |f_n(x)| + 1 \quad \text{for these n's.}$$

Hence f is *bounded*. Moreover, if $\varepsilon > 0$ is given, we shall have

$$|f_n(x) - f(x)| \leq \varepsilon \quad \text{for all } x \in I$$

as soon as n *is large enough*, namely for all $n \geq N = N_\varepsilon$. In particular,

$$|f_N(x) - f(x)| \leq \varepsilon \quad \text{for all } x \in I.$$

Hence

$$|f(y) - f(x)| \le |f(y) - f_N(y)| + |f_N(y) - f_N(x)| + |f_N(x) - f(x)|$$
$$\le 2\varepsilon + |f_N(y) - f_N(x)| .$$

But the function f_N is continuous, i.e. continuous at each point $x \in I$ and there exists a positive $\delta = \delta_{x,\varepsilon,N} > 0$ for which

$$|f_N(y) - f_N(x)| < \varepsilon \quad \text{as soon as} \quad |y - x| < \delta.$$

Hence

$$|f(y) - f(x)| < 3\varepsilon \quad \text{as soon as} \quad |y - x| < \delta.$$

This proves continuity of f at x, and a) is established. For b), simply observe that

$$\left|\int_\Pi f_n(x)dx - \int_\Pi f(x)dx\right| = \left|\int_\Pi \left(f_n(x) - f(x)\right)dx\right| \le$$
$$\le \int_\Pi \mathrm{Sup}\left|f_n(x) - f(x)\right| dx = \|f_n - f\| \int_\Pi dx \longrightarrow 0 \quad (n \to \infty).$$

Here, dx represents the element of volume dV in $\mathbb{R}^n$ (or more precisely the measure $|dx_1 \wedge \ldots \wedge dx_n|$ as in 12.2, however, the crucial point here is already interesting when n = 1 and $\Pi = J$ is a compact sub-interval of I). ■

It may be interesting to reformulate the assertions of the basic theorem in terms of *limits*. Still under the same assumptions we have

$$\text{a)} \lim_{y \to x} \lim_{n \to \infty} f_n(y) = \lim_{n \to \infty} \lim_{y \to x} f_n(y),$$

$$\text{b)} \lim_{n \to \infty} \int_\Pi f_n(x)dx = \int_\Pi \lim_{n \to \infty} f_n(x)dx.$$

For a), observe that the left hand side is $\lim_{y \to x} f(y)$, whereas the right hand side is $\lim_{n \to \infty} f_n(x)$ (since f_n is continuous) and is thus equal to f(x). Recall now that $\lim_{y \to x} f(y) = f(x)$ precisely characterizes continuity of f at x. Thus, the statement of the basic theorem is a *justification* for *interchanging limits* (resp. *limits and integrals*). The two examples preceding the notion of uniform continuity show how both properties fail in general for simple convergence. Let us add a more physical **example**. Consider the operation of filling up a container (unit volume for simplicity) with water, using several pipes. Denote by $V_n(t)$ the volume of water in the container at time t when the procedure starts at time t = 0 and the pipe has a flow of n liters per hour. Thus $V_n(t) = 0$ for $t \le 0$, is linear between t = 0 and t = 1/n and $V_n(t) = 1$ $(t \ge 1/n)$

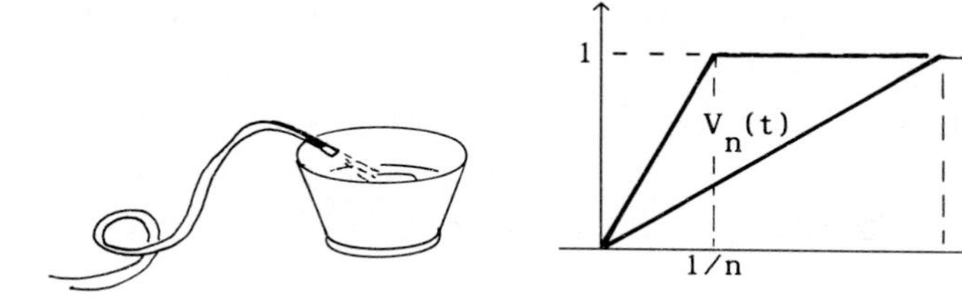

Fig.15.4

as in the figure. Obviously

$$V_n(t) \longrightarrow V(t) = \begin{cases} 0 \text{ for } t \leq 0 \\ 1 \text{ for } t > 0 \end{cases}$$

and e.g.on the interval I = [0,1]

$$\lim_{t \to 0} V_n(t) = 0 \quad \text{for any } n,$$

so that

$$\lim_{n \to \infty} \lim_{t \to 0} V_n(t) = 0.$$

But for any $t > 0$,

$$\lim_{n \to \infty} V_n(t) = 1 \quad (V_n(t) = 1 \text{ for } n \geq 1/t)$$

and so

$$\lim_{t \to 0} \lim_{n \to \infty} V_n(t) = 1.$$

15.3 APPROXIMATION OF THE SQUARE ROOT

Let us now turn to a more technical lemma having an interesting application.

Lemma of Dini. Let $f_n : I \longrightarrow \mathbb{R}$ be a *monotonous* sequence of continuous functions which converges *pointwise* to the *continuous* function f. Then, f_n converges *uniformly* to f *on any compact subset* $J \subset I$. □

Proof. Without loss of generality, we can assume that the sequence (f_n) is increasing, with $f_n(x) \nearrow f(x)$. By continuity assumption on f (and all f_n), the $g_n = f - f_n$ form a *decreasing* sequence of *continuous* functions with $g_n(x) \searrow 0$ $(x \in I)$. Let $\varepsilon > 0$ be *fixed* and consider the open sets

$$U_n = \{ x \in I : g_n(x) < \varepsilon \}.$$

Since $g_{n+1} \leq g_n$, we have $U_{n+1} \supset U_n$ and $g_n(x) \longrightarrow 0$ implies that the union of all U_n is the whole of I. In particular, any compact subset $J \subset I$ is *covered* by the family (U_n). The Lebesgue property shows that J is already covered by a *finite family* of U_n, and by one U_N since these open sets form an increasing sequence. This means that $g_N(x) < \varepsilon$ for all $x \in J \subset U_N$ and *a fortiori*

$$g_n(x) < \varepsilon \quad \text{for all } x \in J \quad \text{and all } n \geq N.$$

With the norm of uniform convergence on J, $\|g_n\| < \varepsilon$ $(n \geq N)$. This proves uniform convergence on J. ■

The **application** we have in mind is the following (it will have a far-reaching theoretical consequence).

Proposition. There exists a sequence (p_n) of *polynomials* which converges uniformly on the interval $I = [0,1]$ to the square root function. □

Proof. Let us define by induction the sequence (p_n) by

$$p_o = 0,$$
$$p_{n+1} = p_n + (x - p_n^2)/2 \ (n \geq 0).$$

For example

$$p_1(x) = x/2,$$
$$p_2(x) = x - x^2/8,$$
$$p_3(x) = 3x/2 - 5x^2/8 + x^3/8 - x^4/128 \quad \text{etc.}$$

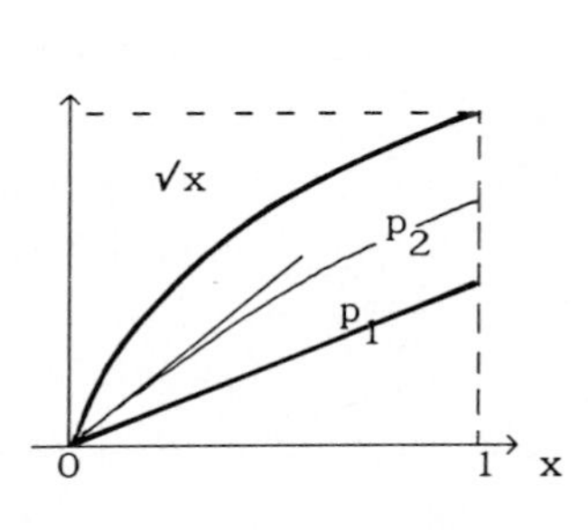

Fig.15.5

By induction, it is obvious that all p_n's are polynomials and that $p_n(0) = 0$. Still by induction, let us show that $p_n(x) \leq \sqrt{x}$, for $x \in I$, this being obvious for $n = 0$. To estimate

$$\sqrt{x} - p_{n+1}(x) = \sqrt{x} - p_n - (x - p_n^2)/2 =$$
$$= (\sqrt{x} - p_n)\Big(1 - (\sqrt{x} + p_n)/2\Big)$$

we use the induction assumption $p_n \leq \sqrt{x}$ hence

$$(\sqrt{x} + p_n) \leq 2\sqrt{x}, \quad 1 - (\sqrt{x} + p_n)/2 \geq 1 - \sqrt{x} \geq 0 \quad (0 \leq x \leq 1)$$

and finally

$$\sqrt{x} - p_n \geq 0 \quad \Longrightarrow \quad \sqrt{x} - p_{n+1} \geq 0.$$

As a consequence, we infer

$$p_{n+1} \geq p_n \qquad \text{(still for } 0 \leq x \leq 1).$$

For each x in the given interval I, we have

$$0 = p_o(x) \leq \ldots \leq p_n(x) \leq \ldots \leq \sqrt{x}.$$

By a well known property of the real number system,

$$\ell_x = \lim_n p_n(x) \text{ exists, and } \ell_x = \sup_n p_n(x) \leq \sqrt{x}.$$

Going to the limit (which exists as we have just seen) in the relation

$$p_{n+1}(x) = p_n(x) + (x - p_n(x)^2)/2,$$

we obtain

$$\ell_x = \ell_x + (x - \ell_x^2)/2$$

whence

$$x - \ell_x^2 = 0, \quad \ell_x^2 = x, \quad \ell_x = \sqrt{x} \text{ (since } \ell_x \geq p_o(x) = 0).$$

This proves that $p_n(x) \longrightarrow \sqrt{x}$ for $x \in I$ (pointwise convergence). Since the convergence is *monotonous* with all p_n *and* the square root function *continuous*, the Dini theorem implies that the convergence is *uniform* on $I = [0,1]$. ∎

Remark 1. Perhaps it is not superfluous to draw the attention of the reader to the fact that the square root function $\sqrt{x}$ is *not a polynomial...* Polynomials are by definition linear combinations of positive integral powers of x (in mathematics, linear combinations are *always assumed to be finite* linear combinations!). In particular, polynomial functions are indefinitely differentiable. But the square root function is *not differentiable at* $x = 0$. The tangent to the graph of $y = \sqrt{x}$ is *vertical* at $x = 0$, and the reader will check (as an exercise) that $p_n'(0) = n/2$ diverges for $n \to \infty$.

Remark 2. Although the preceding proof is neat, one should not get the impression that the sequence (p_n) just constructed is of much value in numerical analysis to estimate square roots ! In fact, there is a polynomial of the third degree, say p_3, which approximates the square root *uniformly up to* 10^{-3} on the interval $[1/4,1]$ (roughly speaking, giving the correct value of $\sqrt{a}$ up to the third decimal when $1/4 \leq a \leq 1$). From this approximate value x_1 of the positive root of the equation $x^2 - a = 0$, each application of Newton's method will roughly double the number of correct decimals (two applications will ensure the 8 or 10 decimals of pocket calculators). Finally, observe that a knowledge of the square root (in the decimal system) of numbers in the interval $[10^{-2},1]$ is sufficient for all practical purposes. Similarly, in the binary system, it is enough to determine the square root of numbers in the interval $[2^{-2},1] = [1/4,1]$.

15.4 ALGEBRAS OF CONTINUOUS FUNCTIONS

From now on, our objective will be the proof of the approximation theorem of Weierstrass. It states that if I is a compact (i.e. closed and bounded) interval, *any continuous function on* I *is a uniform limit of polynomials.* This statement is remarkable since it shows that all continuous non differentiable functions are uniform limits of $\mathcal{C}^\infty$ functions. A similar statement holds for functions of several variables : if Π is a compact subset in some $\mathbb{R}^n$ and f is any continuous numerical function on Π, then there exists a sequence of polynomials converging uniformly on Π to f. It happens that a further generalization is useful and no more difficult to prove... In fact, the general proof given will be solely based on the explicit construction of a sequence of polynomials approximating the square root on [0,1].

For the generalization we have in view, the notion of polynomial has to be replaced by an abstract notion which we now introduce.

Definition. A **subalgebra** A of $\mathcal{C}_b(I)$ is a vector subspace which is stable under products. ■

By definition $A \subset \mathcal{C}_b(I)$ is a subalgebra if the following condition holds

$$f \text{ and } g \in A,\ \lambda \in \mathbb{R} \text{ (or } \mathbb{C}) \Longrightarrow \lambda f,\ f + g \text{ and } fg \in A.$$

Let us give a few **examples** of subalgebras of $\mathcal{C}([0,1])$.

a) Take for A the subspace consisting of $\mathcal{C}^1$ functions (recall that by our general conventions, f'(0) is the *right* derivative at the origin and f'(1) is the left derivative at the point 1).

b) Take for A the subspace consisting of indefinitely differentiable functions (same conventions as in a) for the extremities).

c) Take for A the subspace consisting of polynomial functions (this subalgebra is contained in the previous two examples).

d) Take for A the subspace consisting of functions vanishing at the point a = 1/2.

e) Take for A the subspace of functions f with f(0) = f(1) (also called *periodic* functions).

Definition. For any subalgebra $A \subset \mathcal{C}_b(I)$ we define the **closure** $\bar{A}$ of A by

$$A \subset \bar{A} = \{f \in \mathcal{C}_b(I) : \exists\ (f_n) \subset A \text{ with } \|f_n - f\| \longrightarrow 0\} \subset \mathcal{C}_b(I).$$

If $A = \bar{A}$ we say that A is **closed**, and if $\bar{A} = \mathcal{C}_b(I)$, we say that A is **dense**. □

In other words, $\bar{A}$ consists of the uniform limits of convergent sequences of A. Obviously

$$f_n \longrightarrow f \quad \text{and} \quad g_n \longrightarrow g \quad \text{(uniformly)}$$

imply

$$\lambda f_n \longrightarrow \lambda f\ (\lambda \in \mathbb{R}),\quad f_n + g_n \longrightarrow f + g \quad \text{and} \quad f_n g_n \longrightarrow fg \quad \text{(uniformly)}$$

and this proves that the closure of any subalgebra of $\mathcal{C}_b(I)$ is also a subalgebra.

Let us prove a sequence of easy lemmas.

Lemma 0. For any subalgebra A, $\bar{A}$ is closed. □

Proof. We have to check that the closure $\bar{\bar{A}} = (\bar{A})^-$ of $\bar{A}$ is equal to $\bar{A}$. It is enough to check the inclusion $\bar{\bar{A}} \subset \bar{A}$. Let $f \in \bar{\bar{A}}$ and choose a sequence $(f_n) \subset \bar{A}$ converging uniformly to f (on I). Define

$$\varepsilon_n = \|f - f_n\| \longrightarrow 0.$$

By definition of $f_n \in \bar{A}$, for each n there must exist a sequence of A converging uniformly to f_n (on I). Hence, for each n, it is possible to choose a $g_n \in A$ such that $\|f_n - g_n\| \le \varepsilon_n$. Then we shall have

$$\|f - g_n\| \le \|f - f_n\| + \|f_n - g_n\| \le 2\varepsilon_n \longrightarrow 0.$$

This proves that the sequence $(g_n) \subset A$ converges uniformly to f and $f \in \bar{A}$. ■

The next properties will be valid for *real valued* functions.

Lemma 1. If A is a subalgebra of $\mathcal{C}_b(I)$ and $f \in A$ is a real valued function, then $|f| \in \bar{A}$. □

Proof. Only the case $f \neq 0$ deserves attention ! Thus we may assume $c = \|f\| \neq 0$ and the function $f^2/c^2 \in A$ takes its values in the interval [0,1] (because f takes *real* values). From 15.3, we know that there exists a sequence (p_n) of polynomials with no constant term ($p_n(0) = 0$) converging uniformly to the (positive) square root function on [0,1]. Hence we shall have

$$p_n(f^2/c^2) \longrightarrow |f|/c \quad \text{and} \quad cp_n(f^2/c^2) \longrightarrow |f|$$

uniformly on I. Since the polynomials p_n have no constant term, the functions $cp_n(f^2/c^2)$ belong to A (even if A does not contain the constant functions !) whence the assertion. ■

Lemma 2. With the notations of lemma 1, for real valued functions

f and $g \in A$ (or $\bar{A}$) $\Longrightarrow$ $\mathrm{Sup}(f,g)$ and $\mathrm{Inf}(f,g) \in \bar{A}$ $(= \bar{\bar{A}})$. □

Proof. The function $\mathrm{Sup}(f,g)$ is defined by $x \longmapsto \mathrm{Sup}(f(x),g(x))$ so that it is bounded, continuous and

$$\mathrm{Sup}(f,g) = (f + g)/2 + |f - g|/2$$

and similarly

$$\mathrm{Inf}(f,g) = (f + g)/2 - |f - g|/2.$$

The announced result follows immediately from lemma 1. ■

A diagram may clarify the concept of supremum between two real functions.

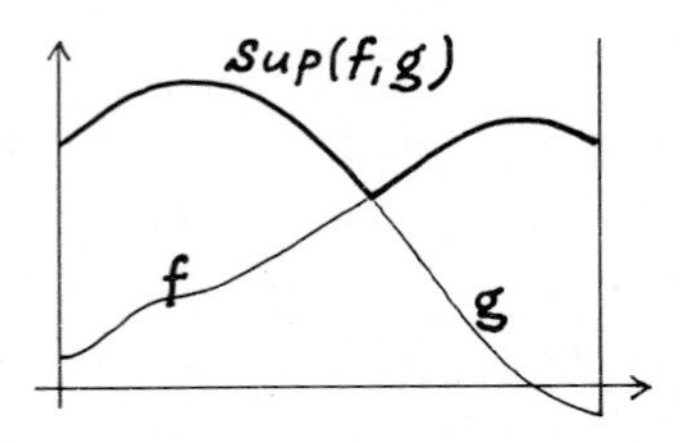

Fig.15.6

Also observe that conversely, the absolute value can be deduced from the Sup operator by $|f| = \mathrm{Sup}(f,-f)$.

Definition. We say that a subalgebra $A \subset \mathcal{C}_b(I)$ **separates points** when the following condition holds :

for each couple of distinct points $a \neq b \in I$,

there exists a function $f \in A$ with $f(a) \neq f(b)$. ■

Observe that in this definition, "the" function f can depend on the couple (a,b). We are *not* requiring that the subalgebra A contains an *injective* function. For example, if I is the square $[0,1] \times [0,1]$ in $\mathbb{R}^2$, there is no real valued continuous function f which is injective. But if the points $a = (a_1,a_2)$ and $b = (b_1,b_2)$ are distinct, either $a_1 \neq a_2$ or $b_1 \neq b_2$. In the first case, we can take the x function to separate a and b, and in the second case, we can take the y function.

Lemma 3. If a subalgebra $A \subset \mathcal{C}_b(I)$ contains the constant 1 (hence all constant functions) and separates points, then

$\forall\ a \neq b \in I$, $\forall\ \alpha$ and $\beta \in \mathbb{R}$, $\exists\ f \in A$ with $f(a) = \alpha$ and $f(b) = \beta$.■

Proof. This *interpolation lemma* follows almost immediately from the definition of the property of separating points. Choose indeed a function $g \in A$ with $g(a) \neq g(b)$. Put $g(a) = a'$, $g(b) = b'$ and consider the function

$$f = \alpha + (\beta - \alpha)\frac{g - a'}{b' - a'} = \left(\alpha - a'\frac{\beta - \alpha}{b' - a'}\right)\cdot 1 + \frac{\beta - \alpha}{b' - a'}\cdot g = \lambda\cdot 1 + \mu g.$$

One checks immediately that $f \in A$ (since $1 \in A$) and f interpolates the required values. ■

15.5 THE STONE-WEIERSTRASS THEOREM

Theorem of Stone-Weierstrass (real case). Let I be a compact space and $\mathcal{C}(I) = \mathcal{C}_b(I)$ the algebra of real valued continuous functions on I. Then all subalgebras $A \subset \mathcal{C}(I)$ containing the constant 1 and separating points are dense in $\mathcal{C}(I)$. □

Proof. Let $A \subset \mathcal{C}(I)$ be a subalgebra containing the constants and separating points. We have to show that any $f \in \mathcal{C}(I)$ belongs to the closure $\bar{A}$ of A. Let us fix a continuous $f : I \longrightarrow \mathbb{R}$. It will be enough to show that for any given $\varepsilon > 0$, there is a $f_\varepsilon \in \bar{A}$ with $\|f - f_\varepsilon\| \leq \varepsilon$: in this case, taking successively $\varepsilon = 1/n$, we will be able to produce a sequence $f_{1/n} \in \bar{A}$ converging uniformly to f, proving that $f \in \bar{\bar{A}} = \bar{A}$ as required. In a *first step*, for each $a \in I$, we show that there exists a function $g = g_a \in \bar{A}$ satisfying

$$g(a) = f(a) \quad \text{and} \quad g \leq f + \varepsilon.$$

For this purpose, invoking the interpolation lemma 3, we see that for each $x \in I$ there exists a $h_x \in A$ with

$$h_x(a) = f(a) \quad \text{and} \quad h_x(x) = f(x)$$

(if $x = a$, we can simply take the constant $h_x = f(a)\cdot 1$). Consequently, the function $h_x - f$ vanishes at the points a and x. By continuity

$$(h_x - f)(y) < \varepsilon \quad \text{or equivalently} \quad h_x(y) < f(y) + \varepsilon$$

will still hold for all y in an open neighborhood V_x of x. In this way, we get an open covering (V_x) of the compact space I. It is possible to extract a *finite* covering from this one, say

$$V_i = V_{x_i} \ (1 \leq i \leq m) \quad \text{(corresponding to functions } h_i = h_{x_i}).$$

The function $g = \mathrm{Inf}(h_1,\dots,h_m)$ *belongs to* $\bar{A}$ (as an easy induction based on lemma 2 shows) and satisfies

▷ $g(a) = f(a)$ (since all $h_x(a) = f(a)$),

▷ $g(y) \leq h_i(y) < f(y) + \varepsilon$ for $y \in V_i$.

Since the neighborhoods V_i cover I, the second property implies

▷ $g \leq f + \varepsilon$ (in I).

In a *second step*, let us show how the preceding $g = g_a$ can be patched together (one further *bed of Procrustes procedure!).* The difference $g_a - f$ vanishes at a, hence

$$(g_a - f)(y) > -\varepsilon \quad \text{or equivalently} \quad g(y) > f(y) - \varepsilon$$

will still hold in an open neighborhood W_a of a. From the open covering (W_a) of I we again extract a finite sub-covering, say

$$W_j = W_{a_j} \ (1 \le j \le m) \quad \text{(corresponding to functions } g_j = g_{a_j}\text{)}.$$

We finally define

$$f_\varepsilon = \mathrm{Sup}(g_1,\ldots,g_m) \in \bar{A} \ .$$

Since all $g_j \le f + \varepsilon$, we certainly have $f_\varepsilon \le f + \varepsilon$. Moreover,

$$f_\varepsilon(y) \ge g_j(y) > f(y) - \varepsilon \quad \text{for } y \in W_j$$

implies that $f_\varepsilon \ge f - \varepsilon$ on I (since the W_j cover I). Altogether, we have

$$f_\varepsilon \in \bar{A}, \quad f - \varepsilon \le f_\varepsilon \le f + \varepsilon \quad \text{namely} \quad \|f - f_\varepsilon\| \le \varepsilon$$

as expected. ■

There is a version of the preceding approximation theorem which is valid for complex valued functions and which we now explain. To be able to deduce this version from the real case, we have to consider subalgebras $A \subset \mathcal{C}(I,\mathbb{C})$ which are **stable under conjugation**

$$f \in A \implies \bar{f} \text{ (complex conjugate of f)} \in A.$$

Theorem of Stone-Weierstrass (complex case). Let I be a compact space and $\mathcal{C}(I,\mathbb{C}) = \mathcal{C}_b(I,\mathbb{C})$ the algebra of complex valued continuous functions on I. Then all $\mathbb{C}$-subalgebras $A \subset \mathcal{C}(I,\mathbb{C})$

- ▷ containing the constant 1 (hence all constants),
- ▷ separating points of I,
- ▷ stable under conjugation,

are dense in $\mathcal{C}(I)$. □

Proof. Introduce the real subalgebra $A_r \subset \mathcal{C}(I)$ defined by

$$A_r = \{ f \in A : f \text{ takes only real values } \}.$$

Since A contains all constants, A_r contains all real constants. If $f \in A$, we know that $\bar{f} \in A$ hence

$$g = \mathrm{Re}(f) = (f + \bar{f})/2 \quad \text{and} \quad h = \mathrm{Im}(f) = (f - \bar{f})/2i$$

also belong to A :

$$f = g + ih \in A \implies g \text{ and } h \in A_r.$$

Moreover, if f takes different values at two points, either g or h take distinct values at these points. This proves that

$$A \text{ separates points} \implies A_r \text{ separates points.}$$

The assumptions of the real Stone-Weierstrass theorem are satisfied by the subalgebra A_r of $\mathcal{C}(I)$: A_r is dense in $\mathcal{C}(I)$. Now, if $f \in \mathcal{C}(I,\mathbb{C})$ is arbitrary, decompose $f = g + ih$ into real and imaginary parts. Both g and h are uniform limits of elements of A_r and consequently f itself is a uniform limit of elements of $A_r + iA_r \subset A$ (in fact, equality holds here). ■

It is important to know that the complex version of the Stone-Weierstrass theorem would *fail without the extra assumption* concerning stability of the subalgebra A with respect to conjugation (the reader is expected to be able to distinguish when a bar in superscript position refers to a complex conjugation or to a closure!).

For **example**, let I denote the unit circle $|z| = 1$ in $\mathbb{C}$ and A the complex subalgebra of $\mathcal{C}(I,\mathbb{C})$ consisting of the complex polynomials in $z = x+iy$. A typical element of A is thus

$$p(z) = a_0 + a_1 z + \dots + a_n z^n.$$

This subalgebra is *not dense* in $\mathcal{C}(I,\mathbb{C})$ since uniformly convergent sequences (p_n) of polynomials will never converge to the function $f(z) = \bar{z}$:

▷ $\oint_{|z|=1} p(z)dz = 0$ for any polynomial in z (cf.13.3);

▷ If (p_n) is a sequence of polynomials in z, 15.1 shows that
$p_n \longrightarrow f$ uniformly on I $\Longrightarrow$ $\oint f(z)dz = \lim \oint p_n(z)dz = 0$;

▷ But $\oint_{|z|=1} \bar{z}\, dz = \int_0^{2\pi} e^{-it} ie^{it} dt = 2\pi i \neq 0$.

15.6 APPLICATIONS

The Stone-Weierstrass theorem has several important corollaries that we list now. First of all, let us formulate its original form, due to Weierstrass.

The Weierstrass approximation theorem. Let I be a compact (i.e. closed and bounded) subset of $\mathbb{R}^n$ and $f : I \longrightarrow \mathbb{R}$ (or $\mathbb{C}$) a continuous numerical function on I. Then, there exists a sequence of polynomials which converges uniformly to f on I. □

Proof. Let $A \subset \mathcal{C}(I)$ denote the subalgebra consisting of polynomial functions (in n variables in $\mathbb{R}^n$). This subalgebra contains the constants, separates points (choose coordinate functions x_i for this purpose) and is stable under complex conjugation in the complex case. All assumptions of the Stone-Weierstrass theorem are satisfied.■

In the simplest case $n = 1$, the compact set I would typically be an interval of the form $[a,b]$. It is already quite remarkable that any continuous function on such an interval is a uniform limit of polynomials. Such an approximation has been explicitely constructed in the case of the square root on $[0,1]$ (cf.15.3). It is also quite surprising that the general result has been based on this unique example !

In the case $n = 2$, the compact set could be a product $I \times J$ of two compact intervals (a compact *rectangle*), and any continuous function on such a set is again a uniform limit of poynomials in x and y. Still with $n = 2$, we could take for I the compact unit disc $x^2 + y^2 \leq 1$, or the *unit circle* $x^2 + y^2 = 1$. This last example is so important that it deserves a special mention. Let us recall that a **T-periodic** function f on the real line $\mathbb{R}$ is characterized by the property $f(t + T) = f(t)$ $(t \in \mathbb{R})$. Examples of 2π-periodic functions are given by the exponentials $t \longmapsto e^{ikt}$ $(k \in \mathbb{Z})$ and their finite linear combinations

$$t \longmapsto \sum_{k \in F} c_k e^{ikt} \quad (c_k \in \mathbb{C},\ F \text{ finite} \subset \mathbb{Z})$$

also called **trigonometric polynomials** for obvious reasons.

Corollary. Let $f : \mathbb{R} \longrightarrow \mathbb{C}$ be continuous and 2π-periodic. Then f is a uniform limit of a sequence of *trigonometric polynomials*. □

Proof. If we let $z = x + iy = e^{it}$ $(t \in \mathbb{R})$, then $z \in \mathbb{C}$ will lie on the unit circle $|z| = 1$ or $x^2 + y^2 = 1$. Any 2π-periodic function f of $t \in \mathbb{R}$ can be identified to a function F of z on this unit circle according to the identification

$$f(t) = F(e^{it}) \ (= F(z) = F(x,y) \ ).$$

Let us consider the algebra A of polynomial functions in z *and* $\bar{z}$ on the unit circle. (Since $x = (z + \bar{z})/2$, $y = (z - \bar{z})/2i$, the polynomials in z and $\bar{z}$ are precisely the polynomials in x and y.) These polynomials are the finite linear combinations of *monomials* $z^i \bar{z}^j = z^{i-j} = z^k$ (on the unit circle $z \cdot \bar{z} = 1$), where i and $j \in \mathbb{N}$, i.e. $k \in \mathbb{Z}$. This algebra contains the constants, separates the points (the complex function $z \longmapsto z$ is *injective* on the circle!) and is stable under conjugation. ■

The above applications are *existence results*. They do not really say how to find polynomials uniformly near a given continuous function f on a compact interval. No more do they say how to choose the coefficients of trigonometric polynomials to approximate uniformly periodic functions. We shall come back to these points in 22.3.

Theorem of Fubini. Let I and J be compact sets in $\mathbb{R}^n$ resp. $\mathbb{R}^m$. For any continuous $f : I \times J \longrightarrow \mathbb{R}$ (or $\mathbb{C}$) we have

$$\int_I \left(\int_J f(x,y)dy \right) dx = \int_J \left(\int_I f(x,y)dx \right) dy$$

$$\left(= \iint_{I \times J} f(x,y)dxdy \right) . \qquad \square$$

Proof. The set $K = I \times J$ is compact (contained in $\mathbb{R}^{n+m}$) and we can consider the subalgebra A of $\mathcal{C}(K)$ consisting of the finite sums of *decomposable functions* $(x,y) \longmapsto \varphi(x)\psi(y)$ on K. Thus typical elements of A are finite sums $\sum \varphi_i(x)\psi_i(y)$. The statement of the theorem obviously holds for decomposable functions

$$\int_I \left(\int_J \varphi(x)\psi(y)dy \right) dx = \left(\int_I \varphi(x)dx \right) \left(\int_J \psi(y)dy \right)$$

and hence for functions in A by linearity. Since A contains the constants, separates points and is stable under complex conjugation, it is dense in $\mathcal{C}(K)$. Let $f \in \mathcal{C}(K)$ be an arbitrary continuous function and select a sequence $(f_n) \subset A$ which converges uniformly to f (on K). Then

$$\int_I \left(\int_J f_n(x,y)dy \right) dx = \int_J \left(\int_I f_n(x,y)dx \right) dy$$

since $f_n \in A$. Letting $n \longrightarrow \infty$ we shall obtain a similar equality for f in place of the f_n's (by the basic theorem of 15.2). ■

CHAPTER 16
UNIFORM CONVERGENCE: PRACTICE

16.1 CRITERIA FOR NORMAL CONVERGENCE

All criteria of uniform convergence are based on the following *completeness* result.

For any set $I \subset \mathbb{R}^n$, the vector space $\mathcal{C}_b(I)$ of bounded continuous functions $f : I \longrightarrow \mathbb{R}$ (or $\mathbb{C}$) is a *metric space* with respect to the distance function (introduced in 15.2)

$$d(f,g) = \|f - g\| = \sup_{x \in I} |f(x) - g(x)|.$$

Theorem. The metric space $\mathcal{C}_b(I)$ is **complete**, i.e. for a sequence (f_n) to converge in $\mathcal{C}_b(I)$ it is necessary and sufficient that it is a *Cauchy sequence*, namely, that $d(f_n, f_m) \longrightarrow 0$ when $n, m \to \infty$. □

Proof. Let $(f_n) \subset \mathcal{C}_b(I)$ be a Cauchy sequence. For fixed $x \in I$,

$$|f_n(x) - f_m(x)| \le \|f_n - f_m\|$$

proves that the numerical sequence $(f_n(x)) \subset \mathbb{R}$ (or $\mathbb{C}$) is a Cauchy sequence, hence converges (both $\mathbb{R}$ and $\mathbb{C}$ are *complete* fields). Let $f(x)$ denote its limit. In this way, we define a numerical function $f : I \longrightarrow \mathbb{R}$. We intend to prove that f_n converges uniformly to f, hence f is bounded and continuous (basic theorem of 15.2 again) and $f_n \longrightarrow f$ in the metric space $\mathcal{C}_b(I)$. For fixed $\varepsilon > 0$, there is an $N = N_\varepsilon$ with $\|f_n - f_m\| \le \varepsilon$ when n and $m \ge N$. For these large n and m, we have *a fortiori*

$$|f_n(x) - f_m(x)| \le \|f_n - f_m\| \le \varepsilon \qquad \text{(for all } x \in I).$$

Letting $m \to \infty$ we get

$$|f_n(x) - f(x)| \le \varepsilon \qquad \text{(for all } x \in I),$$

hence

$$\|f_n - f\| = \sup_{x \in I} |f_n(x) - f(x)| \le \varepsilon \qquad \text{for } n \ge N. \qquad ■$$

A *normed* vector space which is *complete* is called **Banach space**. Hence $\mathcal{C}_b(I)$ is always a Banach space.

Convergence of a *series* $\sum u_k$ is defined by the convergence of the corresponding *sequence of partial sums* $f_n = \sum_{k \le n} u_k$. (Conversely, convergence of a sequence (f_n) can be reduced to the convergence of the series $f_o + (f_1 - f_o) + (f_2 - f_1) + \ldots$ having as a general term $u_k = f_k - f_{k-1}$ for $k \ge 1$.) The convergence *criterium* which follows from the completeness of the space $\mathcal{C}_b(I)$ now reads

$$\sum_k u_k \textit{ converges uniformly (in } \mathcal{C}_b(I))$$
$$\textit{if and only if}$$
$$\|\sum_{n<k\le m} u_k\| \longrightarrow 0 \quad \textit{for } m > n \longrightarrow \infty.$$

Since

$$\|\sum_{n<k\le m} u_k\| \le \sum_{n<k\le m} \|u_k\|,$$

the series will certainly converge (uniformly) when $\sum_k \|u_k\| < \infty$. In practice, when $\|u_k\|$ is difficult to estimate, it will be enough to find a convergent series $\sum_k c_k < \infty$ with

$$|u_k(x)| \le c_k \quad \text{(for all } x \in I).$$

Such a series has positive terms and is called **numerical majorant** of the series $\sum_k u_k$. This method of proof of uniform convergence of series is so important that we give it a special name.

Definition. A series of functions $\sum_k u_k$ is said to converge **normally** (or **in norm**) when $\sum_k \|u_k\|$ converges (i.e. $\sum_k \|u_k\| < \infty$). ■

We have seen that

normal convergence *implies* *uniform convergence.*

The following "tests" give *sufficient conditions* for *normal convergence.*

Practical test of d'Alembert. Assume that a sequence $(u_n) \subset \mathcal{C}_b(I)$ satisfies $\|u_{k+1}/u_k\| \longrightarrow r < 1$. Then the series $\sum_k u_k$ converges normally (hence uniformly on I). □

Proof. Select any number $R = r + \varepsilon$ between r and $1 : r < R < 1$, so that $\varepsilon > 0$. Thus there exists a rank $N = N_\varepsilon$ for which

$$\|u_{k+1}/u_k\| \le r + \varepsilon = R \quad \text{for } k \ge N.$$

This implies

$$|u_{k+1}(x)/u_k(x)| \le R \quad \text{for } k \ge N \textit{ and all } x \in I.$$

Hence

$$|u_{N+1}(x)| \le R|u_N(x)|,$$
$$|u_{N+2}(x)| \le R|u_{N+1}(x)| \le R^2|u_N(x)|, \ \dots$$

and by induction

$$|u_{N+i}(x)| \le R^i|u_N(x)| \quad (i \in \mathbb{N}).$$

The series $\sum_{k\ge N} u_k = \sum_{i\ge 0} u_{N+i}$ admits the numerical majorant

$$\sum \operatorname{Sup}_{x\in I}|u_N(x)| \, R^i = \|u_N\| \sum R^i$$

which is a convergent geometrical series since $0 < R < 1$. ■

As an **application** of this test, let us consider the series

$$1 + \frac{\alpha\cdot\beta}{\gamma}\frac{x}{1!} + \frac{\alpha(\alpha+1)\cdot\beta(\beta+1)}{\gamma(\gamma+1)}\frac{x^2}{2!} + \dots$$

where the parameters α, β and γ are any complex numbers *but* γ *is not a negative integer* : $-\gamma \notin \mathbb{N}$. The general term of this series is

$$u_{k+1}(x) = \frac{x^{k+1}}{(k+1)!} \frac{\alpha(\alpha+1)\dots(\alpha+k)\cdot\beta(\beta+1)\dots(\beta+k)}{\gamma(\gamma+1)\dots(\gamma+k)}$$

so that

$$u_{k+1}(x)/u_k(x) = \frac{x}{k+1} \frac{(\alpha+k)(\beta+k)}{\gamma+k} = x \frac{(1+\alpha/k)(1+\beta/k)}{(1+1/k)(1+\gamma/k)} .$$

On an interval $I_r = [-r,r]$, we have

$$\|u_{k+1}/u_k\| = r\left|\frac{(1+\alpha/k)(1+\beta/k)}{(1+1/k)(1+\gamma/k)}\right| \longrightarrow r \quad \text{when } k \longrightarrow \infty$$

and normal convergence is assured if $0 \le r < 1$ (normal convergence is also assured in discs $|z| < 1$ in $\mathbb{C}$ for the same reason). The sum of this series is the Gauss' **hypergeometric function** also denoted by

$$F\left({\alpha\ \beta \atop \gamma}\middle| x\right) = {}_2F_1\left({\alpha\ \beta \atop \gamma}\middle| x\right) \quad (-\gamma \notin \mathbb{N} \quad \text{and} \quad |x| < 1).$$

The basic theorem of 15.2 shows that this function is continuous on all intervals I_r $(0 < r < 1)$ hence on $]-1,1[$ (or on the open disc $|z| < 1$ in $\mathbb{C}$) since any point of this *open interval* is an interior point of some I_r of the preceding type.

Practical test of Cauchy-Hadamard. Assume that a sequence (u_n) of $\mathcal{C}_b(I)$ satisfies $\limsup_{k \to \infty} \|u_k\|^{1/k} = r < 1$. Then the series $\sum_k u_k$ converges normally (hence uniformly on I). □

Before we give the proof of this test, let us recall the definition of the lim sup (sometimes denoted by $\overline{\lim}$) of a sequence of positive numbers $a_n \ge 0$. Quite generally, we have

$$\dots \ge \sup_{i\ge k} a_i \ge \sup_{i\ge k+1} a_i \ge \dots \ge 0$$

where we put $\operatorname{Sup} a_i = \infty$ if the sequence is unbounded. Then, either all tails $(a_i)_{i\ge k}$ are unbounded and all the Sup are ∞, or the sequence of Sup is decreasing and has a limit. In all cases, the **definition** is

$$\limsup_{k \to \infty} a_k = \lim_{k\to\infty} \left(\sup_{i\ge k} a_i \right) .$$

The lim sup of any sequence of positive numbers *exists* and

$$0 \le \limsup_{k \to \infty} a_k \le \infty.$$

Here is a *picture* of the process involved. Imagine that the a_i represent the lengths of a sequence of sticks. When we move along the line, looking ahead, the horizon will be blocked at the height $\sup_{i\ge k} a_i$. As we pass the highest sticks, the horizon line can only go down and the limit is the lim sup of the sequence.

Proof. Choose again a number $R = r + \varepsilon$ between r and $1 : r < R < 1$ and a rank $N = N_\varepsilon$ with

$$\sup_{i \geq N} \|u_i\|^{1/i} \leq R.$$

Hence

$$\|u_i\|^{1/i} \leq R\ , \quad \|u_i\| \leq R^i \quad (\text{for } i \geq N)$$

and the proposed series $\sum_{i \geq N} u_i$ admits the numerical majorant $\sum R^i$ which is a convergent geometric series since $0 \leq R < 1$. ■

An advantage of the Cauchy-Hadamard test is that the lim sup of the $\|u_k\|^{1/k} \geq 0$ always exists (may be ∞) whereas the lim of the $\|u_{k+1}/u_k\|$ does not in general. For example, for this norm to exist u_{k+1} has to vanish at all points where u_k vanishes (u_{k+1}/u_k has to be bounded). Such a condition is obviously unnatural since zeros of the u_k's certainly do not prevent convergence... (An example of this unnatural feature of the test of d'Alembert is given in the exercise 5.)

Here is a basic **application** of the Cauchy-Hadamard test. Consider a power series, namely a series $\sum u_k$ with $u_k(x) = a_k x^k$. Normal convergence will occur on an interval $I_r = [-r,r]$ when

$$\limsup \|a_k x^k\|^{1/k} = \limsup \|x\|_{I_r} |a_k|^{1/k} = r \limsup |a_k|^{1/k} < 1.$$

If we put

$$0 \leq \rho = 1/\limsup |a_k|^{1/k} \leq \infty\ ,$$

normal (hence uniform) convergence will occur on all closed intervals $I_r = [-r,r]$ for $r < \rho$ and the sum will define a continuous function on the union $]-\rho,\rho[$ of the preceding intervals (any point in this open interval is an *interior point* of one I_r). The number ρ is the **radius of convergence** of the power series $\sum a_k x^k$ and

$$\rho = 1/\limsup |a_k|^{1/k}$$

is **Hadamard's formula** for this convergence radius. Observe that more generally, the series $\sum a_k(z-a)^k$ will converge in the disc $|z-a| < \rho$ in $\mathbb{C}$ whence the terminology.

Let us finally observe that when the limit in the d'Alembert test, resp. the lim sup in the Cauchy-Hadamard test, is 1, then both convergence or divergence can occur. The sequences with $u_k(x) = 1/k^2$ or $= 1/k$ (constant functions) are examples of these cases. It is also obvious that if the lim sup $\|u_k\|^{1/k} = r > 1$, then $\|u_k\|$ tends to infinity and $\sum u_k$ cannot converge uniformly.

16.2 ABEL'S TEST FOR UNIFORM CONVERGENCE

Certain series $\sum u_k$ converge *uniformly* (on some set $I \subset \mathbb{R}^n$) *without converging normally*. Let us give two examples of this situation.

Example 1. Let $u_k = (-1)^k/k$ (constant function on a set I). Then

$$\|\sum_{n\le k\le m} u_k\| = |\sum_{n\le k\le m} u_k| = |1/n - 1/(n+1) + \ldots| \le 1/n \longrightarrow 0$$

and the series converges uniformly. But $\|u_k\| = 1/k$ so that

$$\sum \|u_k\| = \sum 1/k \text{ (harmonic series)}$$

diverges.

Example 2. Let us construct a sequence $(u_k) \subset \mathcal{C}([0,1])$ with

$$\|u_k\| = 1/k \quad \text{(hence } \sum \|u_k\| \text{ diverges as before)}$$

but with $u_k \ge 0$ *and* $\sum u_k$ converges *uniformly*. For this purpose, we define (cf.Fig.16.1) —for $k \ge 1$—

$$u_k(x) = \begin{cases} 0 \text{ outside } [1/(k+1),1/k] \subset [0,1] \\ \text{positive peak of height } 1/k \text{ in } [1/(k+1),1/k] \end{cases}.$$

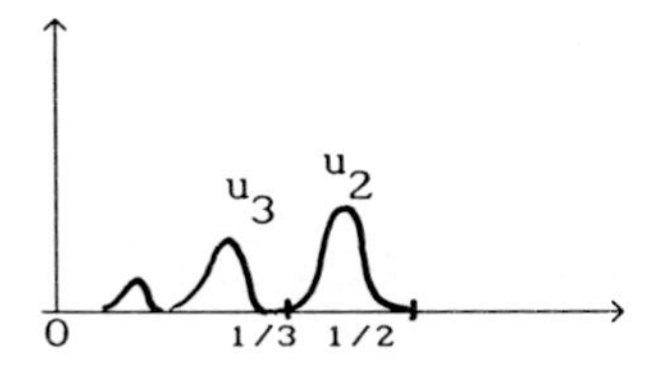

Fig.16.1

Obviously, the maximum of $\sum_{n\le k\le m} u_k \ge 0$ is the maximum of u_n and

$$\|\sum_{n\le k\le m} u_k\| = \|u_n\| = 1/n \longrightarrow 0 \quad \text{for } m \ge n \longrightarrow \infty.$$

This proves that the series $\sum u_k$ converges uniformly. Its sum will be continuous if all u_k are continuous. The figure shows that

$$\sum u_k = \operatorname*{Sup}_k u_k$$

(at a given $x \in [0,1]$, at most one u_k is not 0).

Let us first give a test for numerical series.

Abel's test for numerical series. Let $(a_n) \subset \mathbb{R}$ (or $\mathbb{C}$) be a sequence which can be written in the form $a_n = \varepsilon_n b_n$ with

▷ $\varepsilon_n \searrow 0 \quad (n \to \infty)$,

▷ the partial sums $|\sum_{n\le N} b_n|$ are bounded say $\le \sigma$.

Then the series $\sum a_n$ is convergent and its tails can be estimated by

$$|\sum_{n\ge N} a_n| \le 2\sigma\varepsilon_N \; . \qquad \square$$

Proof. Let us introduce the family of partial sums

$$s_p^q = b_{p+1} + \dots + b_q \qquad (q > p).$$

This is a bounded family since

$$s_p^q = \sum_{n\le q} b_n - \sum_{n\le p} b_n \quad \text{implies} \quad |s_p^q| \le 2\sigma$$

by assumption. We also have

$$b_{p+1} = s_p^{p+1},\ b_{p+2} = s_p^{p+2} - s_p^{p+1},\ \dots,\ b_q = s_p^q - s_p^{q-1}.$$

Consequently

$$\varepsilon_{p+1} b_{p+1} + \dots + \varepsilon_q b_q =$$

$$= \varepsilon_{p+1} s_p^{p+1} + \varepsilon_{p+2}(s_p^{p+2} - s_p^{p+1}) + \dots + \varepsilon_q(s_p^q - s_p^{q-1}) =$$

$$= s_p^{p+1}(\varepsilon_{p+1} - \varepsilon_{p+2}) + \dots + s_p^{q-1}(\varepsilon_{q-1} - \varepsilon_q) + s_p^q \cdot \varepsilon_q$$

(the last step is *Abel's transformation*). By assumption, all differences $\varepsilon_k - \varepsilon_{k+1}$ are positive and we have

$$|\sum_{p<k\le q} a_k| = |\varepsilon_{p+1} b_{p+1} + \dots + \varepsilon_q b_q| \le$$

$$\le |s_p^{p+1}|(\varepsilon_{p+1} - \varepsilon_{p+2}) + \dots + |s_p^{q-1}|(\varepsilon_{q-1} - \varepsilon_q) + |s_p^q|\varepsilon_q \le$$

$$\le 2\sigma(\varepsilon_{p+1} - \varepsilon_{p+2}) + \dots + 2\sigma(\varepsilon_{q-1} - \varepsilon_q) + 2\sigma\cdot\varepsilon_q =$$

$$= 2\sigma\cdot\varepsilon_{p+1} \longrightarrow 0 \quad \text{when} \quad q > p \longrightarrow \infty .$$

This proves that the sequence of partial sums $\sum_{k\le n} a_k$ is a Cauchy sequence and hence converges ($\mathbb{R}$ and $\mathbb{C}$ are *complete*). In the previous inequality

$$|\sum_{p<k\le q} a_k| \le 2\sigma\cdot\varepsilon_{p+1} ,$$

it is now legitimate to let $q \longrightarrow \infty$ and we get $|\sum_{k>p} a_k| \le 2\sigma\cdot\varepsilon_{p+1}$ as expected. ■

For series of functions $a_k(x) = \varepsilon_k(x)u_k(x)$, we can apply the preceding result if the numbers $\varepsilon_k(x)$ tend to zero monotonously (for each $x \in I$) and if the partial sums $|\sum_{n\le N} u_n(x)|$ are bounded (still at each $x \in I$), say

$$|\sum_{n\le N} u_n(x)| \le \sigma(x).$$

In this case, we obtain the estimate

$$|\sum_{n\ge N} \varepsilon_n(x)u_n(x)| \le 2\sigma(x)\varepsilon_N(x)$$

for the tails of the series. These show immediately that uniform convergence will occur if

- ▷ $\sigma(x)$ is bounded, say $\sigma(x) \le \|\sigma\| < \infty$,
- ▷ $\|\varepsilon_N\| \longrightarrow 0 \quad$ (for $N \longrightarrow \infty$).

c the scalar $c = (x|y)/(x|x)$. For this choice, we infer that the vector $y - cx \in W$, hence can be written

$$y - cx = y^1\mathbf{e}_1 + y^2\mathbf{e}_2 + \ldots + y^m\mathbf{e}_m \quad \text{in } W.$$

This proves that

$$y = y^1\mathbf{e}_1 + y^2\mathbf{e}_2 + \ldots + y^m\mathbf{e}_m + cx.$$

Thus $\mathbf{e}_1,\ldots,\mathbf{e}_m$, x *generate* V and constitute an orthogonal basis of this space. ■

Theorem. Let $(\mathbf{e}_i)_{1\le i\le n}$ be an orthogonal basis of a finite dimensional inner product space V and let $v_i = \sum a_i^j\mathbf{e}_j$ $(1 \le i \le n)$ be a system of n vectors in V. The following conditions are equivalent

i) the v_i are linearly independent (constitute a basis of V),

ii) $\det(a_j^i) \ne 0$,

iii) $\mathrm{Gram}(v_1,\ldots,v_n) = \det(v_i|v_j) > 0$. □

Proof. The equivalence i) $\Longleftrightarrow$ ii) is well known and we omit its proof. For the equivalence with iii), let us observe that

$$(v_i|v_j) = (\sum a_i^k\mathbf{e}_k|\sum a_j^\ell\mathbf{e}_\ell) = \sum \bar{a}_i^k a_j^\ell\,(\mathbf{e}_k|\,\mathbf{e}_\ell) = \sum_k \bar{a}_i^k a_j^k$$

is the element located in the row i and column j of the matrix product of

$$A^*A \quad (\,A^* = {}^t\bar{A}\,).$$

This proves that

$$\mathrm{Gram}(v_1,\ldots,v_n) = \det(A^*A) = |\det A|^2$$

and the equivalence ii) $\Longleftrightarrow$ iii) is established (cf.also lemma in 9.1). ■

Practically, the construction of orthonormal systems in inner product spaces (orthonormal bases when the dimension is finite) can be made according to a simple procedure.

Gram-Schmidt orthogonalization. Let $(v_i)_{i\ge 0}$ be a system of linearly independent vectors in an inner product space V. Then, there exists an orthogonal system $(\mathbf{e}_i)_{i\ge 0}$ such that

$\mathbf{e}_o, \mathbf{e}_1, \ldots, \mathbf{e}_m$ generates the same subspace

$V_m \subset V$ as $v_o, v_1, \ldots, v_m$ (for each $m \ge 0$). □

Proof. We use induction on m, noting that for $m = 0$, there is nothing to prove (take $\mathbf{e}_o = v_o$). Assume then that the vectors $\mathbf{e}_i$ have been constructed for $i < m$ and define $\mathbf{e}_m$ by the determinant

$$\mathbf{e}_m = \begin{vmatrix} (v_o|v_o) & \ldots & (v_o|v_m) \\ \vdots & & \vdots \\ (v_{m-1}|v_o) & \ldots & (v_{m-1}|v_m) \\ v_o & \ldots & v_m \end{vmatrix}.$$

This determinant is to be expanded according to its last row, and thus is equal to

$$\mathbf{e}_m = G_m v_m + (\text{linear combination of } v_o, \ldots, v_{m-1})$$

where G_m denotes the Gram determinant formed with the first m vectors $v_o, \ldots, v_{m-1}$. By the preceding theorem, we know that this determinant $G_m > 0$ does not vanish (linear independence assumption). From this expression, we infer that

$$v_o, \ldots, v_{m-1} \text{ and } v_m$$

generate the same subspace as

$$v_o, \ldots, v_{m-1} \text{ and } \mathbf{e}_m .$$

Using the induction assumption, we can even replace $v_o, \ldots, v_{m-1}$ by $\mathbf{e}_o, \ldots, \mathbf{e}_{m-1}$ in the second family. Moreover, distributing scalar products in the last line, we see that the computation of the scalar product $(\mathbf{e}_m|v_i)$ involves a determinant with two equal lines if $i < m$: $\mathbf{e}_m \perp v_i$ for all $i < m$. This proves that the vector $\mathbf{e}_m$ is orthogonal to the subspace V_{m-1} generated by the v_i for $i < m$. Since V_{m-1} contains the $\mathbf{e}_i$ for $i < m$ (it is also generated by these vectors by induction hypothesis), it proves that $\mathbf{e}_m \perp \mathbf{e}_i$ for all $i < m$. ■

17.4 BEST APPROXIMATION THEOREM

Theorem. Let $(\mathbf{e}_i)$ be a *finite* orthonormal system in an inner product space V. Then for each $x \in V$, the linear combination of the $\mathbf{e}_i$ that best approximates x is the one which has for coefficients the scalar products $(\mathbf{e}_i|x)$:

$$\left\| x - \sum a_i \mathbf{e}_i \right\| \text{ is minimum when } a_i = (\mathbf{e}_i|x). \qquad \square$$

Proof. Let us indeed call $x_i = (\mathbf{e}_i|x)$ these scalar products. An obvious computation shows that all $\mathbf{e}_j$ are orthogonal to $x - \sum x_i \mathbf{e}_i$ and we can use the Pythagoras theorem (17.2) to the finite family of orthogonal vectors

$$x - \sum x_i \mathbf{e}_i \ , \ (x_j - a_j)\mathbf{e}_j .$$

We obtain

$$\left\| x - \sum a_i \mathbf{e}_i \right\|^2 = \left\| x - \sum x_i \mathbf{e}_i + \sum (x_i - a_i)\mathbf{e}_i \right\|^2 =$$

$$= \left\| x - \sum x_i \mathbf{e}_i \right\|^2 + \sum \left\| (x_i - a_i)\mathbf{e}_i \right\|^2 =$$

$$= \left\| x - \sum x_i \mathbf{e}_i \right\|^2 + \sum |x_i - a_i|^2 \geq \left\| x - \sum x_i \mathbf{e}_i \right\|^2$$

with equality precisely when $\sum |x_i - a_i|^2 = 0$. Thus the minimal value holds exactly when $a_i = x_i$ for all i . ■

It is important to realize the geometrical content and interpretation of the preceding theorem. Its statement concerns a vector $x \in V$ and linear combinations $\sum a_i\mathbf{e}_i$ formed with a fixed finite orthonormal set. Hence it only concerns the finite dimensional space generated by the $\mathbf{e}_i$ and x. Since we proved that the difference $x - \sum x_i\mathbf{e}_i$ is orthogonal to all $\mathbf{e}_i$, it proves that the *best* linear combination $\sum x_i\mathbf{e}_i$ is the **orthogonal projection** of x on the subspace generated by the $\mathbf{e}_i$'s. The vector x is an orthogonal sum

$$x = \sum x_i\mathbf{e}_i + (x - \sum x_i\mathbf{e}_i)$$

of its projection and an orthogonal *remainder*. The following picture illustrates the situation where two orthonormal vectors only, $\mathbf{e}_1$ and $\mathbf{e}_2$ are considered.

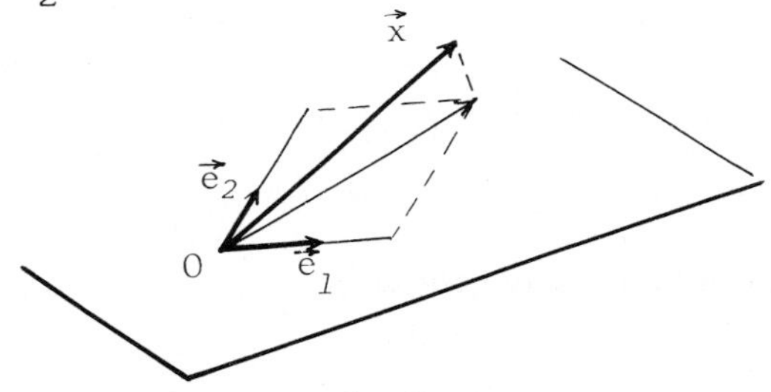

Fig.17.1

17.5 CLASSICAL INEQUALITIES

Theorem (Bessel's inequality). Let $(\mathbf{e}_i)_I$ be any orthonormal system of vectors in an inner product space V. Then, for each $x \in V$, the following series $\sum_I |(\mathbf{e}_i|x)|^2$ converges and its sum satisfies

$$\sum_I |(\mathbf{e}_i|x)|^2 \le \|x\|^2 . \qquad \square$$

Proof. Let $J \subset I$ be a finite subset. As we have seen in the best approximation theorem 17.4, if we introduce the *components* x_i of x by the definition $x_i = (\mathbf{e}_i|x)$,

$$x - \sum_J x_i\mathbf{e}_i \perp \mathbf{e}_j \quad \text{for all } j \in J.$$

The Pythagoras theorem leads to

$$\|x\|^2 = \left\|x - \sum_J x_i\mathbf{e}_i + \sum_J x_i\mathbf{e}_i\right\|^2 =$$

$$= \left\|x - \sum_J x_i\mathbf{e}_i\right\|^2 + \left\|\sum_J x_i\mathbf{e}_i\right\|^2 = \left\|x - \sum_J x_i\mathbf{e}_i\right\|^2 + \sum_J |x_i|^2,$$

and in particular

$$\sum_J |x_i|^2 \le \|x\|^2 .$$

This inequality is true for *all finite* subsets $J \subset I$. Hence

$$\sup_{\substack{J \subset I \\ J \text{ finite}}} \sum_J |x_i|^2 \le \|x\|^2 .$$

But by definition, this Sup value is the sum of the family of (positive) terms $|x_i|^2$. □

One should observe that from the preceding inequality

$$|x_i|^2 \ge 1$$

can only occur for finitely many indices $i \in I$. Similarly,

$$|x_i|^2 \ge 1/n$$

can only occur for finitely many indices, say $i \in J_n$, for all integers $n > 0$. Consequently,

$$x_i \ne 0 \Longrightarrow |x_i|^2 \ge 1/n \text{ for some } n > 0 \Longrightarrow i \in J_n \text{ for some } n > 0$$

and

$$x_i \ne 0 \quad \Longrightarrow \quad i \in I_x = \text{Union of the } J_n \text{ for } n > 0.$$

This union I_x is at most countable and in fact, the series $\sum |x_i|^2$ contains at most countably many non zero terms (for each x). It is a convergent series of positive terms.

Corollary 1 (Cauchy-Schwarz inequality). Let V be an inner product space. Then

$$|(x|y)| \le \|x\| \cdot \|y\| \quad \text{for all } x, y \in V,$$

and equality holds only if x and y are proportional. □

Proof. If $y = 0$, there is nothing to prove (both $(x|y)$ and $\|y\|$ are zero). Otherwise, we can define $\mathbf{e} = y/\|y\|$ and this normed vector constitutes an orthonormal system by itself. The application of the Bessel inequality to this system leads to

$$|(\mathbf{e}|x)|^2 \le \|x\|^2.$$

Expliciting the value of $\mathbf{e}$ we find

$$\left| \|y\|^{-1}(y|x) \right| \le \|x\|$$

hence the result if we multiply both sides by $\|y\|$. Finally, assume that equality holds in the preceding inequality, e.g. assume that $\|x\| = |c| = |(\mathbf{e}|x)|$. In the orthogonal sum decomposition

$$x = ce + (x - ce)$$

we have

$$\|x\|^2 = |c|^2 + \|x - ce\|^2$$

and the assumption implies $\|x - ce\|^2 = 0$, hence $x = ce$. ■

Corollary 2 (Minkowski inequality). In any inner product space V, the following inequality holds

$$\|x + y\| \leq \|x\| + \|y\| \quad (x, y \in V).$$

Consequently, $x \longmapsto \|x\| = \sqrt{(x|x)}$ is a *norm* on V. □

Proof. Let us simply compute the square of the left hand side

$$\|x + y\|^2 = (x + y|x + y) = (x|x) + (x|y) + (y|x) + (y|y) =$$
$$= \|x\|^2 + 2\mathrm{Re}(x|y) + \|y\|^2 \leq \|x\|^2 + 2|(x|y)| + \|y\|^2 \leq$$
$$\leq \|x\|^2 + 2\|x\|\cdot\|y\| + \|y\|^2 = (\|x\| + \|y\|)^2.$$ ■

Remarks. 1) Any inner product space V is automatically considered as a *normed* space with $\|x\| = \sqrt{(x|x)}$ since this expression satisfies the characteristic properties of a norm

$$\|x + y\| \leq \|x\| + \|y\| \quad (x,y \in V),$$
$$\|\lambda x\| = |\lambda|\,\|x\| \quad (\lambda \in \mathbb{C},\ x \in V),$$
$$\|x\| > 0 \text{ if } 0 \neq x \in V.$$

2) In any *normed* space V, the following inequality holds

$$\|x + y\|^2 \leq 2\|x\|^2 + 2\|y\|^2.$$

Indeed since $(\|x\| - \|y\|)^2 \geq 0$, we have $\|x\|^2 + \|y\|^2 \geq 2\|x\|\|y\|$ and $\|x + y\|^2 \leq (\|x\| + \|y\|)^2 = \|x\|^2 + \|y\|^2 + 2\|x\|\|y\| \leq 2\|x\|^2 + 2\|y\|^2$.

3) Any normed space (and in particular any inner product space) is canonically considered as a *metric* space with distance defined by $d(x,y) = \|x - y\|$. The characteristic properties of a *distance* are indeed immediately verified

$$d(x,y) = d(y,x) \quad (x,y \in V),$$
$$d(x,y) \geq 0 \quad \text{and} \quad (d(x,y) = 0 \implies x = y),$$
$$d(x,z) \leq d(x,y) + d(y,z) \quad (x,y,z \in V).$$

Definition. An inner product space V (over $\mathbb{C}$) is called **Hilbert space** when it is complete (with respect to the canonical distance deduced from its scalar product). ■

Proposition. Any finite dimensional inner product space V is complete : finite dimensional complex inner product spaces are Hilbert spaces. □

Proof. Fix an orthonormal basis $(\mathbf{e}_i)_{1\leq i\leq n}$ in V (e.g.constructed as in 17.3) so that for $x = \sum x_i\mathbf{e}_i \in V$

$$\|x\|^2 = \sum|x_i|^2 .$$

In particular, the $i^{\underline{th}}$ component of x satisfies $|x_i| \leq \|x\|$. Consequently, if $(x_n)_{n\geq 0} \subset V$ is a Cauchy sequence and i is fixed, the ith components $x_{n,i}$ will also form a Cauchy sequence in $\mathbb{C}$. Since this field is complete, there is a limit $\xi_i = \lim_{n\to\infty} x_{n,i} \in \mathbb{C}$ and we can put $x = \sum_{1\leq i\leq n} \xi_i\mathbf{e}_i$. Obviously

$$\|x - x_n\|^2 = \|\sum (\xi_i - x_{n,i})e_i\|^2 = \sum|\xi_i - x_{n,i}|^2 \longrightarrow 0 \text{ for } n \longrightarrow \infty$$

and so the Cauchy sequence (x_n) converges to x. ∎

17.6 SQUARE SUMMABLE FUNCTIONS

One of the most interesting inner product spaces that we shall have to consider is

$$V = \mathcal{C}([0,1]) \quad \text{with} \quad (f|g) = \int_0^1 \overline{f(x)}g(x)dx.$$

Let us review a few results obtained in this particular case. First, the *norm* of a function —also called **quadratic norm**— is given by

$$\|f\|^2 = \int_0^1 |f(x)|^2dx.$$

To distinguish it from the Sup norm (or uniform convergence norm), we should write $\|f\|_2$ for the quadratic norm and e.g.

$$\|f\|_u = \|f\|_\infty = \operatorname{Sup}_x |f(x)|.$$

In this section however, only the quadratic norm will be considered and we may dispense with the extra-indices... The Cauchy-Schwarz inequality for the quadratic norm reads

$$\left|\int_0^1 \overline{f(x)}g(x)dx\right|^2 \le \int_0^1 |f(x)|^2dx\cdot\int_0^1 |g(x)|^2dx.$$

Examples of orthogonal functions in this space V abound. Consider for example the complex exponentials $\mathbf{e}_k(x) = e^{2\pi ikx}$. The scalar product of two of them is immediately determined

$$(\mathbf{e}_k|\mathbf{e}_\ell) = \int_0^1 \overline{\mathbf{e}_k(x)}\mathbf{e}_\ell(x)\ dx = \int_0^1 e^{2\pi i(\ell-k)x}dx =$$

$$= \left[\frac{e^{2\pi i(\ell-k)x}}{2\pi i(\ell-k)}\right]_0^1 = 0 \quad \text{if} \quad \ell \ne k\ .$$

For $\ell = k$, $\overline{\mathbf{e}_k(x)}\mathbf{e}_k(x) = e^{2\pi i(k-k)x} = 1$ and $\|\mathbf{e}_k\|^2 = 1$. This shows that $(\mathbf{e}_k)_{k\in\mathbb{Z}}$ is an orthonormal system in $V = \mathcal{C}([0,1])$. We shall study this system in detail in the chapter on Fourier series.

It is also interesting to apply the Gram-Schmidt orthogonalization procedure to the system of independent functions

$$1,\ x,\ x^2,\ \dots,\ x^n,\ \dots$$

The interested reader will determine a corresponding orthogonal basis consisting of polynomials (more on this topic in Ch.18).

Finally, it is essential to observe that the space

$$V = \mathcal{C}([0,1]) \quad \text{with} \quad (f|g) = \int_0^1 \overline{f(x)}g(x)dx$$

is *not complete*.

Proposition. For $n \geq 1$, let $f_n : [0,1] \longrightarrow \mathbb{R} \subset \mathbb{C}$ denote the continuous function

$$f_n(x) = \begin{cases} 0 & \text{for } 0 \leq x \leq 1/2 \\ \text{linear} & \text{for } 1/2 \leq x \leq 1/2 + 1/n \ . \\ 1 & \text{for } 1/2 + 1/n \leq x \leq 1 \end{cases}$$

Then (f_n) is a Cauchy sequence for the quadratic norm but has no limit in the space $\mathcal{C}([0,1])$. □

Proof. If n and $m \geq 1$, it is obvious that $|f_n(x) - f_m(x)| \leq 1$ and

$f_n - f_m$ vanishes identically on $[0,1/2]$
and on $[1/2 + 1/n,1]$ if $m \geq n$.

This shows that for $m \geq n$,

$$\|f_n - f_m\|^2 = \int_0^1 |f_n(x) - f_m(x)|^2 dx = \int_{1/2}^{1/2 + 1/n} |f_n - f_m|^2 dx \leq$$

$$\leq \int_{1/2}^{1/2 + 1/n} dx = 1/n \ .$$

In particular, $\|f_n - f_m\| \longrightarrow 0$ for $m > n \longrightarrow \infty$ and (f_n) is a Cauchy sequence. Let us prove that this sequence has no limit in $\mathcal{C}([0,1])$ (this may be "obvious" to you, but as you might be tempted to say that *the limit* is not continuous, hence is *not a limit*, I prefer to show you a *direct proof* of it !). It will be enough to show that for each continuous f, there exists a positive $\varepsilon = \varepsilon_f > 0$ with $\|f - f_n\| \geq \varepsilon$ (all $n \geq n_f \geq 1$). If $f(1/2) \geq 1/2$ there exists a $\delta > 0$ with

$$1/2 - \delta \leq x \leq 1/2 \quad \Longrightarrow \quad f(x) - f_n(x) = f(x) \geq 1/4$$

and consequently $\|f - f_n\|^2 \geq \delta(1/4)^2$ (for all $n \geq 1$). If, on the contrary $f(1/2) < 1/2$, there exists $\delta > 0$ with

$$1/2 \leq x \leq 1/2 + \delta \quad \Longrightarrow \quad f(x) \leq 3/4.$$

But for n sufficiently large ($1/n \leq \delta/2$, namely $n \geq n_f = 2/\delta$ will do) we shall have

$$f_n(x) - f(x) = 1 - f(x) \geq 1/4 \quad \text{for} \quad 1/2 + \delta/2 \leq x \leq 1/2 + \delta$$

and consequently

$$\|f - f_n\|^2 \geq (\delta/2)(1/4)^2 \quad (n \geq n_f).$$ ■

It is known that any metric space can be embedded in a *complete* metric space. The paradigm of this situation is the embedding of the rational field $\mathbb{Q}$ in the real field $\mathbb{R}$: a real number can be defined as an equivalence class of Cauchy sequences in $\mathbb{Q}$. A similar construction works for all metric spaces and in particular also for $V = \mathcal{C}([0,1])$! Cauchy sequences $(f_n) \subset V$ are well defined and an equivalence relation on them can be introduced by $(f_n) \simeq (g_n)$ whenever $\|f_n - g_n\| \longrightarrow 0$.

Definition. The space $L^2(0,1)$ of square summable *functions* is by definition the completion of the space $\mathcal{C}([0,1])$ for the metric

$$d(f,g) = d_2(f,g) = \sqrt{\int_0^1 |f - g|^2 dx}$$ ■

As the italics are supposed to emphasize, elements of the completion are not really *functions* but *equivalence classes of Cauchy sequences of* V... But fortunately, *discontinuous* functions f for which the integral of $|f|^2$ is well defined and finite can often be viewed as *limits of a Cauchy sequence of* V. For example, the discontinuous function

$$f(x) = \begin{cases} 0 & \text{for } 0 \le x \le 1/2 \\ 1 & \text{for } 1/2 < x \le 1 \end{cases}$$

is a limit of the Cauchy sequence given in the above proposition. Similarly, the functions

$$f_\alpha(x) = \begin{cases} 1/x^\alpha & \text{for } 0 < x \le 1 \\ 0 & \text{for } x = 0 \end{cases}$$

are square summable functions on $[0,1]$ when $2\alpha < 1$. If this condition is satisfied, f_α is for example a limit of the Cauchy sequence

$$f_n(x) = \begin{cases} 1/x^\alpha & \text{for } 1/n \le x \le 1 \\ n^\alpha & \text{for } 0 \le x \le 1/n \end{cases}.$$

More generally, *regulated functions* f (i.e. uniform limits of step functions) for which $\int |f|^2 dx < \infty$ define elements in the completion $L^2(0,1)$. (More on this topic in Ch.19 below!)

CHAPTER 18
SYSTEMS OF POLYNOMIALS

18.1 GENERAL NOTIONS

Most systems of polynomials which play an important role in numerical analysis or physics are real, and thus we limit our discussion to the real case. Let us introduce some notations which will be in constant use in this chapter

Π : real vector space of polynomials in one variable,

$\Pi_{\leq n}$: subspace of polynomials of degree at most n,

$\Pi_{<n}$: subspace of polynomials of degree < n.

To say that $p \in \Pi$ means that there is $n \in \mathbb{N}$ and (real) constants $a_o, \ldots, a_n$ with $p(x) = a_o + a_1 x + \ldots + a_n x^n$. In this case, if $a_n \neq 0$, $\deg(p) = n$ and $p \in \Pi_{\leq n} - \Pi_{<n}$. In particular, we have

$$\dim \Pi_{<n} = n \quad \text{and} \quad \dim \Pi_{\leq n} = n+1.$$

The vector space Π has basis $1, x, \ldots, x^n, \ldots$ hence is infinite dimensional.

Definition. A **system of polynomials** is a sequence $(p_n)_{n\geq 0} \subset \Pi$ such that $\deg p_n = n$. ∎

It follows from the preceding definition that all systems of polynomials are bases of Π.

On the vector space Π, there are *several* important scalar products (since we are working over the reals in this chapter, these scalar products are symmetric and each $\Pi_{\leq n}$ is a Euclidean space for any of them). The only ones that we shall consider are constructed in the following way.

Let I be a closed interval in $\mathbb{R}$ (bounded or not), w a continuous positive function on I (playing the role of a density)

$$w(x) > 0 \text{ for all interior points } x \in I,$$

such that, if I is unbounded,

$$\int_I x^j w(x)dx \text{ converges (absolutely) for all } j \geq 0.$$

Then we define

$$(p|q) = (p|q)_w = \int_I p(x)q(x)w(x)dx \ .$$

All properties of scalar products : *bilinearity* (over $\mathbb{R}$), *symmetry* and *positivity* are immediately verified (for the last one cf.lemma 1 of 14.1). The Gram-Schmidt orthogonalization procedure applied to the sequence $(x^n)_{n\geq 0}$ of powers (cf.17.3) leads to an orthogonal system of polynomials in the inner product space V with $(.|.)_w$. Dividing each polynomial by its norm, one gets an orthonormal system of polynomials. These systems are characterized by

$$\deg(p_n) = n \quad \text{and} \quad (p_n|p_m) = \int p_n(x)p_m(x)w(x)dx = \delta_{nm} .$$

However, it is often more convenient to *normalize* an orthogonal system of polynomials by a different condition. For example, one can prescribe a value $p_n(a) = c_n$ (for some $a \in \mathbb{R}$, usually $a \in I$), or fix the value of the highest coefficient of p_n, etc.

The definition of the scalar products $(p|q)_w$ makes it obvious that

$$(xp|q)_w = (p|xq)_w \quad (p, q \in \Pi).$$

This elementary fact has the following consequence.

Theorem. In any orthogonal system of polynomials, there are recurrence relations of the type

$$p_{n+1} = (A_n + B_n x)p_n - C_n p_{n-1} \quad (B_n \neq 0).$$

for all $n \geq 0$. □

Proof. Since $\deg(p_n) = n$, xp_n has degree n+1 and belongs to Π_{n+1}. Using the basis $(p_i)_{i\leq n+1}$ of this last space, we can write

$$xp_n = \sum_{i\leq n+1} c_i p_i .$$

Since the p_i are mutually orthogonal, the coefficients are proportional to the scalar products $(p_i|xp_n)$ (for an orthonormal system, the coefficients are *exactly* the scalar products, cf.17.4). But

$$(p_i|xp_n) = (xp_i|p_n) = 0 \quad \text{if} \quad xp_i \in \Pi_{<n}$$

since p_n is orthogonal to all polynomials of strictly smaller degree. This proves that $c_i = 0$ if $i+1 < n$, namely if $i < n-1$. The only possibly non-zero coefficients are c_{n+1}, c_n and c_{n-1} and we have a relation of the form

$$xp_n = \alpha_n p_{n+1} + \beta_n p_n + \gamma_n p_{n-1}$$

where $\alpha_n \neq 0$ by degree consideration. Solving for p_{n+1}, we get a relation of the announced type. The name and sign choice for A_n, B_n and C_n is made in accordance with the classical tables on the subject ! ■

18.2 ZEROS OF ORTHOGONAL SYSTEMS OF POLYNOMIALS

One of the most surprising results of the theory is the following general result on the location of zeros of polynomials in any orthogonal system.

Theorem. In any orthogonal system of polynomials (with respect to a scalar product of the form $(.|.)_w$), p_n has exactly n distinct zeros *in* the interior of I. □

Proof. Let indeed $x_1, \ldots, x_m$ be the interior points of I where p_n changes sign (a priori, m can be zero if p keeps a constant sign on the interior of I). These points are roots of p_n (they are precisely the roots with *odd multiplicity*). This implies

$$m \leq n = \deg(p_n).$$

Consider the polynomial p defined by product

$$p(x) = \Pi_{1 \leq i \leq m}(x - x_i)$$

(so that p = 1 if m = 0 by convention). The point is that p changes sign at the same points as p_n (in the interior of I) and the product $p \cdot p_n$ must keep a constant sign on the interior of I. Assume e.g. that $p \cdot p_n \geq 0$ on the interior of I (moreover $\deg(p \cdot p_n)$ = m+n proves that $p \cdot p_n \neq 0$). This implies

$$(p|p_n) = \int_I p \cdot p_n \cdot w dx > 0 \quad \text{(by lemma 1 of 14.1).}$$

This implies

$$p_n \text{ not orthogonal to } p \quad \text{and thus} \quad p \notin \Pi_{<n} ,$$

and proves $m = \deg(p) \geq n$. Comparing with the opposite inequality, we see that m = n. *A posteriori*, we see that p_n did vanish and change sign exactly n times in the interior of I (if n > 0). ■

The preceding proof is so simple and informative that any student interested in mathematics *should enjoy it !* Its usefulness will be explained presently. Let us first review the **Lagrange interpolation method**.

When n+1 distinct points $x_o , \ldots, x_n \in \mathbb{R}$ are fixed, the **Lagrange polynomials** $(\ell_i)_{0 \leq i \leq n}$ are defined by

$$\ell_i(x) = c_i \Pi_{j \neq i}(x - x_j)$$

with a normalization constant c_i chosen such that $\ell_i(x_i) = 1$, i.e.

$$c_i = 1/\Pi_{j \neq i}(x_i - x_j).$$

These polynomials are obviously characterized by the conditions

$$\deg(\ell_i) = n \quad \text{and} \quad \ell_i(x_j) = \delta_{ij}.$$

I claim that these polynomials ℓ_i $(0 \le i \le n)$ *constitute a basis of* $\Pi_{\le n}$ and more precisely, if $\deg(p) \le n$,

$$p = \sum_{0\le i\le n} p(x_i)\, \ell_i$$

(the components of a $p \in \Pi_{\le n}$ in the basis (ℓ_i) are precisely the values $p(x_i)$). To *prove* this assertion, simply consider the polynomial $p - \sum_{0\le i\le n} p(x_i)\, \ell_i \in \Pi_{\le n}$: it vanishes at the n+1 points x_i hence must be identically zero (a polynomial $p \ne 0$ of degree $\le n$ can have at most n roots !)

18.3 GAUSSIAN QUADRATURE

Let now $(p_n)_{n\ge 0}$ be an orthogonal system of polynomials (with respect to a scalar product $(.|.)_w$ as in 18.1). Fix an integer $n \ge 0$ and take for base points $x_o, \ldots, x_n$ the n+1 distinct roots of p_{n+1} (in the interior of the basic interval I over which w is defined by 18.2). The corresponding Lagrange interpolation polynomials will still be denoted ℓ_i : they make up a basis of $\Pi_{\le n}$. For any function f defined at the points x_i, let L_f be the polynomial

$$L_f = \sum_{0\le i\le n} f(x_i)\ell_i \in \Pi_{\le n} .$$

Since all ℓ_i for $i \ne j$ vanish at x_j , it follows that $L_f(x_j) = f(x_j)$ and we call L_f the **interpolation polynomial** corresponding to the values $f(x_j)$. As we have seen, $L_f = f$ if $f = p$ is a polynomial of degree $\le n$. Although $L_f \ne f$ for all polynomials f of degree > n, we can prove that L_f and f may still lead to equal integrals in certain cases.

Theorem (Gaussian quadrature). With the preceding notations, we have

$$\int_I f(x)w(x)dx = \int_I L_f(x)w(x)dx$$

for all polynomials $f \in \Pi_{2n+1}$ of degree $\le 2n+1$. □

Proof. By definition, the polynomial $L_f - f$ vanishes at the n+1 points x_j. Hence this polynomial must be divisible by

$$x - x_o,\ x - x_1,\ \ldots,\ x - x_n$$

and by their product $\Pi_j(x - x_j)$ since the x_j are distinct. This product is proportional to p_{n+1} and we can find a polynomial q

$$L_f - f = p_{n+1}\cdot q .$$

Obviously

$$\deg(f) \le 2n+1 \implies (n+1) + \deg(q) = \deg(L_f - f) \le 2n + 1$$

and in this case $\deg(q) \le n$. But now is the moment to remember that $p_{n+1} \perp \Pi_{\leq n}$ and in particular $p_{n+1} \perp q$. In extenso

$$\int_I (L_f - f)w dx = \int_I p_{n+1} q \cdot w dx = (p_{n+1}|q)_w = 0 ,$$

i.e. $\int_I L_f \cdot w dx = \int_I f \cdot w dx$. ∎

The content of this theorem means that if we are solely interested in the computation of integrals $\int_I f w dx$ for various functions f determined experimentally, we should measure the values $f(x_i)$ at well chosen points x_i. If we take for (x_i) the zeros of the $(n+1)$th polynomial in the orthogonal system determined by $(.|.)_w$, we shall get as good results as with any interpolation of degree $\leq 2n+1$.

Let us explicit the method a little bit further.

Still for polynomials $f \in \Pi_{\leq 2n+1}$ we can write

$$\int_I f(x)w(x)dx = \int_I L_f(x)w(x)dx = \int_I \sum f(x_i)\ell_i(x)w(x)dx = \sum f(x_i)\lambda_i$$

where the constants

$$\lambda_i = \int_I \ell_i(x)w(x)dx$$

are the **Christoffel constants**. These constants depend on the degree of approximation n chosen through the definition of the polynomials ℓ_i (based on the zeros of p_{n+1} in the orthogonal system determined by w). To compute these constants once and for all, we can use the above formula with $f(x) = x^j$

$$\sum_{0 \leq i \leq n} x_i^j \lambda_i = \int_I x^j w(x)dx = (x^j|1) = c_j \quad (0 \leq j \leq n).$$

These constants c_j are the **moments of the measure** $w(x)dx$. If they are known, the Christoffel constants $\lambda_i = \lambda_i^{(n)}$ can be determined by inversion of the matrix having the x_i^j as entries. The determinant of the square system is the well-known *Vandermonde determinant*

$$\begin{vmatrix} 1 & 1 & 1 & \dots \\ x_o & x_1 & x_2 & \dots \\ x_o^2 & x_1^2 & x_2^2 & \dots \\ & \dots & & \dots \end{vmatrix} = \Pi_{i>j} (x_i - x_j) \neq 0 .$$

Hence the system for the c_i is uniquely solvable.

18.4 CLASSICAL DENSITIES

The **classical systems of orthogonal polynomials** are determined by the choice of a few **classical densities** that we explicit now.

There are three kinds of *closed* intervals $I \subset \mathbb{R}$:

▷ $I = [a,b]$ is compact,

▷ $I = [a,\infty[$ is illimited in one direction,

▷ $I = \mathbb{R}$ is illimited in both directions.

Using a linear change of variable, we can assume that I is of the following typical form $[-1,1]$, $[0,\infty[$ or $]-\infty,\infty[= \mathbb{R}$. Here is a list of the classical densities in these cases.

	I	w	Normalization
$P_n^{(\alpha,\beta)}$ Jacobi	$[-1,1]$	$(1-x)^{\alpha}(1+x)^{\beta}$ $\alpha > -1, \ \beta > -1$	$P_n^{(\alpha,\beta)}(1) = \binom{n+\alpha}{n}$
$L_n^{(\alpha)}$ Laguerre	$[0,\infty[$	$x^{\alpha}e^{-x}$ $\alpha > -1$	$L_n^{(\alpha)}(0) = \binom{n+\alpha}{n}$
H_n Hermite	$]-\infty,\infty[$	e^{-x^2}	$H_n(x) = 2^n x^n + \ldots$

Classically, the coefficients of p_n in any orthogonal system of polynomials, are denoted by k_n, k'_n, k''_n, ...

$$p_n(x) = k_n x^n + k'_n x^{n-1} + k''_n x^{n-2} + \ldots$$

In principle, it is possible to compute all these systems of orthogonal polynomials by the Gram-Schmidt orthogonalization procedure. We could also proceed by indeterminate coefficients. For example, the determination of H_7 (Hermite) is given by

$$(H_7|x^i) = 0 \quad \text{for } 0 \leq i \leq 6 \quad \text{and} \quad k_7 = 2^7.$$

However, the preceding system of 7 equations in the 7 unknown k'_7, k''_7, ... is tedious. The reader should also determine the first few **Legendre polynomials** $P_n = P_n^{(0,0)}$ (corresponding to the density $w = 1$ on $I = [-1,1]$). However, there is a faster method to determine the polynomials in classical orthogonal systems.

18.5 RODRIGUES' FORMULA

Theorem (Rodrigues' formula). The classical polynomials as defined in the preceding table are also given by the following formula

$$p_n(x) = \frac{1}{K_n w(x)} (d/dx)^n \left(w(x)p(x)^n\right)$$

where p(x) is resp. given by

$$(1 - x^2) \text{ (Jacobi)} , \ x \text{ (Laguerre)} , \ 1 \text{ (Hermite).}$$

The constant K_n respecting the required normalization has to be

$$(-2)^n n! \text{ (Jacobi)} , \ n! \text{ (Laguerre)} , \ (-1)^n \text{ (Hermite).} \qquad \square$$

Proof. Let p_n be defined by the formula of the theorem. Inspection of the three classes shows that p_n is a polynomial of degree n. Furthermore, let us check that p_n is orthogonal to x^i for $i < n$. For this purpose, we have to estimate the scalar products

$$K_n(p_n|x^i) = \int w^{-1}D^n(wp^n)x^i w dx = \int D^n(wp^n)x^i dx \ .$$

Integrating $i+1 \le n$ times by parts, we get

$$K_n(p_n|x^i) = \left[D^{n-1}(wp^n)x^i - \ldots \pm D^{n-i-1}(wp^n)D^i x^i\right]_{\partial I}$$

since $D^{i+1}x^i = 0$ (the last integral disappears). These integrated terms certainly vanish at all possible infinite extremities of the interval I because the classical densities w decrease exponentially there. These terms are also zero at the extremities of I which are finite (as an easy verification shows). For example, for the interval $I = [0,\infty[$ (case of Laguerre polynomials), the lowest power of x coming in the bracket comes from the lowest power of x in

$$D^{n-j}(x^{\alpha+n}e^{-x})D^{j-1}(x^i) \qquad (1 \le j \le i+1) \ .$$

This power is obtained by derivating n-j times the term $x^{\alpha+n}$ and is thus proportional to

$$x^{\alpha+n-(n-j)}e^{-x}x^{i-(j-1)} = x^{\alpha+n-n+j+i-j+1}e^{-x} = x^{\alpha+i+1}e^{-x} \ .$$

Since we are assuming $\alpha > -1$, we have $\alpha+1 > 0$ and $\alpha+i+1 > 0$ $(i \ge 0)$. Thus all terms vanish at $x = 0$. A similar proof holds for the Jacobi polynomials and $x = \pm 1$. The proof would be completed by the verification that the given values of K_n correspond to the prescribed normalizations. We shall do this only in one case (the other ones being treated similarly as exercises!). Take the Laguerre polynomials as defined by Rodrigues' formula

$$L_n^{(\alpha)}(x) = K_n^{-1}x^{-\alpha}e^x D^n(x^{n+\alpha}e^{-x}) =$$

$$= K_n^{-1}x^{-\alpha}e^x \sum_{0 \le i \le n} \binom{n}{i} D^i(x^{n+\alpha})D^{n-i}(e^{-x}) \ .$$

The constant term (lowest power of x coming in the result) comes from the term i = n in the sum :

$$\binom{n+\alpha}{n} = L_n^{(\alpha)}(0) = K_n^{-1}x^{-\alpha}e^xD^n(x^{n+\alpha})e^{-x} =$$

$$= (n+\alpha)(n-1+\alpha)\ \dots\ (1+\alpha)/K_n\ .$$

To respect the prescribed normalization, K_n has to be n!. ■

As one more **example** of computation with classical polynomials let us prove the following formulas for Laguerre polynomials $L_n^{(\alpha)}$

$$k_n = (-1)^n/n!\ ,\ \|L_n^{(\alpha)}\|^2 = (n+\alpha)!/n!\ .$$

Coming back to the above proof, we have to determine the highest power of x coming in the expression of $L_n^{(\alpha)}$: it is given by the term i = 0

$$k_nx^n = K_n^{-1}x^{-\alpha}e^xx^{n+\alpha}D^n(e^{-x}) = (-1)^nx^n/K_n = (-1)^nx^n/n!\ .$$

Thus $k_n = (-1)^n/n!$. To compute the square h_n of the norm of $L_n^{(\alpha)}$, we write

$$h_n = (L_n^{(\alpha)}|L_n^{(\alpha)}) = (L_n^{(\alpha)}|k_nx^n + k'_nx^{n-1} + \dots) = k_n(L_n^{(\alpha)}|x^n)$$

since $L_n^{(\alpha)}$ is orthogonal to all polynomials of degrees < n. This scalar product is computed by integration by parts (all integrated terms vanish at the extremities)

$$(L_n^{(\alpha)}|x^n) = K_n^{-1}\int_0^\infty D^n(x^{n+\alpha}e^{-x})x^ndx = (-1)^nK_n^{-1}\int_0^\infty x^{n+\alpha}e^{-x}D^nx^ndx =$$

$$= (-1)^nn!K_n^{-1}\int_0^\infty x^{n+\alpha}e^{-x}dx\ .$$

But we have

$$\int_0^\infty x^te^{-x}dt = t!\ (\ = \Gamma(t+1)\ )\qquad \text{for } t > -1$$

and thus

$$h_n = \|L_n^{(\alpha)}\|^2 = k_n(-1)^nn!K_n^{-1}(n+\alpha)! = (n+\alpha)!/n!$$

from the known values of k_n and K_n .

18.6 LEGENDRE POLYNOMIALS

Definition. The **Legendre polynomials** P_n ($n \geq 0$) are defined by the following conditions

▷ $\int_{-1}^{1} P_n(x)P_m(x)dx = 0$ if $n \neq m$ ($\in \mathbb{N}$),

▷ $\deg(P_n) = n$ and $P_n(1) = 1$ ($n \in \mathbb{N}$). ■

In particular, we recognize that $(P_n)_{n\geq 0}$ is the orthogonal system of polynomials corresponding to the uniform density w = 1 on the interval I = [-1,1]. This is the case $\alpha = \beta = 0$ of the

CHAPTER 19

INTRODUCTION TO L^p-SPACES

19.1 HÖLDER INEQUALITY IN C(I)

Let us start with any closed interval $I \subset \mathbb{R}$ and denote by $C_c(I)$ the complex vector space of continuous functions $f : I \longrightarrow \mathbb{C}$ *vanishing outside a bounded set* (depending on f). These functions can always be integrated (cf.Fig.19.1).

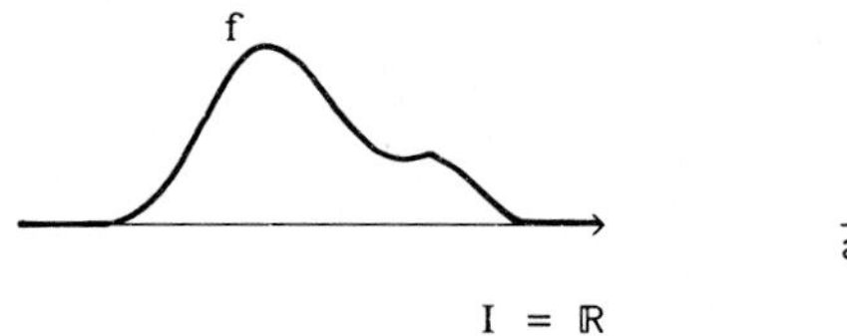

$I = \mathbb{R}$

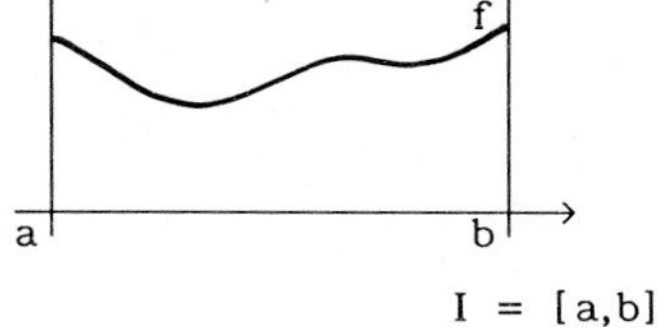

$I = [a,b]$

Fig.19.1

For $1 \leq p < \infty$ let us define

$$\|f\|_p = \left(\int_I |f(x)|^p dx\right)^{1/p} \qquad (f \in C_c(I)) .$$

If $p = 2$, this is the quadratic norm already used in Ch.17. We are going to show that quite generally,

$$f \longmapsto \|f\|_p$$

is always a *norm* on $C_c(I)$, called **norm of the convergence in mean of order p** (for $p = 1$, this is simply the **mean norm,** especially if the interval I is bounded and we divide by its length for a better normalization...) We have

$\triangleright$ $\|f\|_p > 0$ if $f \neq 0$ (cf. lemma 1 of 14.1),

$\triangleright$ $\|\lambda f\|_p = |\lambda| \|f\|_p$ for $\lambda \in \mathbb{C}$.

It only remains to prove

$\triangleright$ $\|f + g\|_p \leq \|f\|_p + \|g\|_p$ $(f, g \in C_c(I))$.

If $p = 1$, the triangle inequality in $\mathbb{C}$

$$|f(x) + g(x)| \leq |f(x)| + |g(x)|$$

can be integrated over I and immediately gives

$$\|f + g\|_1 \leq \|f\|_1 + \|g\|_1 .$$

From now on, we shall assume $p > 1$.

Theorem (Hölder inequality). For $p > 1$, f and $g \in C_c(I)$ we have

$$\|fg\|_1 = \int_I |f(x)g(x)|dx \le \|f\|_p \|g\|_q \quad \text{where} \quad q = p/(p-1) > 1 \quad \square$$

Proof. Observe that q can equivalently be defined by $1/p + 1/q = 1$ and the inequality is obviously true if either f or g is 0 (both sides are then 0). Thus, without loss of generality, we assume that both f and g are $\neq 0$ so that $\|f\|_p \neq 0$, $\|g\|_q \neq 0$. Dividing f and g by their respective norms, we see that it is enough to establish

$$\int_I |f(x)|\,|g(x)|dx \le 1 \quad \text{when} \quad \|f\|_p = 1 \text{ and } \|g\|_q = 1 .$$

Replacing f by $|f|$ and g by $|g|$, we can assume f and $g \ge 0$. Recall now that a function $\varphi : I \longrightarrow \mathbb{R}$ is called **convex** when

$$\varphi(\alpha a + \beta b) \le \alpha\varphi(a) + \beta\varphi(b)$$

for all a, $b \in I$, and α, β are positive real numbers with sum 1 : the graph of φ between two points is *below* the straight line linking them.

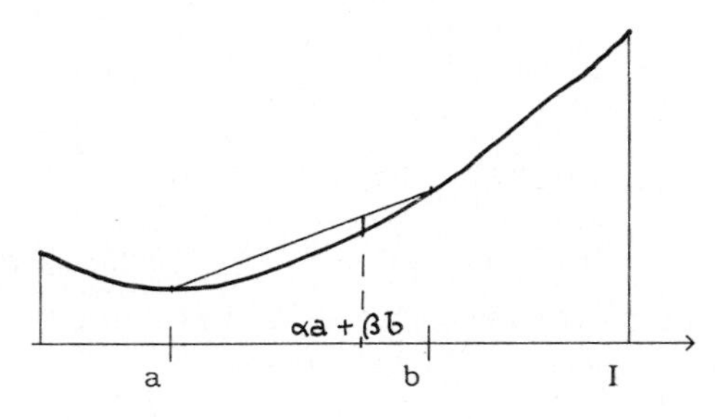

Fig.19.2

A sufficient condition for convexity of φ is $\varphi \in \mathcal{C}^2$ and $\varphi'' \ge 0$. In particular, the exponential function is convex over $\mathbb{R}$: for $\alpha = 1/p$ and $\beta = 1/q$ (positive of sum 1)

$$e^{a/p + b/q} \le \frac{1}{p} e^a + \frac{1}{q} e^b .$$

At all points $x \in I$ where $f(x)g(x) \neq 0$ we can write

$$0 < f(x) = e^{a/p} \quad \text{(i.e. } a = a(x) = p \log f(x)\text{)},$$
$$0 < g(x) = e^{b/q} \quad \text{(i.e. } b = b(x) = q \log g(x)\text{)},$$

so that

$$f(x)g(x) = e^{a/p + b/q} \le \frac{1}{p} e^a + \frac{1}{q} e^b = \frac{1}{p} f(x)^p + \frac{1}{q} g(x)^q$$

at all these points. The same inequality is also obviously true at the points where $f(x)g(x) = 0$! Integrating over x, we get

$$\int f(x)g(x)dx \le \frac{1}{p} \|f\|_p^p + \frac{1}{q} \|g\|_q^q = \frac{1}{p} + \frac{1}{q} = 1. \quad \blacksquare$$

Although very elegant, the preceding proof remains a bit mysterious and we give a **geometrical interpretation** of its main step:

$$f(x)g(x) \le \frac{1}{p} f(x)^p + \frac{1}{q} g(x)^q .$$

Changing notations slightly, let $a > 0$ and $b > 0$ be two positive real numbers. We intend to prove

$$ab \le \frac{1}{p} a^p + \frac{1}{q} b^q .$$

For this purpose, consider the following figure.

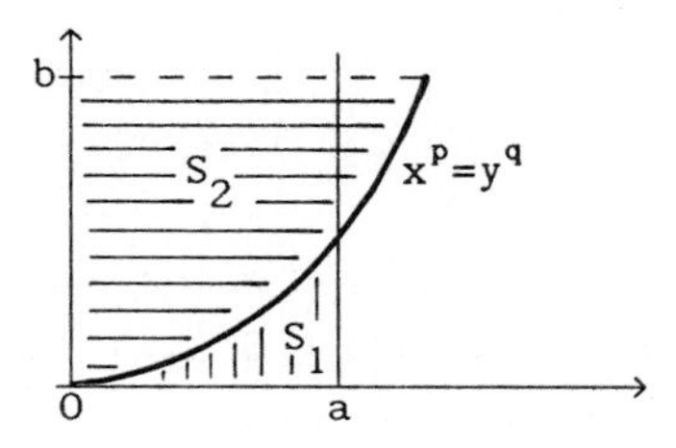

Fig.19.3

The area below the curve $y = x^{p/q}$ and for $x \in [0,a]$ is given by

$$S_1 = \int_0^a x^{p/q}dx = \int_0^a x^{p-1}dx = \left[\frac{x^p}{p}\right]_0^a = a^p/p$$

since $1/q = 1 - 1/p$ hence $p/q = p - 1$. Similarly, the area on the left of the same curve and for $y \in [0,b]$ is

$$S_2 = b^q/q .$$

Now, the picture shows that

$$ab \le S_1 + S_2 .$$

Remark (Legendre transformation). Instead of the curve $x^p = y^q$, we could have considered any monotonously increasing relation

$$\Phi(x,y) = 0 \quad \text{say} \quad y = \eta(x) \quad \text{or} \quad x = \xi(y)$$

(where both η and ξ are monotonous functions). Let us call

$$S(x) = \int_0^x \eta(t)dt , \quad T(y) = xy - S(x) = \int_0^y \xi(s)ds$$

(make a picture similar to Fig.19.3). Then S and T are linked by a Legendre transformation of the same type as the one used in classical mechanics to go from the Lagrange function $\mathscr{L}$ (of a mechanical system) to the Hamilton function $\mathscr{H}$. The same transformation is used systematically in thermodynamics. It belongs to the class of *contact transformations* (for which the interested reader will find an extensive literature linked with tangential coordinates).

19.2 MINKOWSKI'S INEQUALITY

Let us conclude the proof of the fact that $\|f\|_p$ defines a *norm* on $C_c(I)$ (we keep the same notations as in the preceding section).

Theorem (Inequality of Minkowski). For $p \geq 1$ and $f, g \in C_c(I)$,

$$\|f + g\|_p \leq \|f\|_p + \|g\|_p \ . \qquad \square$$

Proof. This inequality is obvious for $p = 1$, so we can assume that $p > 1$. Since $|f + g| \leq |f| + |g|$, we can again replace f by $|f|$ and g by $|g|$, so that it is enough to prove the inequality for positive functions ($f \neq -g$, otherwise there is nothing to prove !). Write

$$(f + g)^p = f(f + g)^{p-1} + g(f + g)^{p-1}$$

(recall that $p > 1$). Apply Hölder's inequality for both terms (writing simply $\int f$ instead of $\int_I f(x)dx$) :

$$\int f(f+g)^{p-1} \leq \|f\|_p \left(\int (f+g)^{(p-1)q}\right)^{1/q},$$

and, exchanging the roles of f and g

$$\int g(f+g)^{p-1} \leq \|g\|_p \left(\int (f+g)^{(p-1)q}\right)^{1/q}.$$

Adding these two inequalities we obtain

$$\int (f+g)^p \leq \left(\|f\|_p + \|g\|_p\right) \cdot C$$

with

$$C = \left(\int (f+g)^{(p-1)q}\right)^{1/q} \quad \text{or} \quad C^q = \int (f+g)^{(p-1)q} \ .$$

But $p + q = pq \implies pq - q = p$ and $C^q = \int (f+g)^p > 0$. Dividing by C both sides of the inequality obtained, we are left with

$$C^{q-1} \leq \|f\|_p + \|g\|_p \ ,$$

i.e. $\|f + g\|_p = \|f + g\|_p^{(p/q)(q-1)} = C^{q-1} \leq \|f\|_p + \|g\|_p$. ■

Since this inequality proves that $\|f\|_p$ is a norm on $C_c(I)$, this vector space is also a *metric space* for the *distance* defined by $d_p(f,g) = \|f - g\|_p$. By definition, the **Lebesgue space** $L^p(I)$ is the completion of $C_c(I)$ for this distance d_p. We have already considered the case $p = 2$ in 17.6 —square summable functions— for which $L^2(I)$ is a Hilbert space. In general, when $1 \leq p \leq \infty$, $L^p(I)$ is a Banach space (cf.16.1). By definition, $f \in L^p(I)$ simply means that there exists a Cauchy sequence $(f_n) \subset C_c(I)$ with $\|f - f_n\|_p \longrightarrow 0$. This information is often sufficient.

19.3 NEGLIGIBLE SETS

Definition. A subset $N \subset \mathbb{R}$ is called **negligible** when for each $\varepsilon > 0$, it is possible to find a sequence $(I_n)_{n\geq 0}$ of open intervals with

$\triangleright$ $N \subset$ union of the I_n ,

$\triangleright$ $\sum_{n\geq 0} \ell_n \leq \varepsilon$ where ℓ_n = length of I_n . ∎

It is obvious that any finite set of real numbers is negligible. It is also true (but less obvious !) that the set of rational numbers $\mathbb{Q} \subset \mathbb{R}$ is negligible. Here is a reason : we know that $\mathbb{Q}$ is countable, so that we can list all rational numbers in a sequence

$$r_o, r_1, r_2, \ldots, r_n \ldots \in \mathbb{Q}$$

and if $\varepsilon > 0$ is given, we take for I_o the open interval of center r_o and length $\varepsilon/2$, for I_1 the open interval of center r_1 and length $\varepsilon/4$, etc. Since each r_i is in an open interval I_i, the set $\mathbb{Q}$ of rational numbers is covered by the union of the intervals I_n and the sum of their lengths is

$$\varepsilon/2 + \varepsilon/4 + \varepsilon/8 + \ldots = \varepsilon(1/2 + 1/4 + 1/8 + \ldots) = \varepsilon .$$

(In particular, the *dense* subset $\mathbb{Q}$ of $\mathbb{R}$ is contained in the *open* set $U = \bigcup_{n\geq 0} I_n \neq \mathbb{R}$; also observe that the I_n's will not be disjoint intervals in general...) The same kind of argument gives a more general result.

Theorem. Let $(N_n)_{n\geq 0} \subset \mathbb{R}$ be a countable family of negligible sets. Then their union $\bigcup_{n\geq 0} N_n$ is also negligible. □

Proof. If $\varepsilon > 0$ is given, and since N_n is negligible, it is possible to find a family of open intervals $(I_{n,m})_{m\geq 0}$ with

$\triangleright$ $\bigcup_{m\geq 0} I_{n,m} \supset N_n$,

$\triangleright$ $\sum_{m\geq 0} \ell_{n,m} \leq \varepsilon/2^{n+1}$ (where $\ell_{n,m}$ = length of $I_{n,m}$).

Since $\mathbb{N} \times \mathbb{N}$ is countable, the double family $(I_{n,m})_{n,m}$ is countable with $\sum_{n,m} \ell_{n,m} \leq \sum_{n\geq 0} \varepsilon/2^{n+1} \leq \varepsilon$. ∎

However, there exists some *uncountable* negligible sets.

The Cantor set. Let us define a decreasing sequence of compact subsets of $K_o = I = [0,1]$ as follows. To obtain K_1, remove the open middle third from K_o

$$K_1 = [0,1/3] \cup [2/3,1].$$

Obtain more generally K_n by removing the open middle thirds of the segments constituting K_{n-1}. In particular, K_n consists of 2^n closed segments of lengths $1/3^n$. The Cantor set K is by definition the intersection of the family of K_n just defined

$$K = \bigcap_{n\geq 0} K_n .$$

This set is compact (and not empty since it contains the points 0 and 1). In fact, K contains the points

$$0,\ 1\ ,\ 1/3,\ 2/3,\ 1/9,\ 2/9,\ 7/9,\ 8/9,\ 1/27,\ \ldots$$

which are called *ends* of K. This already proves that K is *infinite* but we intend to show that K is even *uncountable*. For this purpose, let us *number* the elements of K in the following binary way. The elements of K which lie in K_1 will have first digit 0 if they lie in the first segment and digit 1 if they lie in its second segment. More precisely, the elements of K will have first two digits 00, 01, 10 and resp.11 if they lie in the corresponding

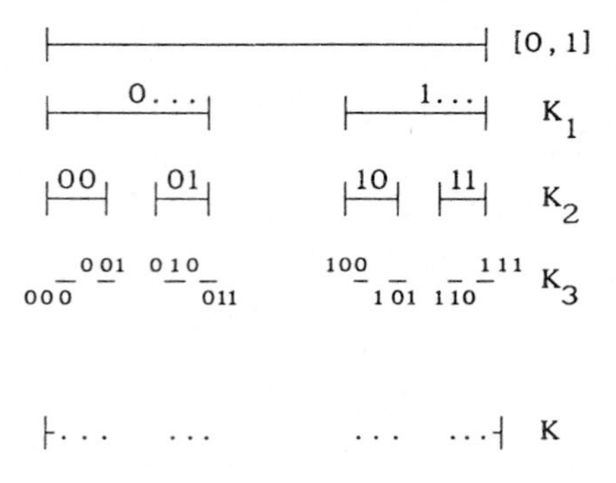

Fig.19.4

interval of K_2. In this way, infinite sequences of 0 and 1's correspond one-to-one to points of K. But the set of such sequences is *not countable*. Furthermore, we claim that K is *negligible*. Since K_n is a union of 2^n closed intervals of lengths $1/3^n$, it is easy to cover it with a union of 2^n open intervals of double length $2/3^n$. The sum of the lengths in this covering of K_n (and *a fortiori* covering K) is $2(2/3)^n$ hence $\leq \varepsilon$ for large n's if $\varepsilon > 0$ is fixed.

Let us observe finally that **negligible** sets in $\mathbb{R}^n$ (or $\mathbb{C}^n$) can be defined similarly, using products of open intervals in place of the I_n and replacing ℓ_n by $\mathrm{vol}(I_n)$.

19.4 CLASSES OF FUNCTIONS

Let us show that the Hölder inequality is still true in the completed spaces. More precisely, let $f \in L^p(I)$, $g \in L^q(I)$ where $p > 1$ and $1/p + 1/q = 1$. By definition, there are Cauchy sequences (f_n) and (g_n) in $C_c(I)$ with $f_n \longrightarrow f$ and $g_n \longrightarrow g$ (more precisely, the completion $L^p(I)$ consists of equivalence classes of Cauchy sequences and we mean f = "equivalence class of (f_n)" or $(f_n) \in f$). These sequences are necessarily bounded, say

$$\|f_n\|_p \le M \quad \text{and} \quad \|g_n\|_q \le M \quad (n \in \mathbb{N}).$$

The trivial case $p = 1$ of the Minkowski inequality gives

$$\|f_n g_n - f_m g_m\|_1 \le \|f_n(g_n - g_m)\|_1 + \|(f_n - f_m)g_m\|_1 .$$

With the Hölder inequality, we shall obtain

$$\|f_n g_n - f_m g_m\|_1 \le \|f_n\|_p \|g_n - g_m\|_q + \|f_n - f_m\|_p \|g_m\|_q \le$$
$$\le M\varepsilon + \varepsilon M = 2M\varepsilon$$

if n and m are large enough. This proves that $(f_n g_n)$ is a Cauchy sequence (in $C_c(I)$) for $\|.\|_1$ and has a limit in the complete space $L^1(I)$. This limit *will be denoted by* fg. Since

$$\|f_n g_n\|_1 \le \|f_n\|_p \cdot \|g_n\|_q$$

for all $n \ge 0$, we can let $n \longrightarrow \infty$ and we obtain

$$\|fg\|_1 \le \|f\|_p \cdot \|g\|_q .$$

The only problem with the preceding derivation is the following. It would be nice to know that the limit fg of the Cauchy sequence $(f_n g_n)$ is still *a function*, and in particular, that it *is* the product of two *functions*. Of course, to be able to identify elements f of the completion $L^p(I)$ of $C_c(I)$ to (classes of) *functions*, the Lebesgue theory is needed. We shall only give an idea of how this theory deals with these problems.

Definition. Let $I \subset \mathbb{R}$ be a closed interval. Two functions

$$f,\ g : I \longrightarrow \mathbb{C}$$

are equal **almost everywhere** when the set of points $x \in I$ where $f(x) \ne g(x)$ is negligible. This situation is abbreviated by

$$f(x) = g(x) \text{ a.e.} \quad (\text{or even } f = g \text{ a.e.}) \qquad \square$$

The relation just introduced is an *equivalence relation*: this means that

▷ $f = f$ a.e.

▷ $f = g$ a.e. $\Longrightarrow$ $g = f$ a.e.

▷ $f = g$ a.e. and $g = h$ a.e. $\Longrightarrow$ $f = h$ a.e.

(the last implication holds since the union of two negligible sets is still negligible : we have proved much more...) In particular, we can form *equivalence classes* with respect to this equivalence relation. The class containing a function f is usually denoted by the same bold letter **f**.

Lemma. Equivalence classes of functions make up a commutative ring if we define

▷ the sum of two classes **f** and **g** to be the class of $f + g$,

▷ the product of two classes **f** and **g** to be the class of $f \cdot g$.□

Proof. Let us only check that the first definition works : if

$$f, f_1 \in \mathbf{f} \quad \text{and} \quad g, g_1 \in \mathbf{g} \quad (\text{i.e. } \mathbf{f_1} = \mathbf{f} \text{ and } \mathbf{g_1} = \mathbf{g})$$

we have to check that $f + g$ and $f_1 + g_1$ define the same class, namely that $f + g = f_1 + g_1$ a.e. But $f(x) = f_1(x)$ outside a negligible set N_f and $g(x) = g_1(x)$ outside another negligible set N_g. The equality $f(x) + g(x) = f_1(x) + g_1(x)$ will hold outside $N_f \cup N_g$ which is negligible. All properties can be similarly verified. In more algebraic terms, the set $\mathcal{N}$ of *negligible functions* $f = 0$ a.e. is an *ideal* in the ring of complex valued functions and we are only considering the *quotient ring...* ■

19.5 COMPLETED SPACES $L^p(I)$

A first basic result of Lebesgue is the following.

Theorem. Let $(f_n) \subset C_c(I)$ be a Cauchy sequence for $\|.\|_p$. Then, there exists a subsequence $(f_{n_i})_{i \geq 0}$ and a negligible set $N \subset I$ such that $f_{n_i}(x)$ converges (for $i \longrightarrow \infty$) for $x \notin N$. If we let **f** be the class (for equality a.e.) of any function f with

$$f(x) = \lim_{i \to \infty} f_{n_i}(x) \qquad (x \notin N),$$

then this class is well defined by the Cauchy sequence (f_n). ■

In this way, elements of the completion $L^p(I)$ are identified to classes **f** of functions (equal a.e.). The first aim of integration theory is to identify which functions f lead to classes in L^p and to explain the meaning of a good answer (*a criterium*)

f must be *measurable* and $\int_I |f|^p dx$ must be finite.

The space $L^p(I)$ is also the completion of the space $\mathcal{S}(I)$ of **step functions** (with respect to the norm $\|.\|_p$). Recall the definition of step functions

$$f \in \mathcal{S}(I) \iff f \text{ is a finite linear combination of characteristic functions of bounded subintervals}$$

Here is a diagram of the situation concerning inclusions between these spaces

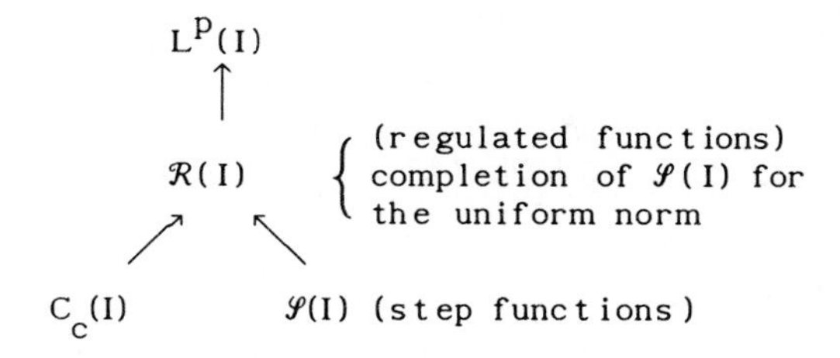

Also observe that the only continuous step functions are constant functions

$$C_c(I) \cap \mathcal{S}(I) = \begin{cases} \{0\} & \text{if } I \text{ is unbounded} \\ \mathbb{C} & \text{if } I \text{ is bounded} \end{cases}$$

19.6 VARIATION OF $\|.\|_p$ WITH p

As in the preceding chapters, I will denote a closed interval of $\mathbb{R}$ (with non empty interior). The norm of uniform convergence on $C_c(I)$ is usually denoted by

$$\|f\|_\infty = \sup_{x\in I} |f(x)|.$$

The reason for this notation is that this supremum is indeed the limit of $\|f\|_p$ for $p \longrightarrow \infty$.

Theorem (Polya). For $f \in C_c(I)$

$$\|f\|_p = \left(\int_I |f(x)|^p dx\right)^{1/p} \longrightarrow \sup_{x\in I} |f(x)| = \|f\|_\infty . \qquad \square$$

Proof. It is enough to prove this for $f \neq 0$ and replacing f by $f/\|f\|_\infty$, we can assume that $\sup|f| = 1$. Since f vanishes outside a compact subset $I(f) \subset I$, this supremum is a maximum, attained at some point $a \in I(f)$. By continuity, for each $\varepsilon > 0$, there is an interval J of positive length (in I(f)) containing a with

$$|f(x)| \geq 1 - \varepsilon \quad \text{for all } x \in J.$$

We are going to show that the contribution of the interval J to the integrals defining $\|f\|_p^p$ is dominant for large p. First, let us observe that $|f| \leq 1$ implies

$$\|f\|_p \leq \ell^{1/p} \quad \text{where } \ell = \ell_{I(f)} \text{ denotes the length of } I(f).$$

But when $p \longrightarrow \infty$, $\ell^{1/p} \longrightarrow 1$ and we deduce $\limsup_{p\to\infty} \|f\|_p \le 1$.

It will then be enough to show that $\|f\|_p \ge 1 - 2\varepsilon$ for large enough p. But for $\varepsilon < 1$

$$\|f\|_p \ge \left(\int_J |f(x)|^p dx\right)^{1/p} \ge \left(\int_J |1-\varepsilon|^p dx\right)^{1/p} = (1-\varepsilon)\,\ell_J^{1/p}$$

and $\ell_J > 0 \implies \ell_J^{1/p} \longrightarrow 1$. Consequently

$$\|f\|_p \ge 1 - 2\varepsilon \quad \text{for large enough } p > 1. \qquad \blacksquare$$

The preceding proof does not inform us on the variation of the norms $\|f\|_p$ for a fixed $f \in C_c(I)$. In fact, the convergence is monotonous. Here is a preliminary observation.

Let us choose a fixed function $f \in C_c(I)$. In the computation of the quadratic norm of f, use the estimate $|f| \le \|f\|_\infty$ in one of the two factors of $|f|^2$:

$$\|f\|_2^2 = \int_I |f|^2 dx \le \|f\|_\infty \int_I |f|\,dx = \|f\|_\infty \|f\|_1 .$$

Taking logarithms, the preceding inequality can be rewritten

$$2\log\|f\|_2 \le \log\|f\|_\infty + \log\|f\|_1$$

or

$$\log\|f\|_2 \le \tfrac{1}{2}\log\|f\|_\infty + \tfrac{1}{2}\log\|f\|_1 .$$

This inequality is a *convexity* property. In fact, we can prove the following quite general result.

Theorem. For fixed $f \in C_c(I)$, the function

$$t = 1/p \longmapsto \varphi(t) = \log\|f\|_p$$

is a *convex* function of $t = 1/p \in [0,1]$ (i.e. $p \in [1,\infty]$). $\square$

Proof. We have to prove that $\varphi(t) = \log\|f\|_{1/t}$ is a convex function on the interval $[0,1]$. For this purpose, we shall use the Hölder inequality. Let $t = 1/\tau$ and $s = 1/\sigma$ be in the basic interval $[0,1]$ and $\alpha + \beta = 1$ with $\alpha \ge 0$, $\beta \ge 0$. We have to estimate $\varphi(\alpha t + \beta s)$. Introduce $\alpha t + \beta s = 1/\rho$ so that

$$\alpha\rho/\tau + \beta\rho/\sigma = 1$$

and $p = \tau/\alpha\rho$ and $q = \sigma/\beta\rho$ are conjugate exponents for which the Hölder inequality applies

$$\varphi(\alpha t + \beta s) = \log \|f\|_\rho$$

and

$$\|f\|_\rho = \left(\int |f|^\rho\right)^{1/\rho} = \left(\int |f|^{(\alpha+\beta)\rho}\right)^{1/\rho} =$$

$$= \left(\int |f|^{\alpha\rho}\cdot|f|^{\beta\rho}\right)^{1/\rho} = \left\| \, |f|^{\alpha\rho}\cdot|f|^{\beta\rho} \right\|_1^{1/\rho} \le$$

$$\leq \left\| |f|^{\alpha\rho} \right\|_p^{1/\rho} \cdot \left\| |f|^{\beta\rho} \right\|_q^{1/\rho} =$$

$$= \left(\int |f|^{\alpha\rho p} \right)^{1/p\rho} \cdot \left(\int |f|^{\beta\rho q} \right)^{1/q\rho} =$$

$$= \left(\int |f|^{\tau} \right)^{\alpha/\tau} \cdot \left(\int |f|^{\sigma} \right)^{\beta/\sigma} = \|f\|_\tau^\alpha \cdot \|f\|_\sigma^\beta .$$

Taking logarithms, this proves

$$\varphi(\alpha t+\beta s) = \varphi(1/\rho) = \log \|f\|_\rho \leq$$
$$\leq \alpha \log \|f\|_\tau + \beta \log \|f\|_\sigma =$$
$$= \alpha\, \varphi(t) + \beta\, \varphi(s) .$$

■

On a *bounded* interval I, the constant 1 is is in $L^2(I)$ and for any $f \in L^2(I)$, the scalar product of 1 and f is defined by an integration of f : f must be integrable and

$$f \in L^2(I) \implies f \in L^1(I).$$

More generally, if $f \in L^p(I)$ and q is the conjugate exponent to p (namely $1/p + 1/q = 1$), $1 \in L^q(I)$ and the Hölder inequality shows that $f = 1 \cdot f \in L^1(I)$:

$$L^p(I) \subset L^1(I) \quad \text{for } p \geq 1 .$$

The convexity property of $1/p \longmapsto \log \|f\|_p$ shows still more generally

$$L^1(I) \supset L^p(I) \supset L^q(I) \quad \text{for } 1 \leq p \leq q \leq \infty .$$

PART FOUR

FOURIER SERIES

CHAPTER 20

TRIGONOMETRIC SERIES AND PERIODIC FUNCTIONS

20.1 TRIGONOMETRIC AND EXPONENTIAL FORM

A **trigonometric polynomial** is any finite linear combination of positive powers of the functions cost and sint. By definition, any trigonometric polynomial can be written in the form

$$\alpha_o + \sum_{1\le k\le n} (\alpha_k \cos^k t + \beta_k \sin^k t).$$

The coefficients α_k and β_k may be real or complex numbers. Since we can express the trigonometric functions with the complex exponential according to the Euler formulas

$$\cos t = (e^{it} + e^{-it})/2 \ , \ \sin t = (e^{it} - e^{-it})/(2i) \ ,$$

we can equivalently write trigonometric polynomials in the following form

$$\sum_{-n\le k\le n} c_k e^{ikt}.$$

In turn, since $e^{ikt} = \cos kt + i\sin kt$, the preceding expression can still be written as

$$a_o + \sum_{1\le k\le n} (a_k \cos kt + b_k \sin kt).$$

To obtain this expression, we have had to group the $\pm k$ indices and the relation between the coefficients is $a_o = c_o$ and

$$\begin{cases} a_k = c_k + c_{-k} \\ b_k = i(c_k - c_{-k}) \end{cases} \quad \text{for } k \ge 1 \ .$$

Conversely, $c_o = a_o$ and

$$\begin{cases} c_k = (a_k - ib_k)/2 \\ c_{-k} = (a_k + ib_k)/2 \end{cases} \quad \text{for } k \ge 1 \ .$$

Using the Tchebycheff polynomials T_k of the first kind (resp. the U_k of the second kind, cf.18.7),

$$a_o + \sum_{1\le k\le n} (a_k \cos kt + b_k \sin kt) =$$

$$= a_o + \sum_{1\le k\le n}(a_k T_k(\cos t) + b_k \sin t \cdot U_{k-1}(\cos t))$$

and we are back to the first defining form of a trigonometric polynomial.

Finally, let us observe that the real case is characterized by

$$a_k \text{ and } b_k \in \mathbb{R} \text{ or } c_{-k} = \bar{c}_k \text{ (complex conjugation).}$$

20.2 ABSOLUTE CONVERGENCE

Trigonometric series will equivalently be represented either by

$$\sum_{k\in\mathbb{Z}} c_k e^{ikt}$$

or by

$$a_o + \sum_{k\geq 1} (a_k \cos kt + b_k \sin kt) .$$

However, convergence is more easily studied in the complex exponential form since

$$|e^{ikt}| = 1 \quad (k \in \mathbb{Z} \quad \text{and} \quad t \in \mathbb{R}).$$

Let us recall that a series $\sum_{k\in\mathbb{Z}} \ldots = \sum_{-\infty}^{\infty} \ldots$ is said to converge when separately $\sum_{k\geq 0} \ldots$ and $\sum_{k<0} \ldots$ converge.

Absolute convergence at one point t of a series $\sum_{k\in\mathbb{Z}} c_k e^{ikt}$ obviously occurs when

$$\sum_{k\in\mathbb{Z}} |c_k e^{ikt}| = \sum_{k\in\mathbb{Z}} |c_k| < \infty .$$

In this case, we have *normal convergence* (in the sense of the norm of uniform convergence) since

$$|c_k| = \operatorname*{Sup}_{t\in\mathbb{R}} |c_k e^{ikt}| = \|c_k e^{ikt}\|_{\infty} .$$

Let us summarize the results obtained.

Theorem. If a series $\sum_{k\in\mathbb{Z}} c_k e^{ikt}$ converges absolutely for one value of t, then $\sum_{k\in\mathbb{Z}} |c_k| < \infty$. When this condition is satisfied, then $\sum_{k\in\mathbb{Z}} c_k e^{ikt}$ converges normally (hence uniformly) on the whole real line $\mathbb{R}$ and its sum represents a continuous function. ■

20.3 APPLICATIONS OF ABEL'S CRITERIUM

A trigonometrical series $\sum c_k e^{ikt}$ can converge without converging absolutely. For example, we have seen in 16.2 that this is the case when the coefficients $c_k > 0$ decrease monotonously (both for $k \longrightarrow +\infty$ and $k \longrightarrow -\infty$). Let us recall that the majoration

$$|\sum_p^q e^{ikt}| \leq \sigma_t = \frac{2}{|1 - e^{it}|}$$

gives a uniform bound

$$|\sum_p^q e^{ikt}| \leq \sigma = \frac{2}{|1 - e^{i\delta}|} \qquad t \in [\delta, 2\pi-\delta] = I_\delta$$

(for each $0 < \delta < \pi$). Hence the series $\sum_{k\geq 0} c_k e^{ikt}$ converges uniformly on all intervals I_δ when the coefficients $c_k > 0$ tend monotonously to 0. In this case, the sum

$$f(t) = \sum_{k\geq 0} c_k e^{ikt}$$

is a *continuous* function on the union $I =]0,2\pi[$ of all I_δ (each point $t \in I$ is an *interior* point of some I_δ).

To show that Abel's criterium is not entirely restricted to the case of positive coefficients, let us consider another typical case. Still assume that the coefficients $c_k > 0$ decrease monotonously to 0. To find uniform convergence domains for the series

$$\sum_{k\geq 0} (-1)^k c_k e^{ikt}$$

we need uniform estimates for $\sum_p^q (-1)^k e^{ikt}$. These finite sums are again geometric sums, but with ratio $-e^{ikt}$. As before, we get

$$\left|\sum_p^q (-1)^k e^{ikt}\right| \leq \frac{2}{|1 + e^{it}|} = \sigma_t .$$

Now, we get uniform bounds on all intervals $J_\delta = [-\pi+\delta, \pi-\delta]$ since

$$\sigma_t \leq \sigma_\delta \quad \text{for} \quad t \in J_\delta.$$

In this case, the sum is continuous on the union $J =]-\pi,\pi[$ of all J_δ. This should not come as a surprise... since $-1 = e^{i\pi}$,

$$(-1)^k e^{ikt} = e^{ik\pi + ikt} = e^{ik(t+\pi)}$$

and our series is simply obtained by a translation (of π) from the original one !

The preceding considerations apply *a fortiori* for the real and imaginary parts of series $\sum_k c_k e^{ikt}$, namely for the series

$$\sum_{k\geq 0} a_k \cos kt \ , \ \sum_{k\geq 1} b_k \sin kt \ .$$

In particular, the series

$$\sum_{k\geq 1} (1/k)\sin kt$$

converges uniformly on all $I_\delta = [\delta, 2\pi-\delta]$ and its sum represents a continuous function on $I =]0,2\pi[$. Here, we should add that the series also converges when t is an integral multiple of π since all its terms are then zero ! The sum vanishes at 0 and 2π and is a continuous function on $]0,2\pi[$ but is *not continuous* on $[0,2\pi]$. (We shall come back to this series in 23.4.)

20.4 PERIODIC FUNCTIONS AND THEIR MEAN VALUE

Definition. A **periodic function of period** $T > 0$ (or more simply **T-periodic function**) is a function $f : \mathbb{R} \longrightarrow \mathbb{C}$ such that

$$f(t + T) = f(t) \quad \text{for all} \quad t \in \mathbb{R}. \qquad \blacksquare$$

For example

- ▷ sint, cost, tan t are periodic of period 2π ,
- ▷ tan t is periodic of period π ,
- ▷ e^{ikt} is periodic of period 2π (if $k \in \mathbb{Z}$) .

As these examples show, a periodic function of period T may have smaller periods (e^{ikt} has also the period $2\pi/k$ if $k > 1$). If we want to produce T-periodic exponentials, we can look at

$$e^{ik\omega t} \quad \text{for} \quad \omega = 2\pi/T \quad (\text{and } k \in \mathbb{Z}).$$

In particular, the exponentials $e^{2\pi ikt}$ are 1-periodic (for $k \in \mathbb{Z}$).
Here is another way of constructing periodic functions. Take any $f \in C_c(\mathbb{R})$ (i.e. f is continuous and vanishes outside a bounded set of $\mathbb{R}$), and consider

$$\phi(t) = \phi_f(t) = \sum_{k\in\mathbb{Z}} f(t-k).$$

If t remains in a bounded set $\Omega \subset \mathbb{R}$, it is obvious that only finitely many terms in the sum are not zero on Ω. Hence the series converges uniformly on all bounded sets $\Omega \subset \mathbb{R}$. The sum $\phi = \phi_f$ is consequently a continuous function. This sum is also obviously a 1-periodic function. (A similar argument would hold if $f \in C(\mathbb{R})$ was only assumed to decrease sufficiently fast for $|t| \longrightarrow \infty$.)

Since continuous periodic functions are bounded, we can look for their *average value*.

Definition. Let f be a continuous T-periodic function. We define the **mean value** of f to be

$$c_o = c_o(f) = \frac{1}{T}\int_0^T f(t)dt \in \mathbb{C}$$ ■

Observe that the mean value of a T-periodic function f can also be computed by any integral

$$c_o(f) = \frac{1}{T}\int_a^{a+T} f(t)dt \qquad (a \in \mathbb{R}).$$

When f takes real values only, the image of f is an interval of $\mathbb{R}$ and it is obvious that $c_o(f)$ belongs to this interval of values. When f takes complex values, we can view it as a *closed* parameterized curve in $\mathbb{C}$, and it would not be difficult to check that $c_o(f)$ belongs to the *convex hull* of this image curve (it is a kind of center of gravity of this curve, but depends on the parametrization).

Let $\omega = 2\pi/T$ and consider in particular the T-periodic exponentials $e^{ik\omega t}$. When $k \neq 0$, their mean value is

$$\frac{1}{T}\int_0^T e^{ik\omega t}dt = \frac{1}{T}\left[\frac{e^{ik\omega t}}{ik\omega}\right]_0^T = 0.$$

When the T-periodic function f is defined by a uniformly convergent series

$$f(t) = \sum_{k\in\mathbb{Z}} c_k e^{ik\omega t},$$

the mean value of f can be computed by an integration term by term and the only contribution will be given by the constant term as we have just seen

$$c_o(f) = c_o$$

(justifying the notation !).

Definition. Let f be a continuous T-periodic function. The **Fourier sequence of** f is the sequence of mean values

$$c_k(f) = c_o(f\cdot e^{-ik\omega t}) = \frac{1}{T}\int_0^T f(t)e^{-ik\omega t}dt \qquad (k \in \mathbb{Z}) \qquad \blacksquare$$

When f is given as above by a uniformly convergent series

$$f(t) = \sum_{\ell\in\mathbb{Z}} c_\ell e^{i\ell\omega t},$$

we deduce uniformly convergent representations

$$f(t)\cdot e^{-ik\omega t} = \sum_{\ell\in\mathbb{Z}} c_\ell e^{i(\ell-k)\omega t},$$

having mean values equal to their constant term

$$c_k(f) = c_o(f\cdot e^{-ik\omega t}) = c_k \qquad (k \in \mathbb{Z}).$$

On the space of continuous T-periodic functions $f : \mathbb{R} \longrightarrow \mathbb{C}$, we shall consider the norms

$$\|f\|_p = \left(\frac{1}{T}\int_0^T |f(t)|^p dt\right)^{1/p} \qquad (1 \le p < \infty)$$

and

$$\|f\|_\infty = \operatorname*{Sup}_{t\in\mathbb{R}}|f(t)| = \operatorname*{Max}_{t\in\mathbb{R}}|f(t)|.$$

The p-norms will mainly be used for $p = 1$ and $p = 2$.

Proposition. The mapping $f \longmapsto (c_k(f))_{k\in\mathbb{Z}}$ is a linear map from the space of T-periodic continuous functions on $\mathbb{R}$ to the space $\mathcal{B}(\mathbb{Z})$ of bounded sequences on $\mathbb{Z}$. More precisely

$$\operatorname*{Sup}_{k\in\mathbb{Z}}|c_k(f)| \le \|f\|_1 \le \|f\|_\infty. \qquad \square$$

Proof. Once stated, the preceding assertion is immediately proved

$$|c_k(f)| = \frac{1}{T}\left|\int_0^T f(t)e^{-ik\omega t}dt\right| \le \frac{1}{T}\int_0^T |f(t)e^{-ik\omega t}|dt =$$

$$= \frac{1}{T}\int_0^T |f(t)|dt = \|f\|_1 \le \|f\|_\infty \frac{1}{T}\int_0^T dt = \|f\|_\infty$$

for all $k \in \mathbb{Z}$. $\blacksquare$

The preceding inequalities will be improved in 23.1 .

20.5 BEHAVIOR OF THE FOURIER SEQUENCE FOR $f \in \mathcal{C}^n$

From now on let us fix the period $T = 2\pi$. The corresponding periodic exponentials are then the e^{ikt} $(k \in \mathbb{Z})$ and the Fourier sequence of a $(2\pi$-)periodic function f is defined by

$$c_k(f) = \frac{1}{2\pi}\int_0^{2\pi} f(t)e^{-ikt}dt \quad (k \in \mathbb{Z}) .$$

Proposition. If f is a periodic $\mathcal{C}^n$ function (for some $n > 0$), we have

$$c_k(f) = (ik)^{-1}c_k(f') = (ik)^{-n}c_k(f^{(n)}) \quad \text{for } k \neq 0$$

and in particular

$$|c_k(f)| \le \|f^{(n)}\|_\infty / |k|^n = \mathcal{O}(1/|k|^n) \quad (k \longrightarrow \infty). \qquad \square$$

Proof. If f is $\mathcal{C}^1$, integration by parts in the defining formula for c_k $(k \neq 0)$ leads to

$$c_k(f) = \left[\frac{e^{-ikt}f}{-ik}\right]_0^{2\pi} - \int_0^{2\pi} \frac{e^{-ikt}}{-ik} f'(t)dt.$$

The term between brackets vanishes by periodicity of f and only

$$c_k(f) = \int_0^{2\pi} \frac{e^{-ikt}}{ik} f'(t)dt = \frac{1}{ik}\, c_k(f')$$

remains. When $n > 1$, the same formula can be used for f' in place of f and the assertion of the proposition follows by induction. ■

Corollary. If $f \in \mathcal{C}^2(\mathbb{R})$ is periodic, the series $\sum_{k\in\mathbb{Z}} c_k(f)e^{ikt}$ converges normally (hence uniformly) on $\mathbb{R}$. ■

Definition. The **Fourier series** of a $(2\pi$-)periodic function f is the series $\sum_{k\in\mathbb{Z}} c_k(f)e^{ikt}$. ■

In this definition, we implicitely assume that f can be integrated against the exponentials (e.g. f continuous or regulated). Fourier series of T-periodic functions f are defined similarly using the basic exponentials $e^{ik\omega t}$ $(\omega = 2\pi/T)$.

20.6 FIRST CONVERGENCE RESULT

Theorem. Let $f \in \mathcal{C}^2(\mathbb{R})$ be $(2\pi$-)periodic. Then the Fourier series of f converges normally (hence uniformly) to f :

$$f(t) = \sum_{k\in\mathbb{Z}} c_k(f)e^{ikt} \quad (t \in \mathbb{R}) \qquad \square$$

Proof. By the preceding section, we already know that the Fourier series of f converges normally. Hence its sum

$$g(t) = \sum_{k\in\mathbb{Z}} c_k(f)e^{ikt}$$

is a continuous periodic function. Moreover, if we let $h = f - g$, the Fourier sequence of h is

$$c_k(h) = c_k(f) - c_k(g) = 0$$

since the computation of the coefficients $c_k(g)$ can be determined by termwise integration. Thus

▷ h is a continuous periodic function,

▷ the Fourier sequence of h is the zero sequence.

These properties will be shown to imply that $h = 0$ vanishes identically. By assumption,

$$\int_0^{2\pi} h(t) \sum_{\text{finite}} a_k e^{ikt}\, dt = 2\pi \sum_{\text{finite}} a_k \cdot c_k(h) = 0 .$$

But there is a sequence of trigonometrical polynomials

$$p_n = \sum_{\text{finite}} a_k(n)e^{ikt} \longrightarrow \bar{h} \quad \text{uniformly on} \quad [0,2\pi]$$

(cf.15.6, Corollary). Now —by the basic theorem (15.2 b)—

$$0 = \int_0^{2\pi} p_n h\, dt \longrightarrow \int_0^{2\pi} \bar{h} \cdot h\, dt = \int_0^{2\pi} |h|^2 dt.$$

Since $|h|^2 \geq 0$ is a continuous function, this proves $h = 0$ (14.1, lemma 1). ∎

When analyzed more closely, the preceding proof reveals that we have only used the $\mathcal{C}^2$ assumption on f to be able to integrate term by term the defining series for g (its coefficients are of the order of $1/k^2$ for $f \in \mathcal{C}^2$). Consequently, we get the following generalization as a corollary of the preceding proof.

Theorem. Let f be a continuous periodic function having an absolutely summable Fourier sequence $\sum_{k\in\mathbb{Z}} |c_k(f)| < \infty$. Then, the Fourier series of f converges normally (hence uniformly) on the whole real line $\mathbb{R}$ and

$$f(t) = \sum_{k\in\mathbb{Z}} c_k(f)e^{ikt} \qquad (t \in \mathbb{R}) . \qquad ∎$$

As we are going to show in the next section, this theorem has a wider field of applications than the preceding one. If it is less attractive (?!), it is because the verification of its hypothesis requires a computation of all Fourier coefficients of f. A picture of the graph of f immediately reveals properties of f but may not inform us about properties of the sequence $(c_k(f))$... (In 23.5, we shall be able to generalize the situation.)

It is important to realize that —in general— the Fourier series of f should be *distinguished* from f (even if it converges), and until equality is proved, one should write $f : \sum_{k\in\mathbb{Z}} c_k(f)e^{ikt}$.

CHAPTER 21
APPLICATIONS

21.1 EXPANSION OF cos(at) AND EULERIAN FORMULA FOR cot(t)

Let $a \in \mathbb{R}$ (or $\mathbb{C}$) and consider the function

$$f_a(t) = \begin{cases} \cos at & \text{for } |t| \le \pi \\ 2\pi - \text{periodic on } \mathbb{R} \end{cases} .$$

The graph of f_a appears as follows (for small real values of a).

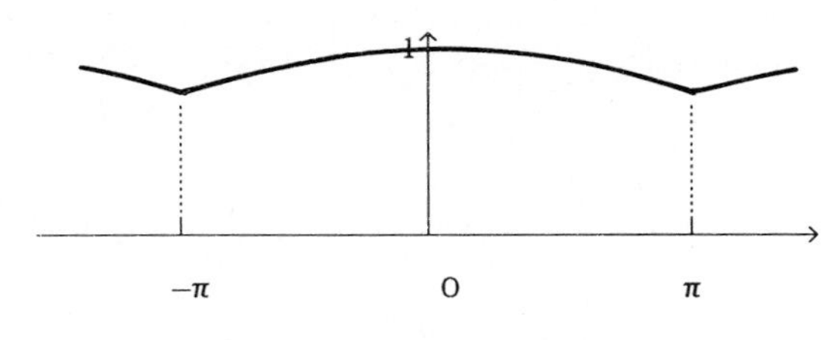

Fig. 21.1

If a is not an integer, f_a is not differentiable and the theory does not tell us that the Fourier sequence of f_a decreases fast. However, we can compute this Fourier sequence.

From now on, let us assume that $a \notin \mathbb{Z}$ (the case $a \in \mathbb{Z}$ is trivial since $f_n(t) = \cos nt = T_n(\cos t)$ is a trigonometric polynomial in that case). The computation of $c_k(f_a)$ is best made using an integral on $[-\pi,\pi]$ of

$$f_a(t) = \cos at = (e^{iat} + e^{-iat})/2 \qquad (-\pi \le t \le \pi) .$$

The first term will lead to an integral proportional to

$$\int_{-\pi}^{\pi} e^{iat} e^{-ikt} dt = \left[\frac{e^{i(a-k)t}}{i(a-k)} \right]_{-\pi}^{\pi} = \frac{2}{a - k} \sin(a-k)\pi .$$

Moreover, $\sin(a-k)\pi = (-1)^k \sin a\pi$, and the other computation is obtained by replacing a by −a. Finally, we see that

$$c_k(f_a) = \frac{(-1)^k}{2\pi} \sin a\pi \left(\frac{1}{a - k} + \frac{1}{a + k} \right) = \frac{(-1)^k a}{\pi} \, \frac{\sin \pi a}{a^2 - k^2} .$$

In this formula we see that $c_k(f_a) = O(1/k^2)$ for $|k| \longrightarrow \infty$ although $f \notin \mathcal{C}^2$. The last theorem of 20.6 is applicable and

$$f_a(t) = \sum_{k \in \mathbb{Z}} c_k(f_a) e^{ikt} \qquad (t \in \mathbb{R})$$

with normal and uniform convergence over $\mathbb{R}$. In particular we have

$$\cos at = \sum_{k\in\mathbb{Z}} c_k(f_a)e^{ikt} = \sum_{k\in\mathbb{Z}} \frac{(-1)^k a}{\pi} \frac{\sin\pi a}{a^2 - k^2} e^{ikt} \quad \text{for } |t| \le \pi.$$

Two famous formulas can be derived from this one.

For $t = 0$, $\cos at = 1$ and dividing throughout by $\sin\pi a$, we get

$$\frac{1}{\sin\pi a} = \sum_{k\in\mathbb{Z}} \frac{(-1)^k a}{\pi} \frac{1}{a^2 - k^2} \qquad (a \notin \mathbb{Z}) .$$

For $t = \pi$, dividing as before by $\sin\pi a$, we get

$$\pi\cot\pi a = \sum_{k\in\mathbb{Z}} \frac{a}{a^2 - k^2} = \frac{1}{a} + \sum_{k\ge1} \frac{2a}{a^2 - k^2} =$$

$$= \frac{1}{a} + \sum_{k\ge1} \left(\frac{1}{a - k} + \frac{1}{a + k}\right) .$$

Both formulas can be found in Euler's complete works ! The second one is even more striking than the first one : we can rewrite it in the form

$$\pi\cot\pi x = \lim_{N\to\infty} \sum_{|k|\le N} \frac{1}{x - k} \qquad (x \notin \mathbb{Z}).$$

As such, the cotangent function appears as a sum of hyperbolas.

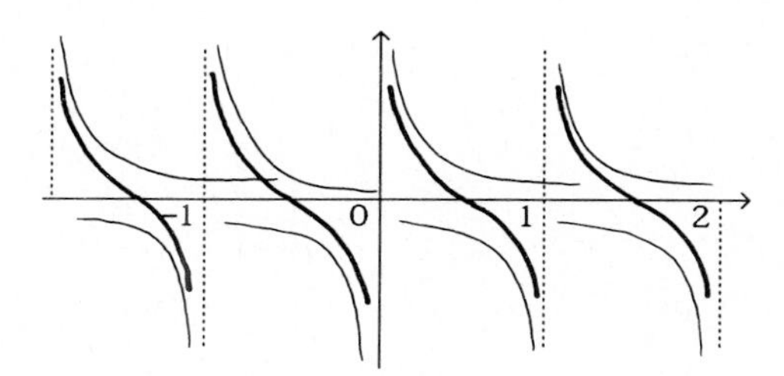

Fig.21.2

However, one should not write a sum $\sum_{-\infty}^{\infty} \frac{1}{x - k}$ without warning : the general term of this series is of the order $1/k$ (for $k \to \infty$) and consequently the series $\sum_{k\ge0} \frac{1}{x - k}$ *diverges*. The limit has to be taken *symmetrically*. A series $\sum_{k\in\mathbb{Z}} u_n$ is sometimes called *semi-convergent* when only $\lim_{N\to\infty} \sum_{|k|\le N} u_k$ exists (but $\sum_{k\ge0} u_k$ or $\sum_{k<0} u_k$ diverges).

21.2 APPLICATION TO THE INTEGRAL OF sinx/x

In 16.6 we gave a proof of the formula $\int_{\mathbb{R}} \sin x/x dx = \pi$.

At that point, the main purpose was to illustrate the use of all convergence theorems. But this integral can be computed more easily with other methods (the fastest one being without any doubt the one based on the calculus of residues, cf.13.5).

Let us base another evaluation of the preceding definite integral on the Eulerian formula just proved. Once the convergence of the integral is established (e.g.by means of Abel's theorem as in 16.6), its computation can be made using a limit with symmetric intervals

$$\int_{-\infty}^{\infty} \frac{\sin x}{x}\, dx = \sum_{k\in\mathbb{Z}} \int_{k\pi}^{(k+1)\pi} \frac{\sin x}{x}\, dx = \lim_{N\to\infty} \sum_{|k|\le N} \cdots$$

But

$$\int_{k\pi}^{(k+1)\pi} \frac{\sin x}{x}\, dx = (-1)^k \int_0^{\pi} \frac{\sin x}{x+k\pi}\, dx .$$

We have to compute

$$\lim_{N\to\infty} \sum_{|k|\le N} (-1)^k \int_0^{\pi} \frac{\sin x}{x+k\pi}\, dx = \lim_{N\to\infty} \int_0^{\pi} \sum_{|k|\le N} (-1)^k \frac{\sin x}{x+k\pi}\, dx .$$

We know that

$$\lim_{N\to\infty} \sum_{|k|\le N} (-1)^k \frac{\sin x}{x+k\pi} = \sin x \left(\frac{1}{x} + \sum_{k\ge1}(-1)^k\Big(\frac{1}{x-k\pi} + \frac{1}{x+k\pi}\Big)\right) =$$

$$= \sin x \left(\frac{1}{x} + \sum_{k\ge1}(-1)^k \frac{2x}{x^2-k^2\pi^2}\right) = \sin x \frac{1}{\sin x} = 1 .$$

In order to be able to permute $\lim_{N\to\infty}$ with $\int_0^{\pi} \ldots$, we have to prove that the preceding convergence is uniform on $[0,\pi]$ (cf.below). It will then follow that

$$\int_{-\infty}^{\infty} \frac{\sin x}{x}\, dx = \int_0^{\pi} \lim_{N\to\infty} \sum_{|k|\le N} (-1)^k \frac{\sin x}{x-k\pi}\, dx = \int_0^{\pi} dx = \pi .$$

The estimates

$$\Big|\sum_{|k|>N} (-1)^k/(x-k\pi)\Big| = \Big|\sum_{k>N} (-1)^k 2x/(x^2-k^2\pi^2)\Big| \le$$

$$\le \sum_{k>N} 2\pi/(k^2\pi^2-\pi^2) = (2/\pi) \sum_{k>N} 1/(k^2-1)$$

are uniform in $x \in [0,\pi]$ and prove our contention since the series $\sum_{k\ge0} 1/(k^2-1)$ converges.

21.3 ZETA VALUES

The Eulerian formulas proved in 21.1 have such attractive applications that we cannot resist giving some of them.

The Riemann zeta function is defined by

$$\zeta(s) = \sum_{k>0} 1/k^s \quad \text{for} \quad \mathrm{Re}(s) > 1 .$$

Indeed, in that half plane,

$$\sum |1/k^s| = \sum 1/k^{\mathrm{Re}(s)} \quad \text{converges} .$$

We intend to show that the values of this function on the

even integers $s = 2n \geq 2$ can be computed in elementary terms. In particular, the following special values will be derived

$$\zeta(2) = \sum_{k>0} 1/k^2 = \pi^2/6 ,$$
$$\zeta(4) = \sum_{k>0} 1/k^4 = \pi^4/90 ,$$
$$\zeta(6) = \sum_{k>0} 1/k^6 = \pi^6/945 .$$

It is obvious that $\zeta(2n) > 1$ (look at the definition) and $\zeta(2n)$ decreases monotonously to 1. In particular, we can form the *generating function*

$$f(z) = \sum_{n\geq 1} \zeta(2n)z^{2n} = \sum_{n\geq 1} \left(\sum_{k\geq 1} k^{-2n}\right) z^{2n}$$

which defines a holomorphic function in the disc $|z| < 1$ in $\mathbb{C}$. For such values, the double sum converges absolutely and we can permute them

$$f(z) = \sum_{k\geq 1}\left(\sum_{n\geq 1}(z^2/k^2)^n\right) = \sum_{k\geq 1} \frac{z^2/k^2}{1 - z^2/k^2} = \sum_{k\geq 1} \frac{z^2}{k^2 - z^2} =$$
$$= -\frac{z}{2} \sum_{k\geq 1} \frac{2z}{z^2 - k^2} = \frac{1}{2} - \frac{z}{2}\left[\sum_{k\geq 1}\left(\frac{1}{z-k} + \frac{1}{z+k}\right) + \frac{1}{z}\right] =$$
$$= \frac{1}{2} - \frac{z}{2}\pi\cot\pi z = \frac{1}{2} - \frac{z}{2} i\pi \frac{e^{2i\pi z} + (-1+2)}{e^{2i\pi z} - 1} =$$
$$= \frac{1}{2} - \frac{i\pi z}{2} - \frac{1}{2}\frac{2i\pi z}{e^{2i\pi z} - 1} .$$

Now is the moment to recall that the Taylor expansion

$$\frac{t}{e^t - 1} = \sum_{k\geq 0} b_k t^k/k!$$

defines the **Bernoulli numbers** $b_k \in \mathbb{Q}$. We have just obtained the expansion of the *even* generating function

$$f(z) = \frac{1}{2} - \frac{i\pi z}{2} - \frac{1}{2}\sum_{\ell\geq 0} b_\ell(2\pi i z)^\ell/\ell! .$$

Since f vanishes at the origin and is even, only the terms with $\ell = 2n \geq 2$ have a non zero coefficient and there remains

$$f(z) = -\frac{1}{2}\sum_{n\geq 1} b_{2n}(2\pi i z)^{2n}/(2n)! .$$

This delivers the explicit value for the coefficients $\zeta(2n)$

$$\zeta(2n) = -\frac{1}{2}(2i\pi)^{2n} b_{2n}/(2n)! = (-1)^{n+1} b_{2n} \cdot \frac{2^{2n-1}}{(2n)!} \cdot \pi^{2n}.$$

This formula shows in particular

$$1 < \zeta(2n) = \sum_{k\geq 1} 1/k^{2n} = |b_{2n}| \cdot \frac{2^{2n-1}}{(2n)!} \cdot \pi^{2n} \in \pi^{2n}\mathbb{Q}$$
$$\mathrm{sgn}(b_{2n}) = (-1)^{n-1} \quad (n \geq 1) .$$

The preceding deduction also shows

$$b_{2n+1} = 0 \quad \text{for} \quad n \geq 1 ,$$

$$|b_{2n}| \sim \frac{(2n)!}{2^{2n-1}\pi^{2n}} \quad \text{for} \quad n \longrightarrow \infty$$

(the notation $a_n \sim b_n$ means $a_n/b_n \longrightarrow 1$).

There remains to give the first few coefficients of the expansion of $t/(e^t - 1)$. They can be determined by the method of indeterminate coefficients, amplifying by the denominator $e^t - 1$ (which has a known expansion !). Here they are :

$$b_1 = -1/2, \; b_{2n+1} = 0 \text{ for } n \geq 1 ,$$

$$b_2 = 1/6, \; b_4 = -1/30, \; b_6 = 1/42, \; b_8 = -1/30, \; \ldots$$

(more values are tabulated in the classical tables, e.g. in the Abramowitz-Stegun "Handbook of Mathematical Functions" Dover 1972 p.810). They lead to the announced values for $\zeta(2n)$, $n = 1, 2, 3$.

Finally, let us observe that no similar computations are available for the $\zeta(2k+1)$... and $\zeta(3)$ is already quite a mysterious number (R.Apery showed in 1978 that $\zeta(3)$ is *irrational*, but there is no reason to expect that this number is a rational multiple of π^3 !).

21.4 FOURIER SERIES OF POLYGONS

When the sum of a Fourier series $\sum_{k\in\mathbb{Z}} c_k e^{ikt} = f(t)$ is a continuous function f of $t \in \mathbb{R}$, we can interpret it as a closed parameterized curve

$$t \longmapsto f(t) \in \mathbb{C}$$

in the complex plane. In general, this closed curve will have multiple points (namely, f is not one-to-one). It may be interesting to determine the Fourier series corresponding to some simple closed curves which are rectifiable. For polygons, we can require to take a uniform parameter (i.e proportional to the *special parameter*) on each side.

Proposition. If $f(t) = \sum_{k\in\mathbb{Z}} c_k e^{ikt}$ is the Fourier series of a polygon in $\mathbb{C}$ —uniformly parameterized— then $c_k = \mathcal{O}(1/k^2)$. □

Proof. By assumption, f is 2π-periodic and *piecewise linear*. If we denote by s_j the vertices of the image polygon, and by t_j the corresponding parameter values, $f(t_j) = s_j \in \mathbb{C}$ and f is affine linear between t_j and t_{j+1} $(0 \leq j < n)$. Say

$$f(t) = s_j + v_j(t - t_j) \quad t_j \leq t \leq t_{j+1} , \; f(t_o) = f(t_n).$$

The Fourier coefficients are then given by

$$2\pi c_k(f) = \int_0^{2\pi} f(t)e^{-ikt}dt = \sum_{0\le j<n} \int_{t_j}^{t_{j+1}} f(t)e^{-ikt}dt .$$

Let us evaluate a typical term by an integration by parts

$$\int_{t_j}^{t_{j+1}} f(t)e^{-ikt}dt = \left[f(t)\frac{e^{-ikt}}{-ik}\right]_{t_j}^{t_{j+1}} - \int_{t_j}^{t_{j+1}} v_j\cdot\frac{e^{-ikt}}{-ik}\, dt.$$

In the sum over j, the integrated terms cancel out two by two (f is continuous and periodic) and we obtain

$$2\pi c_k(f) = \sum_j \int_{t_j}^{t_{j+1}} v_j\cdot\frac{e^{-ikt}}{ik}\, dt = \frac{1}{ik} \sum_j v_j \int_{t_j}^{t_{j+1}} e^{-ikt}dt =$$

$$= k^{-2}\sum_j v_j\left[e^{-ikt}\right]_{t_j}^{t_{j+1}} .$$

For the absolute values, we obtain

$$|2\pi c_k(f)| \le k^{-2}\sum_j 2|v_j| = M/k^2 .$$ ■

Lemma. Let f be a continuous, 2π-periodic function presenting a symmetry of order $n \ge 2$ in the sense

$$f(t + 2\pi/n) = e^{2\pi i/n} f(t) \quad (t \in \mathbb{R}) .$$

Then the Fourier sequence of f satisfies

$c_k(f) = 0$ if $k - 1$ is not a multiple of n,

$c_k(f) = \frac{n}{2\pi}\int_0^{2\pi/n} f(t)e^{-ikt}dt$ if $k - 1$ is a multiple of n. □

Proof. Simply observe

$$2\pi c_k(f) = \int_{2\pi/n}^{2\pi+2\pi/n} f(\tau)e^{-ik\tau}d\tau = \int_0^{2\pi} f(t+2\pi/n)e^{-ikt}e^{-ik2\pi/n}dt =$$

$$= e^{(1-k)2\pi i/n}\cdot 2\pi c_k(f) .$$

This proves

$$c_k(f)\cdot(1 - e^{(1-k)2\pi i/n}) = 0 .$$

For k-1 not an integral multiple of n, $e^{(1-k)2\pi i/n} \ne 1$ and we must have $c_k(f) = 0$. When k-1 is a multiple of n, the integral over the interval $[0,2\pi]$ is a sum of n equal integrals over subintervals of lengths $2\pi/n$. ■

Let us consider more particularly the case of a regular polygon with n sides.

Proposition. Let $s_j = e^{2\pi ij/n}$ $(1 \le j \le n)$ be the nth roots of 1 in $\mathbb{C}$. The Fourier series of the regular polygon having the s_j as vertices —uniformly parameterized— is

$$f_n(t) = C_n \sum_{k\equiv 1 \bmod n} e^{ikt}/k^2 = C_n \sum_{\ell\in\mathbb{Z}} e^{i(1+\ell n)t}/(1+\ell n)^2 .$$

The normalization constant is $C_n = (\pi/n)^{-2}\sin^2(\pi/n)$. □

Proof. According to the result just established in the lemma, we only have to compute the coefficients c_k for $k \equiv 1 \bmod n$ and this can be done by means of the simpler formula

$$c_k = \frac{n}{2\pi}\int_0^{2\pi/n} f(t)e^{-ikt}dt$$

in this case. But for $t \in [0,2\pi/n]$,

$$f(t) = 1 + (s_1 - 1)\frac{nt}{2\pi} .$$

The announced result then follows from an easy computation. ■

Let us observe at this point that $C_n < 1$ tends monotonously to 1 for $n \longrightarrow \infty$ and $f_n \longrightarrow f = e^{it}$ (uniform parametrization of the circle) *uniformly* for $n \longrightarrow \infty$. Since the Fourier sequence of any polygon is $c_k = \mathcal{O}(1/k^2)$, the last convergence theorem of 20.6 is applicable and substituting $t = 0$ in the Fourier expansion of the regular polygon, we obtain

$$\sum_{\ell\in\mathbb{Z}} 1/(1+\ell n)^2 = \left(\frac{\pi/n}{\sin(\pi/n)}\right)^2.$$

In particular, for $n = 2$,

$$1 + 1/3^2 + 1/5^2 + \dots = 1/2 \sum_{\substack{k\in\mathbb{Z} \\ k \text{ odd}}} 1/k^2 = \frac{1}{2}\cdot\frac{\pi^2}{4} = \pi^2/8 .$$

In the following picture, the Fourier series of the pentagon is illustrated : partial sums $\sum_{k\equiv 1(5),|k|\le n}\cdots$ are plotted for $n = 1, 2, 3, 4$ and 8.

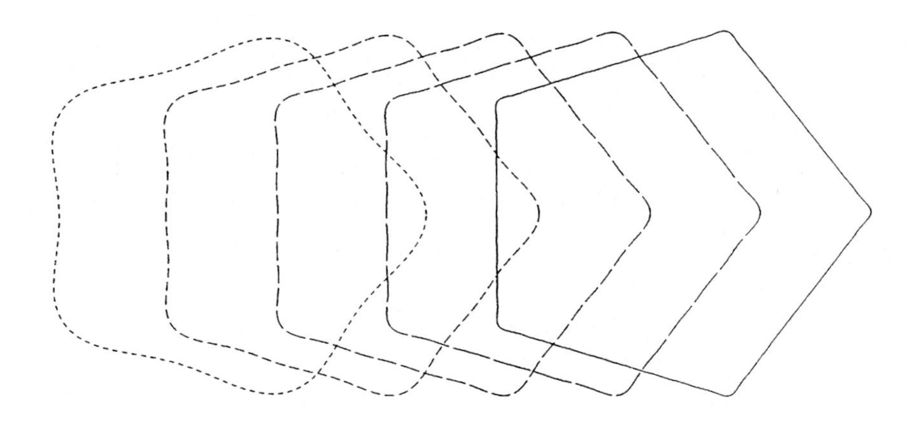

Fig.21.3

21.5 A FRACTAL CURVE

It is easy to give examples of interesting curves using Fourier series. Here is one. Let

$$\varphi(t) = \left(1 + 1/\sqrt{3}\right) e^{2\pi i t} \sum_{n\geq 1} 3^{-n} \, e^{-2\pi i 4^n t/3} + $$

$$+ \left(1 - 1/\sqrt{3}\right) e^{-2\pi i t} \sum_{n\geq 1} 3^{-n} \, e^{+4\pi i 4^n t/3} .$$

This is a closed parameterized curve (in $\mathbb{C}$). It is centered at the origin, the period is $T = 3$ and $\varphi(0) = 1$. It exhibits a symmetry of order 6 (like a snowflake!). The image of φ is given in the following figure.

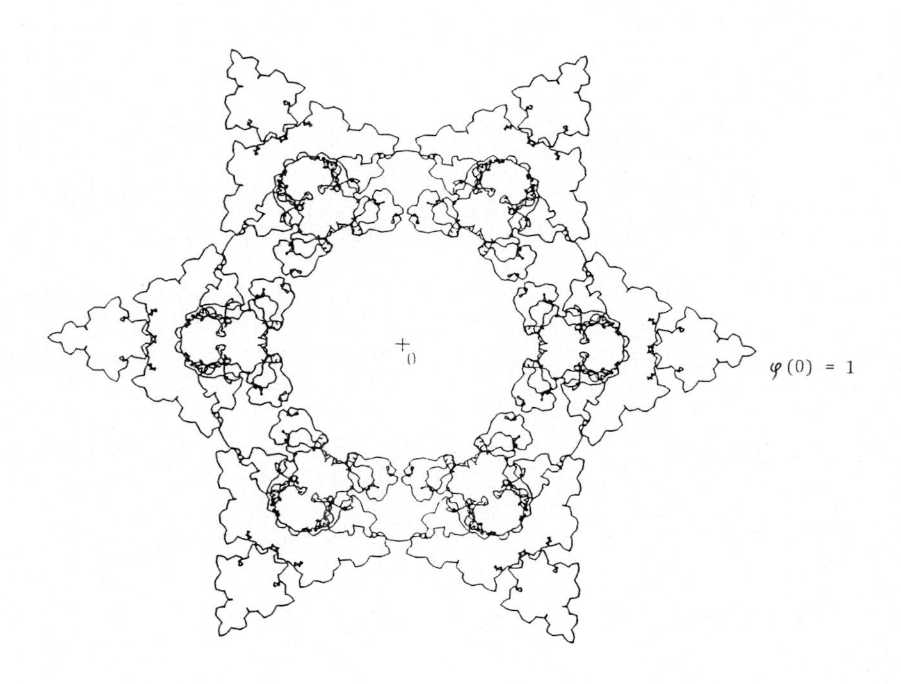

Fig.21.4

CHAPTER 22
DIRICHLET AND FEJER'S RESULTS

22.1 DIRICHLET'S CONVERGENCE THEOREM

Let us start with a 1-periodic function $f : \mathbb{R} \longrightarrow \mathbb{C}$ having Fourier coefficients $c_k(f)$ (in particular, let us assume that f is *integrable* but not necessarily continuous ; precise assumptions will be made in due time !). We still denote by $\mathbf{e}_k(t) = e^{2\pi ikt}$ the normalized 1-periodic exponentials and $\mathbf{e}(t) = \mathbf{e}_1(t) = e^{2\pi it}$.

Let us estimate the symmetric sums

$$S_N(f) = \sum_{|n|\le N} c_n(f)\mathbf{e}_n$$

using the defining expressions for the Fourier coefficients (to simplify notations, we adopt the convention that all integrals $\int$ have to be computed on any interval of unit length). Thus

$$S_N(f)(x) = \sum_{|n|\le N} \mathbf{e}_n(x)\cdot\int \mathbf{e}_n(-y)\ f(y)\ dy =$$

$$= \int f(y) \sum_{|n|\le N} \mathbf{e}_n(x-y)\ dy = \int f(y)\ K_N(x-y)\ dy\ .$$

Let us compute

$$K_N(t) = \sum_{|n|\le N} \mathbf{e}_n(t) = \sum_{|n|\le N} \mathbf{e}(t)^n =$$

$$= \mathbf{e}(t)^{-N}(1 + \mathbf{e}(t) + \dots + \mathbf{e}(t)^{2N}) =$$

$$= \mathbf{e}(t)^{-N}\frac{\mathbf{e}(t)^{2N+1} - 1}{\mathbf{e}(t) - 1} = \frac{\mathbf{e}(t)^{N+1/2} - \mathbf{e}(t)^{-N-1/2}}{\mathbf{e}(t)^{1/2} - \mathbf{e}(t)^{-1/2}} =$$

$$= \sin(N + 1/2)2\pi t \ / \ \sin\pi t\ .$$

This function K_N is called **the Dirichlet kernel**. Its properties are

- ▹ K_N is an even, 1-periodic function,
- ▹ the mean value of K_N is 1 (i.e. $c_o(K_N) = 1$),
- ▹ $K_N(0) = 2N + 1$.

These properties are indeed obvious in the definition

$$K_N = \sum_{|n|\le N} \mathbf{e}_n\ .$$

Theorem (Dirichlet). Let f be a 1-periodic function on $\mathbb{R}$ which is *regulated* on the basic period interval $[0,1]$ ($f \in L^1([0,1])$ would be enough). Let $x \in \mathbb{R}$ and assume moreover that both limits of $[f(x+h) - f(x)]/h$ and of $[f(x-h) - f(x)]/h$ exist when $h \searrow 0$. Then $\sum_{|k|\le N} c_k(f)e^{2\pi ikx} \longrightarrow [f(x+0) + f(x-0)]/2 \quad (N \longrightarrow \infty)$. □

Before we give the proof of this quite interesting result, let us recall that regulated functions are defined as uniform limits of step functions. As such, both limits at *right* or *left* of any point *exist*, and this property even *characterizes* regulated functions (on a compact interval). Thus, $f(x+0)$ denotes the right limit

$$f(x+0) = \lim_{0<h\to 0} f(x+h)$$

(and similarly for $f(x-0)$).

The proof will show that the theorem holds for a limit in $\|.\|_1$ of step functions. But the space $L^1(0,1)$ *can also be characterized as the completion of the space of step functions for the distance* $d_1(f,g) = \|f - g\|_1$, hence the generalization announced in the statement between parentheses.

Proof. We have already estimated the symmetric sums $\sum_{|n|\le N} \cdots$ by

$$S_N(f)(x) = \int f(x+t)\ K_N(t)dt$$

$$= \int_0^{1/2} f(x+t)K_N(t)dt + \int_0^{1/2} f(x-t)K_N(t)dt$$

and we now subtract

$$f(x+0)/2 = f(x+0)\int_{-1/2}^{1/2} K_N(t)dt/2 = f(x+0)\int_0^{1/2} K_N(t)dt$$

and

$$f(x-0)/2 = f(x-0)\int_{-1/2}^{1/2} K_N(t)dt/2 = f(x-0)\int_0^{1/2} K_N(t)dt.$$

This gives

$$S_N f(x) - f(x+0)/2 - f(x-0)/2 =$$

$$= \int_0^{1/2} [f(x+t) - f(x+0)]K_N(t)dt +$$

$$+ \int_0^{1/2} [f(x-t) - f(x-0)]K_N(t)dt .$$

It will be enough to show that both terms tend to zero for $N \longrightarrow \infty$. The first one is

$$\int_0^{1/2} [f(x+t) - f(x+0)]K_N(t)dt =$$

$$= \int_0^{1/2} \frac{f(x+t) - f(x+0)}{\sin\pi t} \sin(2N+1)\pi t\ dt =$$

$$= \int_0^{1/2} \frac{f(x+t) - f(x+0)}{\pi t} \frac{\pi t}{\sin\pi t} \sin(2N+1)\pi t\ dt.$$

By assumption,

$$\frac{f(x+t) - f(x+0)}{t} \text{ is regulated (has limit } f'_+(x) \text{ for } t \searrow 0)$$

and $\pi t/\sin\pi t$ is a continuous function so that

$$F(t) = \frac{f(x+t) - f(x+0)}{\pi t} \frac{\pi t}{\sin \pi t} \text{ is regulated .}$$

(This same function is integrable if we assume f integrable). We see that it will be enough to prove

$$\int_0^\ell F(t)\sin\lambda t dt \longrightarrow 0 \quad \text{for } \lambda \longrightarrow \infty \text{ (if F regulated or integrable).}$$

Replacing $\sin\lambda t$ by $(e^{i\lambda t} - e^{-i\lambda t})/(2i)$, it is enough to prove the following lemma (observe that the second term to be evaluated has a similar form and thus the proof will be completed).

Lemma. Let F be regulated or integrable over $\mathbb{R}$. Then

$$\Phi(\lambda) = \int_0^\ell F(t)e^{i\lambda t}dt \longrightarrow 0 \quad \text{for } |\lambda| \longrightarrow \infty . \qquad \square$$

Proof. When F is the characteristic function of an interval (a,b) (closed, open or semi-open, this does not matter...), the integral is simply

$$\Phi(\lambda) = \int_a^b e^{i\lambda t}dt = (e^{i\lambda b} - e^{i\lambda a})/(i\lambda)$$

so that $|\Phi(\lambda)| \le 2/|\lambda| = O(|\lambda|^{-1})$ in this case. If F is a *step function*, i.e. a (finite) linear combination of functions of the preceding type, we shall still have $|\Phi(\lambda)| = O(|\lambda|^{-1})$. In general, F will be a limit of step functions F_n and with

$$\Phi(\lambda) = \int_0^\ell F(t)e^{i\lambda t}dt \ , \quad \Phi_n(\lambda) = \int_0^\ell F_n(t)e^{i\lambda t}dt \ ,$$

we can estimate

$$|\Phi(\lambda) - \Phi_n(\lambda)| \le \int_0^\ell |F(t) - F_n(t)|dt \le \varepsilon \quad \text{for large n}$$

both if $F_n \longrightarrow F$ uniformly or in L^1-norm (i.e. $\|F - F_n\|_1 \longrightarrow 0$). Hence $\Phi_n \longrightarrow \Phi$ uniformly (in $\lambda \in \mathbb{R}$) . But we already know that each $\Phi_n(\lambda) \longrightarrow 0$ when $|\lambda| \longrightarrow \infty$. This implies that $\Phi(\lambda)$ also tends to 0 when $|\lambda| \longrightarrow \infty$: if $\varepsilon > 0$ is given, say $\|\Phi - \Phi_n\|_\infty \le \varepsilon/2$ for $n \ge N$ and $|\Phi_N(\lambda)| \le \varepsilon/2$ for $|\lambda| \ge M$, we shall have

$$|\Phi(\lambda)| \le |\Phi(\lambda) - \Phi_N(\lambda)| + |\Phi_N(\lambda)| \le \varepsilon/2 + \varepsilon/2 = \varepsilon$$

when $|\lambda| \ge M$. ■■

One should observe that according to the formulas of 20.1, symmetric sums $\sum_{|k|\le N} c_k e^{ikt}$ precisely correspond to partial sums

$$a_o + \sum_{k=1}^{N} (a_k \cos kt + b_k \sin kt) .$$

Hence Dirichlet's theorem simply states that the Fourier series of f in trigonometrical form *converges* to the mean values of the right and left limits of f (under the stated assumptions).

22.2 FEJER AVERAGES

We have already studied the symmetric partial sums of the Fourier series of a function f by

$$S_N = S_N(f) = \sum_{|n|\le N} c_n(f)\cdot \mathbf{e}_n$$

Let us now introduce the **Fejer sums** as averages of the preceding ones

$$A_N = A_N(f) = (S_o + S_1 + \ldots + S_{N-1})/N =$$
$$= \sum_{|n|\le N} \frac{N - |n|}{N} c_n(f)\cdot \mathbf{e}_n \quad .$$

These expressions are trigonometric polynomials and we intend to prove the following basic convergence result.

Theorem (Fejer). Let f be continuous and 1-periodic over $\mathbb{R}$. Then the Fejer averages $A_N(f)$ converge uniformly to the function f. □

Proof. An explicit computation of these averages is at hand. We have indeed obtained

$$S_n f(x) = \int f(x+t)K_n(t)dt$$

with

$$K_n(t) = \sin(2n+1)\pi t \ / \ \sin\pi t \ .$$

Hence

$$A_N(f)(x) = \int f(x+t)\ N^{-1}\sum_{0\le n<N} \frac{\sin(2n+1)\pi t}{\sin\pi t}\ dt$$

and we still have to evaluate

$$\frac{1}{N}\sum_{0\le n<N} \sin\ (2n + 1)\pi t.$$

For this purpose, we observe that the sum is the imaginary part of

$$\mathbf{e}(t/2)\ (1 + \mathbf{e}(t) + \ldots + \mathbf{e}(t)^{N-1}) =$$

$$= \mathbf{e}(t/2)\ \frac{\mathbf{e}(t)^N - 1}{\mathbf{e}(t) - 1} = \frac{\mathbf{e}(t)^N - 1}{2i}\ \frac{1}{\sin\ \pi t} \ .$$

But now,

$$\text{Im}\ (\mathbf{e}(t)^N - 1)/2i = \text{Re}\ (1 - \mathbf{e}(t)^N)/2 = (1 - \cos 2N\pi t)/2 =$$
$$= \sin^2 N\pi t$$

and the sum of the geometric series is

$$\sin^2 N\pi t \ / \ \sin\pi t$$

whereas

$$K_0(t) + \ldots + K_{N-1}(t) = \sin^2 N\pi t \ / \ \sin^2\pi t.$$

Altogether, we have found

$$A_N(f)(x) = \int f(y)\ F_N(x-y)dy = \int f(x-t)F_N(t)dt$$

where

▷ $F_N(t) = \frac{1}{N} \sin^2 N\pi t \;/\; \sin^2 \pi t$,

▷ F_N is an even, 1-periodic *positive* function ,

▷ $F_N(0) = N$,

▷ $\int F_N(t)dt = 1$.

(The last property is a consequence of the definition of F_N as average of the K_n $(0 \le n < N)$, each of which has constant term 1). In particular, the last property implies

$$f(x) = f(x)\int F_N(t)dt = \int f(x)F_N(t)dt$$

and hence

$$A_N(f)(x) - f(x) = \int [f(x-t) - f(x)]\, F_N(t)dt \; .$$

We have to show that this is uniformly small (for large N). The integral can be broken in two

$$\int_{-\delta}^{\delta} [f(x-t) - f(x)]\, F_N(t)dt$$

and

$$\int_{\delta \le |t| \le 1/2} [f(x-t) - f(x)]\, F_N(t)dt.$$

If $\varepsilon > 0$ is given, we can choose $\delta > 0$ (independent of x since f is *uniformly continuous,*) so that $|f(x-t) - f(x)| \le \varepsilon/2$ for all $|t| \le \delta$. Consequently, since F_N is positive

$$\left|\int_{-\delta}^{\delta} [f(x-t) - f(x)]\, F_N(t)dt\right| \le \int_{-\delta}^{\delta} |f(x-t) - f(x)|\, F_N(t)dt \le$$

$$\le (\varepsilon/2) \int_{-\delta}^{\delta} F_N(t)dt \le (\varepsilon/2) \int_{-1/2}^{1/2} F_N(t)dt = \varepsilon/2.$$

On the other hand,

$$0 \le F_N(t) \le \frac{1}{N\sin^2 \pi\delta} \quad \text{for} \quad |t| \ge \delta$$

implies

$$\left|\int_{\delta \le |t| \le 1/2} [f(x-t) - f(x)]\, F_N(t)dt\right| \le$$

$$\le 2\|f\|_\infty \frac{1}{N\sin^2 \pi\delta} \int_{\delta \le |t| \le 1/2} dt \le \varepsilon/2$$

for large enough N. This concludes the proof. ∎

22.3 BACK TO UNIFORM APPROXIMATION

Let f be a continuous periodic function. The Stone-Weierstrass theorem only gave the existence of a sequence of trigonometric polynomials converging to f (cf.15.6). Now, we have an explicit such sequence

$$A_N(f) = (S_o f + S_1 f + \ldots + S_{N-1} f)/N =$$
$$= c_o(f) + \frac{N-1}{N}\left(c_1(f)\, e^{i\omega t} + c_{-1}(f)\, e^{-i\omega t}\right) + \ldots \longrightarrow f\ .$$

BUT, even for continuous periodic functions f, the symmetric partial sums $S_n f$ may *not converge pointwise* to f .

It is interesting to note that Weierstrass' original result can also be derived from Fejer's one.

Theorem (Weierstrass,cf.15.6**)**. Let $f : [-1,1] \longrightarrow \mathbb{C}$ be continuous. Then there exists a sequence of polynomials which converges uniformly to f. □

Proof. Consider the composite map $\varphi : \mathbb{R} \longrightarrow \mathbb{C}$ defined by

$$t \longmapsto x = \cos t \longmapsto f(x) = f(\cos t)\ (\ = \varphi(t))\ .$$

This is an even, continuous function. The sequence of Fejer sums $\sum_{|k| \le N} \gamma_k \cos kt$ converges uniformly to φ. Since $\cos kt = T_k(\cos t)$ (18.7) we get a uniform convergence of

$$\sum_{|k| \le N} \gamma_k T_k(\cos t) \longrightarrow f(\cos t)$$

namely

$$\sum_{|k| \le N} \gamma_k T_k \longrightarrow f \quad \text{uniformly on } [-1,1]\ . \qquad \blacksquare$$

In contrast with the proof given in chapter 15, this uniform approximation is completely explicitely given.

22.4 SUMMATION FILTERS

Let $f : \mathbb{R} \longrightarrow \mathbb{C}$ be (continuous and) 1-periodic. The partial sums $S_N f = \sum_{|k| \le N} c_k \mathbf{e}_k$ of the Fourier series of f can be viewed as produced by a **filter** cutting all frequences of f greater than N. The characteristic of this filter is easily drawn in a figure.

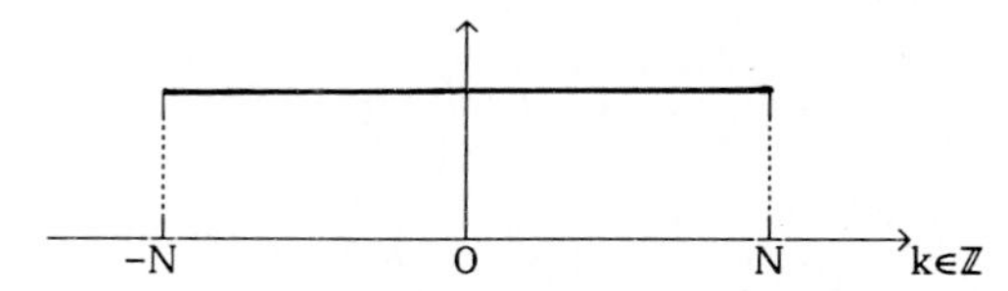

Fig.22.1

Similarly, the Fejer averages

$$A_N = A_N(f) = (S_o + S_1 + \ldots + S_{N-1})/N =$$
$$= \sum_{|n| \le N} \frac{N - |n|}{N}\, c_n(f) \cdot \mathbf{e}_n \quad .$$

can be thought of as produced by a filter cutting linearly the frequences between 0 and N. Here, the characteristic of the filter is represented on the following figure.

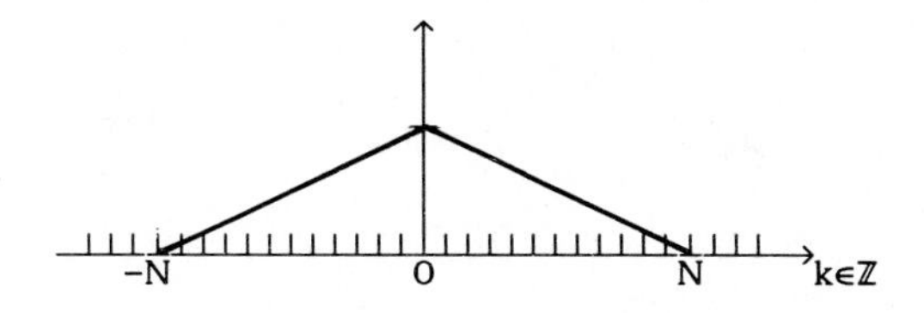

Fig.22.2

Other filters are still very interesting. For example, a filter which attenuates the k-frequency with factors

$$e^{-|k|/n} = r_n^{|k|} \qquad (r_n = e^{-1/n} < 1,\ r_n \longrightarrow 1)\ .$$

The attenuation filter with characteristic

$$r^{|k|} \qquad (\text{for some fixed } 0 \le r < 1)$$

is indeed well known.

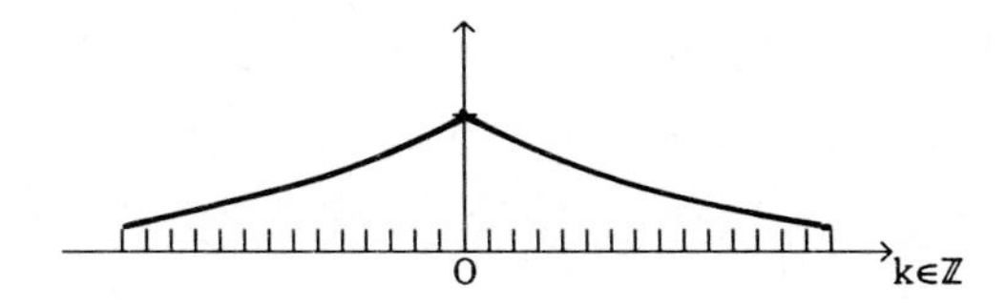

Fig.22.3

This attenuation leads to **Poisson's sums**

$$P_r(f) = \sum_{k\in\mathbb{Z}} r^{|k|} c_k \mathbf{e}_k \ .$$

They play an important role in the study of harmonic functions in the unit disc $|z| < 1$ in $\mathbb{C}$ (say $z = r\mathbf{e}(t) = re^{2\pi it}$).

In the same vein, one could look at the stronger attenuation given by Gaussian factors $\exp(-k^2/n)$ (some large n).

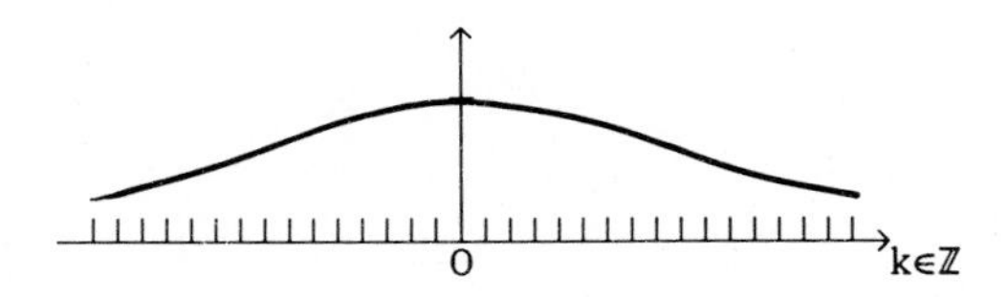

Fig.22.4

Such sums $\sum_k e^{-\varepsilon k^2} c_k \mathbf{e}_k$ were indeed met by Fourier in his study of the heat equation (cf.24.2).

22.5 CONVOLUTION

Let f and g be two continuous (or regulated, or only integrable over a period interval) 1-periodic functions $\mathbb{R} \longrightarrow \mathbb{C}$. We define the **convolution of** f **and** g as the new function

$$F(x) = f*g(x) = \int_0^1 f(y)g(x-y)dy .$$

The change of variable $t = x-y$ ($dt = -dy$) leads to

$$f*g(x) = \int_{x-1}^{x} f(x-t)g(t)dt = \int_0^1 f(x-t)g(t)dt = g*f(x).$$

One checks without difficulty that this convolution is again a 1-periodic function. By 16.5, it is continuous if f and g are continuous (one can show that it is integrable over a period interval if f and g have the same property).

Theorem. The Fourier coefficients of the convolution product f*g of two 1-periodic functions f and g are given by

$$c_k(f*g) = c_k(f)\cdot c_k(g) \quad (k \in \mathbb{Z}). \qquad \square$$

Proof. Let us simply compute these Fourier coefficients according to their definition

$$c_k(f*g) = \int_0^1 dx\ e^{-ikx} \int_0^1 dy\ f(y)g(x-y) =$$

$$= \int_0^1 dy\ f(y) \int_0^1 dx\ e^{-ikx} g(x-y) =$$

$$= \int_0^1 dy\ f(y)e^{-iky} \int_0^1 dx\ e^{-ikx+iky} g(x-y) =$$

$$= \int_0^1 dy\ f(y)e^{-iky}\ c_k(g) = c_k(f)\cdot c_k(g).$$

Notice that we have used Fubini's theorem 16.5 to permute two integrations. ∎

This theorem permits us to say that if an attenuation filter has characteristic attenuation coefficients $(\rho_k)_{k\in\mathbb{Z}}$, then

$\sum_k \rho_k c_k(f)\mathbf{e}_k$ is the Fourier series of

the convolution product $\Phi*f$ where $\Phi = \sum_k \rho_k \mathbf{e}_k$.

In the preceding section, four important cases of this situation have occured. Let us review the corresponding *kernels* Φ .

Dirichlet kernel :

$$K_N(t) = \sum_{|k|\le N} \mathbf{e}_k(t) = \sin(2N + 1)\pi t \ / \ \sin\pi t .$$

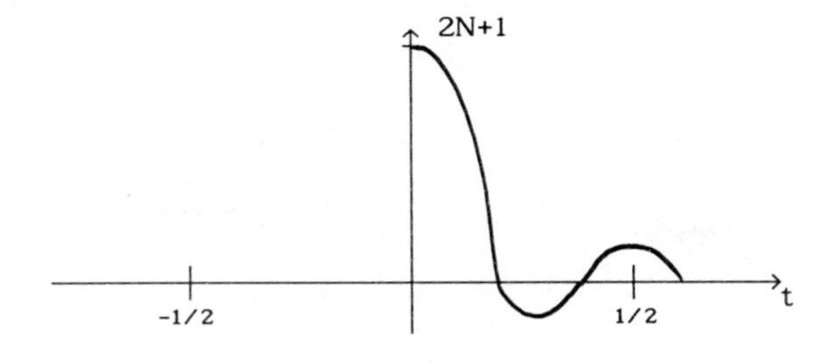

Fig.22.5

Fejer kernel :

$$F_N(t) = \sum_{|k|\le N} \frac{N - |k|}{N} \mathbf{e}_k(t) = \sin^2 N\pi t \,/\, N\sin^2 \pi t \ .$$

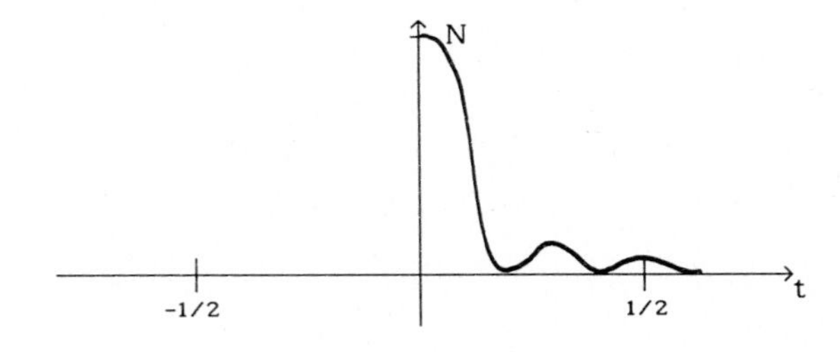

Fig.22.6

Poisson kernel :

$$P_r(t) = \sum_{k\in\mathbb{Z}} r^{|k|} \mathbf{e}_k(t) = (1-r^2)/(1-2r\cos 2\pi t + r^2) =$$

$$= \mathrm{Re}\ (e^{2\pi it} + r)/(e^{2\pi it} - r) \quad (r < 1) \ .$$

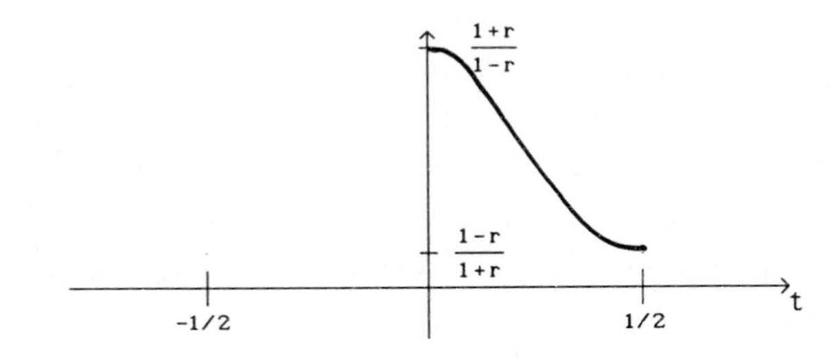

Fig.22.7

Heat equation kernel :

$$\Theta_\varepsilon(t) = \sum_{k\in\mathbb{Z}} \exp(-\varepsilon k^2)\ \mathbf{e}_k(t) \quad (\varepsilon > 0) \ .$$

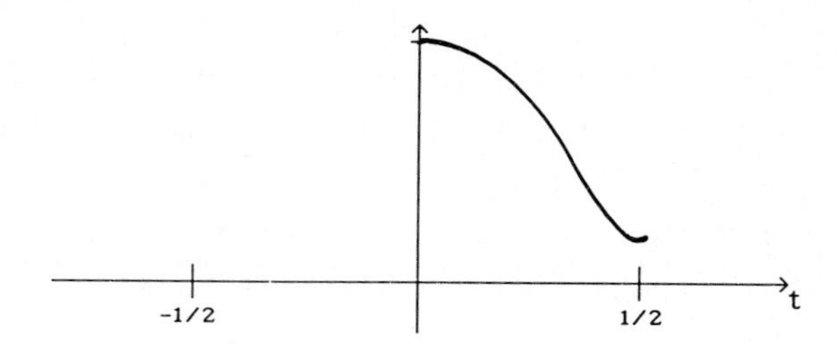

Fig.22.8

In all four cases, the attenuated sums are given by a convolution product.

22.6 A GLIMPSE OF THE FOURIER TRANSFORMATION

Let $f \in \mathcal{C}^2(\mathbb{R})$ be such that $f(x) = \mathcal{O}(x^{-2})$ for $|x| \longrightarrow \infty$ so that $\int_{-\infty}^{\infty} |f|\,dx < \infty$. For $T > 0$, consider the function $\varphi = \varphi_{f,T}$ defined by

$$\varphi(x) = \sum_{n\in\mathbb{Z}} f(x+nT).$$

Since $f(x+nT) = \mathcal{O}((x+nT)^{-2})$, we can find a uniform majoration

$$|f(x+nT)| \le M/n^2 \quad \text{for } n \longrightarrow \pm\infty \quad \text{(uniform for } x \in [0,T]).$$

Hence φ is a continuous, T-periodic function. Its Fourier coefficients can be computed by the usual formulas

$$c_k(\varphi) = \frac{1}{T}\int_0^T \varphi(x)e^{-2\pi ixk/T}dx = \frac{1}{T}\int_0^T \sum_{n\in\mathbb{Z}} f(x+nT)e^{-2\pi ixk/T}dx.$$

By uniform convergence of the sum, we can interchange $\int$ and $\sum$ and

$$c_k(\varphi) = \frac{1}{T}\sum_{n\in\mathbb{Z}} \int_{nT}^{(n+1)T} f(x)e^{-2\pi ixk/T}dx =$$

$$= \frac{1}{T}\int_{-\infty}^{\infty} f(x)e^{-2\pi ixk/T}dx.$$

Under the preceeding assumptions, the **Fourier transform of** f is the function $\hat{f}$ defined by

$$\hat{f}(\xi) = \int_{-\infty}^{\infty} f(x)e^{-2\pi ix\xi}dx.$$

In particular, we have found $c_k(\varphi) = T^{-1}\hat{f}(k/T)$. The Fourier series of φ is

$$\sum_{k\in\mathbb{Z}} T^{-1}\hat{f}(k/T)e^{2\pi ikx/T}.$$

If we can check that $\sum|\hat{f}(k/T)| < \infty$, then we shall have

$$\sum_{n\in\mathbb{Z}} f(x+nT) = \varphi(x) = \sum_{k\in\mathbb{Z}} T^{-1}\hat{f}(k/T)e^{2\pi ikx/T}.$$

In particular, for $T \longrightarrow \infty$, we see that

$$f(x) = \int_{-\infty}^{\infty} \hat{f}(\xi) e^{2\pi i \xi x} d\xi$$

since the right sum is a Riemann sum corresponding to the integral (in fact, this has to be justified more carefully since f does not vanish outside a bounded set...).

Theorem. Assume that $f \in \mathcal{C}^{\infty}(\mathbb{R})$ has fast decreasing derivatives :

$$\text{for all } \ell > 0,\ k > 0 \qquad |x^{\ell} f^{(k)}(x)| \longrightarrow 0 \qquad (|x| \longrightarrow \infty).$$

Then

▷ $\sum_{n\in\mathbb{Z}} f(n) = \sum_{k\in\mathbb{Z}} \hat{f}(k)$ **(Poisson summation formula),**

▷ $f(x) = \int_{-\infty}^{\infty} \hat{f}(\xi) e^{2\pi i \xi x} d\xi$ **(Fourier inversion formula).** □

Proof. As in the case of Fourier series,

$$|\hat{f}(\xi)| \le \int |f(x)| dx = \|f\|_1 \qquad (\xi \in \mathbb{R})$$

so that the Fourier transform of f is a *bounded* function. Moreover an integration by parts (it is valid!) shows that

$$(f')^{\wedge}(\xi) = -2\pi i \xi \cdot \hat{f}(\xi).$$

This proves

$$|\hat{f}(\xi)| \le |2\pi\xi|^{-1} |(f')^{\wedge}(\xi)| \le |2\pi\xi|^{-1} \|f'\|_1.$$

By induction, it is obvious that we can get

$$|\hat{f}(\xi)| \le C_k |\xi|^{-k} \|f^{(k)}\|_1 = O(|\xi|^{-k})$$

so that all integrals considered before the statement of the theorem converge fast. In particular, for T = 1,

$$c_k(\varphi) = c_k(\varphi_{f,1}) = \hat{f}(k)$$

whence $\sum_{k\in\mathbb{Z}} |c_k(\varphi)| < \infty$ and it is legitimate to write (cf.20.6)

$$\sum_{n\in\mathbb{Z}} f(x+n) = \varphi(x) = \sum_{k\in\mathbb{Z}} \hat{f}(k) e^{2\pi i k x}.$$

This equality for x = 0 gives the Poisson summation formula. All steps in the Fourier inversion formula are easily justified with our assumptions on f. ∎

The Fourier transform $\hat{f}$ of f is a *spectral density* of f and f can be recovered from this density by *continuous superposition* of the basic exponentials $x \longmapsto e^{2\pi i \xi x}$ using this density (*spectral synthesis*).

The idea just exploited has been to construct a T-periodic function φ_T from f. If f decreases fast, it is clear that $\varphi_T \longrightarrow f$ (when $T \longrightarrow \infty$) *uniformly on bounded sets*. Another —but similar— idea will be explained in 23.7.

CHAPTER 23
CONVERGENCE IN QUADRATIC MEAN

23.1 ORTHONORMAL SYSTEM OF EXPONENTIALS

Let us consider the scalar product

$$(f|g) = \int_0^1 \overline{f(t)}g(t)dt$$

on the space $C([0,1])$ of continuous functions $f : I = [0,1] \longrightarrow \mathbb{C}$. The completion of this space for the deduced metric

$$d(f,g) = d_2(f,g) = \|f - g\|_2 = (f-g|f-g)^{1/2}$$

is the Hilbert space $L^2(0,1)$ of square summable functions (classes of ..., cf.17.6 and 19.5).

Let us denote by $\mathbf{e}_k$ $(k \in \mathbb{Z})$ the continuous 1-periodic function

$$\mathbf{e}_k(t) = e^{2\pi ikt}$$

(or its restriction to the interval I). We have

$$(\mathbf{e}_k|\mathbf{e}_\ell) = \int_0^1 \mathbf{e}_{\ell-k}(t)dt = \delta_{k\ell} \qquad \text{(Kronecker symbol).}$$

This shows that the system $(\mathbf{e}_k)_{k\in\mathbb{Z}} \subset C(I)$ is an *orthonormal system* (it is also an orthonormal system in $L^2(I)$).

The Fourier sequence of a 1-periodic continuous function f has been defined by

$$c_k(f) = c_o(\mathbf{e}_{-k}\cdot f) = \text{Mean value of } \bar{\mathbf{e}}_k\cdot f\ .$$

We now recognize a scalar product

$$c_k(f) = \int_0^1 \overline{\mathbf{e}_k(t)}f(t)dt = (\mathbf{e}_k|f).$$

In particular, the Fourier sequence $(c_k(f))_{k\in\mathbb{Z}}$ can be *defined* for any $f \in L^2(0,1)$ by the same formulas. The Cauchy-Schwarz inequality now gives

$$|c_k(f)| \le \|\mathbf{e}_k\|_2\cdot\|f\|_2 = \|f\|_2\ .$$

Since

$$\|f\|_1 = \int_0^1 |f(t)|dt = (1|\ |f|\ ) \le \|1\|_2\cdot\|f\|_2 = \|f\|_2$$

we have obtained the following refinement of 20.4.

Proposition. For $f \in C([0,1])$, we have the estimates

$$|c_k(f)| \le \|f\|_1 \le \|f\|_2 \le \|f\|_\infty \qquad (k \in \mathbb{Z})$$

■

The inequality with the quadratic norm can still be improved if we use the Bessel inequality in place of the Cauchy-Schwarz one (cf.17.5). With respect to the orthonormal system $(\mathbf{e}_k)$, this inequality immediately gives

$$\sum_{k\in\mathbb{Z}} |c_k(f)|^2 \le \|f\|_2^2 .$$

We are going to show in 23.3 that *equality holds* in the preceding inequality. Let us only observe now that the result obtained implies that the Fourier sequence always tends to zero (so far, we only have known that it is bounded).

23.2 CONVERGENCE IN QUADRATIC MEAN

Theorem. For any $f \in L^2(0,1)$ (and in particular for any continuous $f : [0,1] \longrightarrow \mathbb{C}$), the Fourier series of f converges to f *in quadratic mean* :

$$\left\| f - \sum_{-m}^{n} c_k(f)\mathbf{e}_k \right\|_2 \longrightarrow 0 \quad (m \text{ and } n \longrightarrow \infty). \qquad \square$$

Proof. This proof is based on the best approximation theorem 17.4, valid in any inner product space. Take $f \in L^2(0,1)$ and $\varepsilon > 0$. By definition, $C([0,1])$ is dense in $L^2(0,1)$, and there exists a continuous f_1 with

$$\|f - f_1\|_2 \le \varepsilon.$$

Let us construct a continuous f_2 with

$$\|f_1 - f_2\|_2 \le \varepsilon \quad \text{and} \quad f_2(0) = f_2(1) .$$

This is easily done as follows : replace the values of f_1 between 0 and some small $\delta > 0$ by an affine linear function interpolating $f_1(1)$ and $f_1(\delta)$ (cf.Fig.23.1).

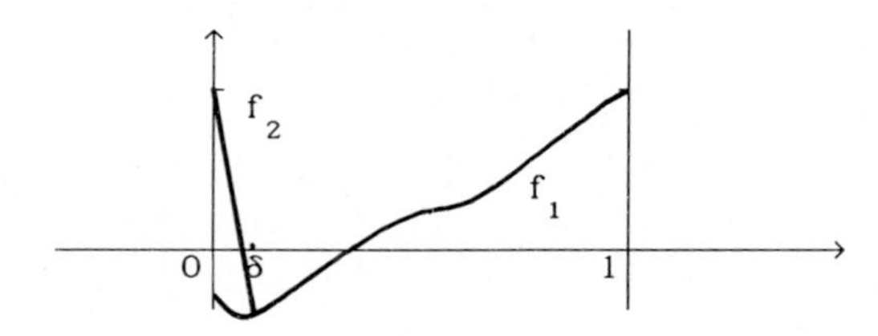

Fig.23.1

Since f_1 and f_2 only differ on $[0,\delta[$ and $|f_1 - f_2| \le 2\mathrm{Sup}|f_1| = M$, it is certainly possible to choose δ small enough to have

$$\|f_1 - f_2\|_2 \le \varepsilon .$$

Now we can extend f_2 continuously by 1-periodicity on the whole real line $\mathbb{R}$. By the corollary to the complex form of the Stone-Weierstrass theorem 15.6, there exists a trigonometric polynomial

$$f_3(t) = p(t) = \sum_{k\in I} a_k e^{2\pi ikt} \quad \text{(for some finite } I \subset \mathbb{Z}\text{)}$$

and uniformly near f_2 to the order ε

$$\|f_2 - f_3\|_\infty \le \varepsilon \ .$$

A fortiori

$$\|f_2 - f_3\|_2 \le \|f_2 - f_3\|_\infty \le \varepsilon$$

(cf.Proposition in 23.1). Gathering all inequalities

$$\|f - f_3\|_2 \le \|f - f_1\|_2 + \|f_1 - f_2\|_2 + \|f_2 - f_3\|_2 \le 3\varepsilon \ .$$

Now, for any finite part $J \supset I$ in $\mathbb{Z}$, the best approximation of f (in quadratic mean) by a linear combination of the $\mathbf{e}_k$ $(k \in J)$ is given by $\sum_{k\in J} c_k(f)\mathbf{e}_k$. Explicitely,

$$\Big\|f - \sum_{k\in J} c_k(f)\mathbf{e}_k\Big\|_2 \le \Big\|f - \sum_{k\in J} \gamma_k \mathbf{e}_k\Big\|_2$$

for *all* families $(\gamma_k)_{k\in J}$. In particular, if we take for (γ_k) the coefficients a_k of f_3 (for $k \in I$ and $\gamma_k = 0$ for $k \in J - I$) we get

$$\Big\|f - \sum_{k\in J} c_k(f)\mathbf{e}_k\Big\|_2 \le \|f - f_3\|_2 \le 3\varepsilon$$

for *all finite sets* J *containing* I (in $\mathbb{Z}$). This proves convergence in quadratic mean of the series $\sum_k c_k(f)\mathbf{e}_k$ towards f. ∎

23.3 FISHER-RIESZ ISOMORPHISM

Theorem. Let f and $g \in L^2(0,1)$. Then

$$(f|g) = \sum_{k\in\mathbb{Z}} \overline{c_k(f)}\cdot c_k(g) \ . \quad \square$$

Proof. To simplify notations, let us abbreviate

$$a_k = c_k(f) \ , \ b_k = c_k(g) \ .$$

For any finite set of integers $J \subset \mathbb{Z}$, the scalar product of

$$f - \sum_J a_k\mathbf{e}_k \quad \text{and} \quad g - \sum_J b_k\mathbf{e}_k$$

is easily determined by distributivity. We find

$$\begin{aligned}
&(f - \textstyle\sum_J a_k\mathbf{e}_k \mid g - \sum_J b_\ell\mathbf{e}_\ell) = \\
&= (f|g) - \textstyle\sum_J b_\ell(f|\mathbf{e}_\ell) - \sum_J \bar{a}_k(\mathbf{e}_k|g) + \sum\sum \bar{a}_k b_\ell\, \delta_{k\ell} = \\
&= (f|g) - \textstyle\sum_J b_\ell\bar{a}_\ell - \sum_J \bar{a}_k b_k + \sum_J \bar{a}_k b_k = \\
&= (f|g) - \textstyle\sum_J b_\ell\bar{a}_\ell \ .
\end{aligned}$$

In particular, we see that

$$\Big|\, (f|g) - \sum_J b_\ell\bar{a}_\ell \,\Big| = \Big|(f - \sum_J a_k\mathbf{e}_k \mid g - \sum_J b_\ell\mathbf{e}_\ell)\Big| \le$$

$$\le \Big\|f - \sum_J a_k\mathbf{e}_k\Big\|\cdot\Big\|g - \sum_J b_\ell\mathbf{e}_\ell\Big\|$$

is arbitrarily small if J is large enough (by 23.2). This proves the theorem. ∎

Corollary (Parseval's identity). For $f \in L^2(0,1)$,

$$\|f\|^2 = \sum_{k\in\mathbb{Z}} |c_k(f)|^2 \ . \quad ∎$$

The above theorem shows that the Fourier map $f \longmapsto (c_k(f))_{\mathbb{Z}}$ defines a linear isometry

$$\mathcal{F} : L^2(0,1) \longrightarrow \ell^2(\mathbb{Z})$$

between the space of square summable *functions* and the space of square summable sequences $(a_k)_{\mathbb{Z}} \subset \mathbb{C}$. This mapping is also onto since for any $(a_k)_{\mathbb{Z}} \subset \mathbb{C}$ satisfying $\sum_{\mathbb{Z}} |a_k|^2 < \infty$, the series $\sum_{\mathbb{Z}} a_k \mathbf{e}_k$ converges in quadratic mean :the space $L^2(0,1)$ is complete and the Cauchy criterium applies since

$$\left|\sum_p^q a_k \mathbf{e}_k\right|^2 = \sum_p^q |a_k|^2 \longrightarrow 0 \quad \text{for p and q} \longrightarrow \infty .$$

In other words, the composite of the two linear maps

$$\begin{array}{ccccc} L^2(0,1) & \longrightarrow & \ell^2(\mathbb{Z}) & \longrightarrow & L^2(0,1) \\ f & \longmapsto & (c_k(f))_{\mathbb{Z}} & & \\ & & (a_k)_{\mathbb{Z}} & \longmapsto & \sum_{\mathbb{Z}} a_k \mathbf{e}_k \end{array}$$

is the identity (on $L^2(0,1)$). This result is known as the **Fisher-Riesz theorem**.

23.4 BACK TO $\sum_{k\geq 1}$ sinkt /k AND $\sum_{k\geq 1}$ coskt /k

Let us group the two proposed series $\sum_{k\geq 1} {}^{\sin}_{\cos})kt \,/\, k$ in a complex one

$$\sum_{k\geq 1} \cos kt \,/\, k + i \sum_{k\geq 1} \sin kt \,/\, k = \sum_{k\geq 1} e^{ikt}/k .$$

To *guess* the sum, let us remember that

$$\sum_{k\geq 1} z^k/k = - \mathrm{Log}(1-z) \quad \text{for } |z| < 1 .$$

If we are lucky (!), this result will still be true for $z = e^{it}$ $(\neq 1)$. We thus expect that

$$\sum_{k\geq 1} e^{ikt}/k = - \mathrm{Log}(1 - e^{it}) \quad \text{for} \quad t \notin 2\pi i\mathbb{Z} .$$

Separating real and imaginary parts,

$$\sum_{k\geq 1} \cos kt/k = - \log|1 - e^{it}| = \log 1/|1 - e^{it}| ,$$

$$\sum_{k\geq 1} \sin kt/k = - \mathrm{Arg}(1 - e^{it}) = \mathrm{Arg}(1 - e^{-it}) =$$

$$= \mathrm{Arg}\,[e^{-it/2} 2i\sin(t/2)] = \mathrm{Arg}\,[ie^{-it/2}] = \frac{\pi}{2} - \frac{t}{2} .$$

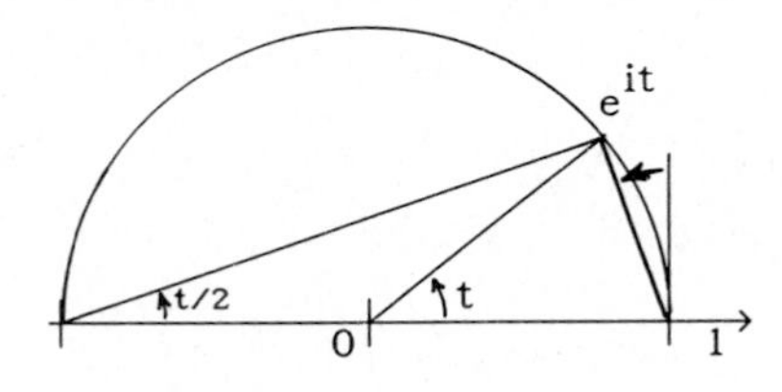

Fig.23.1

This is indeed the case. Both functions are in L^2 : write the first one as

$$-\log|1 - e^{it}| = (-1/2)\log\,|e^{it} - 1|^2 =$$

$$= (-1/2)\log\left[(\cos t - 1)^2 + \sin^2 t\right]$$

$$= (-1/2)\log(t^2 + \mathcal{O}(t^3)) = -\log t + \mathcal{O}(1)$$

for $t \to 0$ (here $\mathcal{O}(1)$ represents a continuous, bounded function). In particular, all powers of $\log|1 - e^{it}|$ are integrable. Their Fourier coefficients are easily determined : for $1 > r_n \to 1$, $t \mapsto \log|1 - r_n e^{it}|$ makes up a Cauchy sequence in $L^1(0,2\pi)$ and it is enough to pass to the limit for the respective Fourier coefficients.

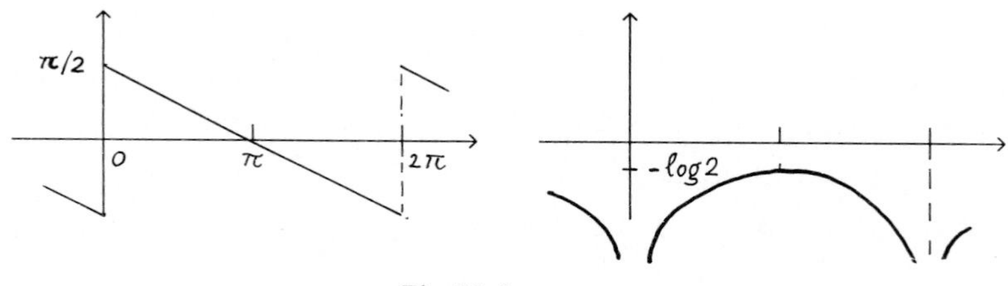

Fig.23.2

It may be interesting to plot a few partial sums of the Fourier series

$$\sin t + \frac{1}{2}\sin 2t + \frac{1}{3}\sin 3t + \ldots$$

and to see how convergence occurs. The next figure exhibits a typical phenomenon near the origin : the partial sums always go much beyond the value $\pi/2$. This is the **Gibbs phenomenon**.

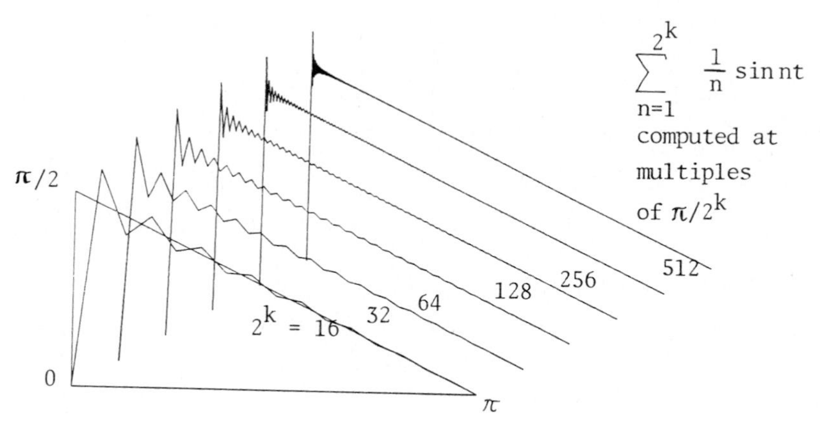

Fig.23.3

23.5 ANOTHER UNIFORM CONVERGENCE RESULT

The first uniform convergence result of 20.6 was valid for $\mathcal{C}^2$ periodic functions. As was already observed there, the same result holds for periodic functions f with $\sum_k |c_k(f)|^2 < \infty$. This last condition is often verified (e.g. it was verified in the application given in 21.1) even when f is not $\mathcal{C}^2$. We can now give an explanation of this feature.

When the continuous 1-periodic function f has a square integrable derivative $f' \in L^2(0,1)$, its Fourier sequence $(c_k(f'))$ is square summable. Since

$$c_k(f) = (ik)^{-1} c_k(f')$$

the Fourier sequence of f appears as product of the two square summable sequences

$$(1/ik)_{k\in\mathbb{Z}} \in \ell^2(\mathbb{Z}) \quad \text{and} \quad (c_k(f')) \in \ell^2(\mathbb{Z}).$$

As such, this defines a summable sequence

$$(c_k(f))_{k\in\mathbb{Z}} \in \ell^1(\mathbb{Z}), \text{ i.e. } \sum_{k\in\mathbb{Z}} |c_k(f)| < \infty.$$

Namely, the conditions of the first uniform convergence theorem (improved version at the end of 20.6) are satisfied. Let us state this result in a simple case (avoiding the L^2-theory).

Theorem. Let f be a continuous 2π-periodic function on $\mathbb{R}$. Assume that f is a primitive of a regulated function (i.e. f' is regulated). Then, the Fourier series of f converges normally and uniformly to f. ∎

23.6 INTERPRETATION OF $\sum k|c_k|^2$

Let $f : \mathbb{R} \longrightarrow \mathbb{C}$ be a continuous 2π-periodic function. It will be considered here as a parameterized closed curve $\gamma \subset \mathbb{C}$ in the complex plane. Let

$$c_k = c_k(f) = \frac{1}{2\pi} \int_0^{2\pi} f(t) e^{-ikt} dt$$

denote the Fourier coefficients of f. We shall also assume that f is piecewise differentiable with $f' \in L^2(0,2\pi)$ (e.g. f' makes finitely many jumps). In this case,

$$c_k(f') = \frac{1}{2\pi} \int_0^{2\pi} f'(t) e^{-ikt} dt = ikc_k .$$

The scalar product of f and f' can be computed by means of the Fisher-Riesz theorem 23.3

$$(f|f') = \sum_{k\in\mathbb{Z}} \bar{c}_k\, ikc_k = \sum_{k\in\mathbb{Z}} ik|c_k|^2.$$

This proves that

$$\sum_{k\in\mathbb{Z}} k|c_k|^2 = \frac{1}{i}(f|f') = \frac{1}{2\pi i}\int_0^{2\pi} \overline{f(t)}f'(t)dt =$$

$$= \frac{1}{2\pi i}\oint_\gamma \bar{z}\, dz\ .$$

In case f is one-to-one into, the closed curve γ is a *simple* closed curve and $\gamma = \partial\Pi$ is the oriented boundary of a piece Π (the Jordan closed curve theorem might be needed here if the full generality of the result was needed...) Assuming that the orientation of γ is the positive one (so that Π inherits the canonical orientation of $\mathbb{R}^2 \cong \mathbb{C}$), Stokes' thorem gives

$$\sum_{k\in\mathbb{Z}} k|c_k|^2 = \frac{1}{2\pi i}\oint_\gamma \bar{z}\, dz = \frac{1}{2\pi i}\iint_\Pi d\bar{z}\wedge dz\ .$$

But now,

$$d\bar{z}\wedge dz = (dx - idy)\wedge(dx + idy) = 2idx\wedge dy$$

and

$$\pi\sum_{k\in\mathbb{Z}} k|c_k|^2 = \frac{1}{2i}\iint_\Pi d\bar{z}\wedge dz = \iint_\Pi dx\wedge dy = \mathrm{Area}(\Pi)\ .$$

It is interesting to observe that in the summation, the coefficient $c_o = c_o(f)$ does not appear : this is reasonable since it represents the center of gravity of the curve (at least if the parameter is the special one).

The interested reader will be able to apply this area formula to the Fourier series of the regular polygons (cf.21.4).

23.7 THEOREM OF KOTELNIKOV AND SHANNON

Let $I = [-T/2,T/2]$ and $f \in L^2(I)$ be a square summable function. Extend f by 0 outside the interval I and compute the Fourier transform of f. According to the definition given in 22.6

$$\hat{f}(\xi) = \int_{-\infty}^{\infty} f(t)e^{-2\pi i\xi t}dt = \int_{-T/2}^{T/2} f(t)e^{-2\pi i\xi t}dt\ .$$

Hence

$$\hat{f}(\xi)/T = \frac{1}{T}\int_{-T/2}^{T/2} e^{-2\pi i\xi t}f(t)dt = (e^{2\pi i\xi t}|f)_I = \sum_{k\in\mathbb{Z}} \bar{\gamma}_k c_k$$

(Fisher-Riesz isomorphism 23.3) where

$$\gamma_k = T^{-1}\int_I e^{-2\pi ikt/T}\ e^{2\pi i\xi t}dt = T^{-1}\left[\frac{e^{2\pi i(\xi-k/T)t}}{2\pi i(\xi-k/T)}\right]_{-T/2}^{T/2} =$$

$$= \frac{\sin \pi T(\xi-k/T)}{\pi T(\xi - k/T)} \quad (\text{if } \xi \notin T^{-1}\mathbb{Z}) .$$

and

$$c_k = T^{-1}\int_I e^{-2\pi ikt/T}\ f(t)dt = T^{-1}\int_{-\infty}^{\infty} e^{-2\pi ikt/T}f(t)dt = \hat{f}(k/T)/T.$$

Altogether, this gives

$$\hat{f}(\xi) = \sum_{k\in\mathbb{Z}} \hat{f}(k/T)\ \frac{\sin \pi T(\xi-k/T)}{\pi T(\xi - k/T)} .$$

Here, it is understood that

$$\frac{\sin \pi T(\xi-k/T)}{\pi T(\xi - k/T)} = \delta_{k\ell} \quad \text{for } \xi = \ell/T \ (\ell \in \mathbb{Z}).$$

This shows that the Fourier transform of f, initially defined by a continuous integral, can also be computed by means of a series (under our assumptions). This observation applied to $\hat{f}$ instead of f (when applicable...) leads to an interesting result.

Theorem (Kotelnikov-Shannon). Let $f \in \mathcal{C}^{\infty}(\mathbb{R})$ have fast decreasing derivatives of all orders. Assume that $\hat{f}$ vanishes outside a bounded interval $I = [-T/2,T/2]$. Then,

$$f(\xi) = \sum_{k\in\mathbb{Z}} \frac{\sin \pi T(\xi-k/T)}{\pi T(\xi - k/T)}\cdot f(k/T) . \qquad \blacksquare$$

This result should be viewed as an *interpolation* result: if f has a compact spectrum in $I = [-T/2,T/2]$ (i.e. $\hat{f}$ vanishes outside I), then the values $f(\xi)$ ($\xi \in \mathbb{R}$) are *entirely determined by the special values of the* $f(k/T)$ ($k \in \mathbb{Z}$). In other words, if f is *synthetized* by a certain procedure which can only produce frequences in a range of a certain width T, f is completely determined when it is measured on a *sample grid of mesh* 1/T.

CHAPTER 24
HISTORICAL PERSPECTIVE

24.1 EULER AND PERIODIC FUNCTIONS

It is interesting to note that already in the 18th.century, Euler (1707–1783) had been led to study trigonometrical series.

Around 1750, Euler wished to solve the functional equation

$$f(x + 1) = f(x)$$

characterizing periodic functions (in order to pursue his studies in astronomy) He tried to solve the equation in general and find *all* periodic functions f (with period 1). For this purpose, he used the Taylor's formula (without bothering to mention the analyticity assumption !)

$$f(x + 1) = f(x) + f'(x) + f''(x)/2! + \dots$$

and thus got the *differential equation* (!)

$$f' + f''/2! + f'''/3! + \dots = 0 .$$

As he had also published an article on constant coefficients linear differential equations, he used the same technique here. This means that he looked for special solutions $f(x) = e^{\lambda x}$ where the parameter λ in the exponent has to satisfy the characteristic equation

$$\lambda + \lambda^2/2! + \lambda^3/3! + \dots = e^{\lambda} - 1 = 0 .$$

The solutions of this equation are the complex numbers $\lambda = k\cdot 2\pi i$ ($k \in \mathbb{Z}$). Did he already know the relation $e^{i\pi} = -1$? He had definitely discovered it in 1777 when it appears in writing for the first time. But if he knew it earlier, in his 1750-53 papers he prefered to remain in the real domain and find another way of proceeding. He writes

$$e^{\lambda} = \lim_{n\to\infty} (1 + \lambda/n)^n$$

and starts by looking at the roots of the polynomial equations

$$(1 + \lambda/n)^n - 1 = 0$$

(hoping that these roots will converge for $n \longrightarrow \infty$ to roots of the proposed equation $e^{\lambda} - 1 = 0$). To simplify, introduce $y = 1 + \lambda/n$ and consider the *cyclotomic equation*

$$y^n - 1 = 0 .$$

Euler also knew that the polynomial $y^n - 1$ is divisible by $y - 1$ and by the *real* quadratic polynomials

$$y^2 - 2y\cos(2\pi k/n) + 1 \quad (= (y - e^{2\pi ik/n})(y - e^{-2\pi ik/n}))$$

(the subject of an earlier joint paper with Cotes). By the duplication formula for the cosine function

$$\cos 2\pi k/n = 1 - 2\sin^2 \pi k/n$$

the quadratic equations reduce to

$$(y - 1)^2 + 4y\sin^2 \pi k/n = 0$$

i.e. to

$$\lambda^2/n^2 + 4(1 + \lambda/n)\sin^2 \pi k/n = 0$$

or

$$\frac{\lambda^2}{4n^2\sin^2 \pi k/n} + 1 + \lambda/n = 0.$$

When $n \longrightarrow \infty$, $\sin^2 \pi k/n \cong (\pi k/n)^2$ and the equation is nearly

$$\frac{\lambda^2}{4\pi^2 k^2} + 1 = 0.$$

But this is the characteristic equation corresponding to the linear differential equation of the second order

$$f'' + 4\pi^2 k^2 f = 0$$

having the general solution

$$a\cos 2\pi kx + b\sin 2\pi kx \quad (a = a_k \text{ , } b = b_k).$$

From this, Euler expects that the general solution of

$$f' + f''/2! + \ldots = 0$$

is a *general infinite linear combination* of these solutions. This means that *any* periodic function of period 1, namely any solution of the functional equation $f(x + 1) = f(x)$ is a series

$$f(x) = \sum_{k\in\mathbb{Z}} (a_k\cos 2\pi kx + b_k\sin 2\pi kx).$$

For some reason, he is looking for special solutions with $f(0) = 1$ and writes them slightly differently

$$f(x) = 1 + \sum_{k\geq 1} \left\{a_k\sin 2\pi kx + A_k(\cos 2\pi kx - 1)\right\} .$$

In another article, Euler gives the following expansions

$$\sum_{n\geq 0} a^n\cos nx = \frac{1 - a\cos x}{1 - 2a\cos x + a^2} ,$$

$$\sum_{n\geq 0} a^n\sin nx = \frac{a\sin x}{1 - 2a\cos x + a^2} .$$

Letting $a = \pm 1$ in the first one he obtains

$$1/2 = 1 + \cos x + \cos 2x + \ldots$$

$$1/2 = 1 - \cos x + \cos 2x - \ldots$$

(in fact both series diverge...). Integrating, he gets

$$(\pi - x)/2 = \sin x + \frac{1}{2}\sin 2x + \frac{1}{3}\sin 3x + \ldots$$

$$x/2 = \sin x - \frac{1}{2}\sin 2x + \frac{1}{3}\sin 3x - \dots$$

$$x^2/4 - \pi^2/12 = -\cos x + \frac{1}{4}\cos 2x - \frac{1}{9}\cos 3x + \dots$$

Of course, Euler does not say that the first expansion is only valid for $0 < x < \pi$ and the next for $-\pi < x < \pi$. It seems that he even *forgets* the periodicity and thinks of a representation valid for *all* x ! He works so formally that he even writes (1739)

$$\sum_{-\infty}^{\infty} x^n = 0$$

probably because

$$\sum_{n\geq 0} x^n = 1/(1 - x) \quad (\text{for } |x| < 1),$$
$$\sum_{n<0} x^n = 1/(x - 1) \quad (\text{for } |x| > 1).$$

Nowadays, it is possible to say that these series converge *in the distribution sense* and $\sum_{-\infty}^{\infty} x^n$ represents the Dirac distribution at the point $x = 1$.

24.2 FOURIER AND THE HEAT EQUATION

In a famous work presented to the Academy of Sciences of Paris in 1807 (in revised form in 1811 and finally published in the Traité Analytique de la Chaleur in 1822), Fourier claims that *any periodic function* can be expanded in a trigonometric series.

His starting point is the **heat equation**

$$\partial^2 T/\partial x^2 = k^2 \partial T/\partial t$$

for a function of two variables $T = T(x,t)$ satisfying

▷ $T(0,t) = T(\ell,t) = 0$ for all $t \geq 0$ (*limit condition*),

▷ $T(x,0) = f(x)$ (*initial condition*).

(We suppose that the initial condition is given by a function f which respects the limit condition : $f(0) = f(\ell) = 0$.)

Fourier looks for special solutions of the partial differential equation (not satisfying the limit and initial conditions !) in the form

$$T(x,t) = \varphi(x)\psi(t) \quad (\textit{separation of the variables}).$$

For these, we must have

$$\varphi''/(k^2\varphi) = \psi'/\psi = \text{const.} = -\lambda^2$$

(the first term is independent of t and equal to the second which is independent of x : both must be independent of x and t, thus constant). Hence

$$\varphi'' + k^2\lambda^2\varphi = 0$$

and $\varphi(x) = a\sin k\lambda x$ (+ $b\cos k\lambda x$). At this point, we try to satisfy the limit condition, imposing $\varphi(0) = \varphi(\ell) = 0$. This will only be possible with ($b = 0$ and) $\lambda = \lambda_n = n\pi/(\ell k)$ ($n \in \mathbb{Z}$). (A priori, the constant $-\lambda^2$ is not necessarily negative and we could take λ complex, e.g. purely imaginary if we wanted $-\lambda^2 > 0$; but the *periodic* limit condition can only be satisfied by periodic functions, imposing $-\lambda^2 < 0$, i.e. λ real.) To simplify computations, let us take $\ell = \pi$. Then, $\lambda_n = n/k$ and $\varphi_n(x) = \sin nx$.

The other differential equation is

$$\psi' = -\lambda_n^2\psi = -(n^2/k^2)\psi$$

whence

$$\psi_n(t) = e^{-(n^2/k^2)t}.$$

Fourier expects thus that the general solution of the heat equation (satisfying the limit condition) is *an infinite linear combination*

$$T(x,t) = \sum_{n\geq 1} c_n e^{-(n^2/k^2)t}\sin nx .$$

The initial condition imposes

$$f(x) = T(x,0) = \sum_{n\geq 1} c_n \sin nx$$

and the problem will be solved if we can determine the coefficients c_n of this expansion of f.

24.3 FOURIER COEFFICIENTS ACCORDING TO FOURIER

The problem is now to find the coefficients c_n of f in a trigonometric expansion $f(x) = \sum_{n\geq 1} c_n \sin nx$ (and prove that such an expansion *always exists!*).

In order to do this, Fourier substitutes the Taylor expansion of $\sin nx$

$$\sin nx = \sum_{\nu\geq 1}(-1)^{\nu-1} \frac{n^{2\nu-1}x^{2\nu-1}}{(2\nu-1)!}$$

in the expansion of f (he also assumes *analyticity* of f...) and compares Taylor coefficients, permuting summations as needed

$$f(x) = \sum_{\nu\geq 1} \frac{(-1)^{\nu-1}}{(2\nu-1)!} \left(\sum_{n\geq 1} n^{2\nu-1}c_n\right) x^{2\nu-1} = \sum_{m\geq 1} f^{(m)}(0)\, x^m/m!.$$

One finds

$$(-1)^{\nu-1}f^{(2\nu-1)}(0) = \sum_{n\geq 1} n^{2\nu-1}c_n .$$

This is a linear system for the unknown numbers c_n which looks like an infinite Cramer system. Fourier hopes that $c_n \longrightarrow 0$ and solves approximate subsystems, ending up in the integral formula. To illustrate his method, let us treat an example (which he also considers). Let us try to find coefficients a,b,c,... in order to have

$$1 = a\cos x + b\cos 3x + c\cos 5x + \dots \quad .$$

Expanding the cosine functions in Taylor series at the origin and equating coefficients of like powers, we find the equations

$$\begin{cases} a + b + c + \dots = 1 \\ a + 3^2 b + 5^2 c + \dots = 0 \\ a + 3^4 b + 5^4 c + \dots = 0 \end{cases} , \dots$$

Forgetting the "...", we can solve this system of three equations in three unknowns a, b and c by elimination (multiply the first two equations by 5^2 and subtract the next

$$\begin{cases} (5^2-1)a + (5^2-3^2)b = 5^2 \\ (5^2-1)a + 3^2(5^2-3^2)b = 0 . \end{cases}$$

Eliminate now b by multiplying the first equation by 3^2 and subtracting the second

$$(3^2-1)(5^2-1)a = 3^2 5^2 .$$

We find

$$a = \frac{3^2}{3^2-1} \frac{5^2}{5^2-1} = \frac{3\cdot 3}{2\cdot 4} \frac{5\cdot 5}{4\cdot 6} .$$

It is clear that if we had taken an extra equation with the next unknown (say d), we would have found a value for a equal to the preceding one multiplied by $7^2/(7^2-1) = \frac{7\cdot 7}{6\cdot 8}$. Going to the limit of infinitely many equations and unknowns, we guess that the solutions are given by infinite products

$$a = \frac{3\cdot 3}{2\cdot 4} \frac{5\cdot 5}{4\cdot 6} \frac{7\cdot 7}{6\cdot 8} \frac{9\cdot 9}{8\cdot 10} \cdots .$$

But Wallis' product formula tells us that this infinite product (converges and) is $4/\pi$! One finds similarly that

$$b = -a/3 , \ c = a/5 , \ \dots$$

and finally,

$$\pi/4 = \cos x - \frac{1}{3}\cos 3x + \frac{1}{5}\cos 5x - \dots \quad .$$

The right hand side indeed converges to the indicated limit when $|x| < \pi/2$ and in fact, the sum of the cosine series is the function sketched in the following figure.

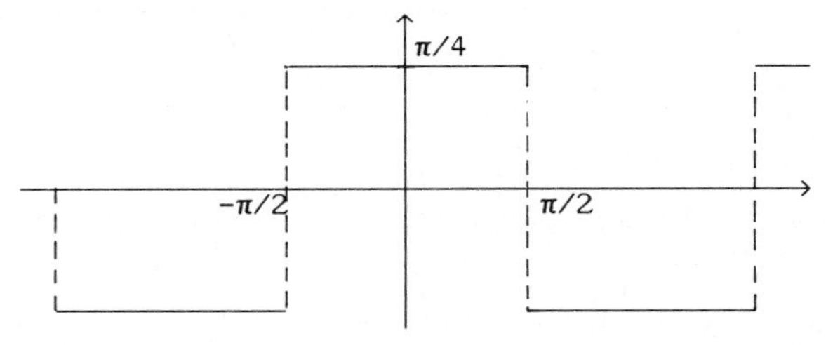

Fig.24.1

With Fourier, let us now turn to the general case of an odd function f

$$a\sin x + b\sin 2x + c\sin 3x + \ldots = Ax + Bx^3/3! + Cx^5/5! + \ldots$$

with $A = f'(0)$, $B = f'''(0)$, $C = f^{(5)}(0)$, ... Proceeding as before, identifying Taylor coefficients at the origin, we now obtain

$$\begin{array}{lll|c|c|} a + 2b + 3c + \ldots & = A & & 3^2 & \\ a + 2^3 b + 3^3 c + \ldots & = -B & & -1 & 3^2 \\ a + 2^5 b + 3^5 c + \ldots & = C & & & -1 \end{array} .$$

Forgetting the "..." and eliminating c according to the indicated scheme we find

$$\begin{array}{ll|c|} (3^2-1)a + 2(3^2-2^2)b = 3^2A + B & & 2^2 \\ (3^2-1)a + 2^3(3^2-2^2)b = -3^2B - C & & -1 \end{array} .$$

Eliminating b as indicated we get

$$(2^2-1)(3^2-1)a = 2^2 3^2 A + (2^2 + 3^2)B + C$$

or

$$2^{-2}(2^2-1)3^{-2}(3^2-1)a = A + (2^{-2} + 3^{-2})B + 2^{-2}3^{-2}C.$$

Looking at bigger square systems, we would find the limit

$$\frac{2^2-1}{2^2}\ \frac{3^2-1}{3^2}\ \frac{\ldots}{\ldots}\ a =$$

$$= A + (2^{-2} + 3^{-2} + \ldots)B + (2^{-2}3^{-2} + 2^{-2}4^{-2} + \ldots)C + \ldots$$

After cancellations, the coefficient of a is recognized to be 1/2. On the other hand, we need to compute the infinite sums : Fourier uses the Eulerian product expansion of sinx

$$\sin x = \begin{cases} x - x^3/3! + x^5/5! - \ldots \\ x(1 - x^2/\pi^2)(1 - x^2/(2\pi)^2)(\ldots) \end{cases}$$

Identifying coefficients of x^3 he gets

$$1/3! = \textstyle\sum_{k\geq 1} k^{-2}\pi^{-2} \qquad \text{hence} \qquad 2^{-2} + 3^{-2} + \ldots = \pi^2/3! - 1.$$

Similarly with the coefficients of x^5 : $1/5! = \sum_{k>\ell\geq 1} k^{-2}\ell^{-2}\,\pi^{-4}$ hence

$$\begin{aligned} \pi^4/5! &= 1^{-2} \textstyle\sum_{k>\ell=1} k^{-2} + \sum_{k>\ell>1} k^{-2}\ell^{-2} \\ &= 1^{-2}\,(\pi^2/3! - 1^{-2}) + \textstyle\sum_{k>\ell>1} k^{-2}\ell^{-2} , \end{aligned}$$

and

$$\textstyle\sum_{k>\ell>1} k^{-2}\ell^{-2} = \pi^4/5! - \pi^2/3!\cdot 1^{-2} + 1^{-4} .$$

Multiplying throughout by π in the equation giving a

$$\pi a/2 = f'(0)\pi + (\pi^3/3! - \pi\cdot 1^{-2})f'''(0) +$$

$$+ (\pi^5/5! - \pi^3/3!1^{-2} + \pi\cdot 1^{-4})f^{(5)}(0) + \ldots =$$

$$= f(\pi) - 1^{-2}f''(\pi) + 1^{-4}f^{(4)}(\pi) - \ldots$$

Finally, the successive coefficients are found to be

$$\pi a/2 = f(\pi) - f''(\pi) + f''''(\pi) - \ldots ,$$

$$\pi b/2 = -\frac{1}{2}\,(f(\pi) - 2^{-2}f''(\pi) + 2^{-4}f''''(\pi) - \ldots) ,$$

$$\pi c/2 = \frac{1}{3}\,((f(\pi) - 3^{-2}f''(\pi) + 3^{-4}f''''(\pi) - \ldots) .$$

One still has to compute the sums

$$f_n(\pi) = f(\pi) - n^{-2}f''(\pi) + n^{-4}f''''(\pi) - \ldots$$

Define $f_n(x)$ correspondingly by

$$f_n(x) = f(x) - n^{-2}f''(x) + n^{-4}f''''(x) - \ldots$$

so that $n^{-2}f_n'' + f_n = f$ and f_n is a solution of the differential equation

$$y'' + n^2y = n^2f .$$

This linear differential equation is easily solved by the method of variation of constants (once the general solution of its homogeneous part is written down). Since f is odd, $f_n(0) = 0$ and

$$f_n(x) = n\sin nx \int_0^x f(t)\cos nt\,dt - n\cos nx \int_0^x f(t)\sin nt\,dt$$

$$+ A\sin nx .$$

In particular

$$f_n(\pi) = -n(-1)^n \int_0^\pi f(t)\sin nt\,dt.$$

Recall that we had discovered that the coefficient of sinnx in the expansion of $(\pi/2)f$ is $(-1)^{n+1}f_n(\pi)/n$. We have now obtained the famous integral for the corresponding coefficient of f

$$\frac{2}{\pi}\int_0^\pi f(t)\sin nt\,dt.$$

The reader will certainly appreciate the dexterity of Fourier compared with the simplicity of today's presentation of the theory ! Fourier only realized later that the orthogonality relations would immediately lead to the same result.

24.4 BESSEL'S SOLUTION TO KEPLER'S PROBLEM

Around 1816, F.W.Bessel —director of the astronomical observatory in Königsberg— was interested in the following well known problem.

The motion of a planet is an ellipse with the sun at a focus (if we neglect the perturbations due to the other planets). How can we determine the *excentric angle* corresponding to the position of the planet (cf.fig.) as a function of time ?

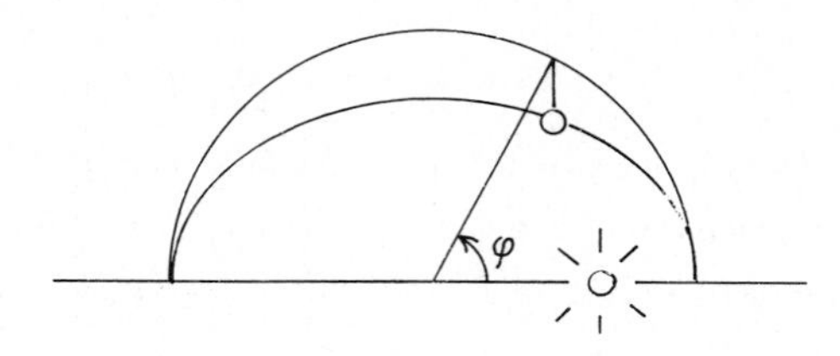

Fig.24.2

If T is the period of rotation, we can also use $\psi = 2\pi t/T = \omega t$ as variable (corresponding to a uniform circular rotation of period T). Denoting by φ the excentric angle, the coordinates of the center P of the planet will be given by

$$x = a\cos\varphi \quad , \quad y = b\sin\varphi$$

where a and b denote as usual the half axis of the ellipse. Kepler's area law can be used and leads to the transcendental equation

$$\varphi - e\sin\varphi = \omega t$$

where $e = c/a$ is the excentricity of the ellipse. In this equation, the unknown is the function $\varphi = \varphi(t)$ and here is Bessel's method for finding it (a mathematical gem !).

Let us change notations slightly : consider an unknown function $y = y(x)$ satisfying the transcendental equation

$$y - \lambda\sin y = x$$

where $\lambda \in [-1,1]$ is a given parameter.

Since $y \longmapsto x = y - \lambda\sin y$ is monotonously increasing, its inverse function must also be monotonously increasing. Let us denote by $y = y(x) = x + f(x)$ this inverse function. Replacing y by $y + 2\pi$, x is replaced by $x + 2\pi$ and conversely : the function f is 2π-periodic. It is also obvious that f is an odd function and consequently, its Fourier expansion has the form

$$f(x) : \sum_{n\geq 1} b_n \sin nx.$$

The Fourier coefficients b_n are given by the integral formula

$$b_n = \frac{2}{\pi}\int_0^{\pi} f(x)\sin nx dx = \frac{2}{\pi}\int_0^{\pi} (y - x)\sin nx dx =$$

$$= \frac{2}{\pi}\left[-(y-x)\frac{\cos nx}{n}\right]_0^{\pi} + \frac{2}{\pi}\int_0^{\pi}(y'-1)\frac{\cos nx}{n}dx =$$

$$= \frac{2}{\pi n}\int_0^{\pi} y'\cos nx dx$$

because $2\int_0^{\pi}\cos nx dx = \int_{-\pi}^{\pi}\cos nx dx = 0$. It is also obvious that

$$x = 0 \Longrightarrow y = 0 \quad \text{and} \quad x = \pi \Longrightarrow y = \pi.$$

As we have seen, when $0 \le x \le \pi$, we also have $0 \le y \le \pi$ and we can use $x \longmapsto y$ as monotonous change of variables (thus *hiding* the unknown function y in an integration variable !) getting

$$b_n = \frac{2}{\pi n}\int_0^{\pi}\cos nx dy = \frac{2}{\pi n}\int_0^{\pi}\cos n(y-\lambda\sin y)\, dy .$$

It is time to remember that Bessel defined functions J_n $(n \in \mathbb{N})$ by

$$J_n(\xi) = \frac{1}{\pi}\int_0^{\pi}\cos(ny - \xi\sin y)dy.$$

With these functions, we have found

$$b_n = \frac{2}{n} J_n(n\lambda).$$

The Fourier expansion of the required function f(x) is thus

$$\sum_{n\ge1} J_n(n\lambda)\frac{\sin nx}{n} .$$

Either we observe a priori that f is a smooth function (indefinitely differentiable), or we prove that $J_n(n\lambda)$ decrease fast so that the Fourier series converges (uniformly) to f and

$$y = y(x) = x + \sum_{n\ge1} J_n(n\lambda)\frac{\sin nx}{n} .$$

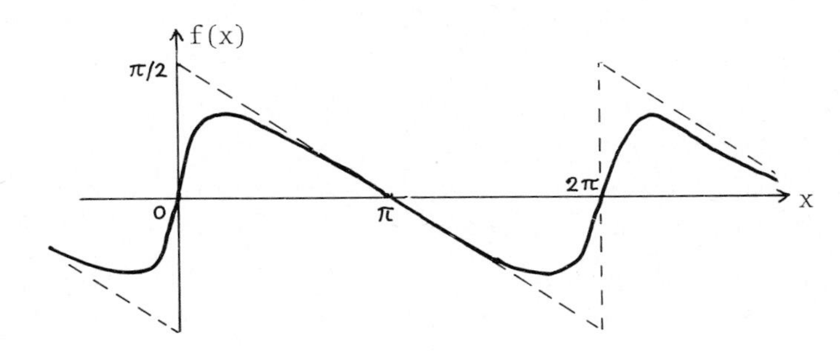

Fig.24.3

For the excentric anomaly, the found formula is

$$\varphi = \varphi(t) = \omega t + \sum_{n\ge1} J_n(ne)\frac{\sin n\omega t}{n} .$$

Similar formulas can be derived for the true anomaly and for the distance between the planet and the sun. Series containing terms

$J_n(nx)$ were systematically studied in the complex domain by Kapteyn (1893) and for example, the following expansions hold

$$\frac{1}{1-x} = 1 + 2 \sum_{n\geq 1} J_n(nx) ,$$

$$(1 - x^2)^{-1/2} = 1 + 2 \sum_{n\geq 1} J_n(nx)^2.$$

24.5 HURWITZ AND THE ISOPERIMETRIC INEQUALITY

The following general deduction of the **isoperimetric inequality** is essentially due to A.Hurwitz (cf.Ann.Ec.Norm.Ser.3, 19, p.392-397, 1902). It makes such a clever use of the theory of Fourier series that I could not resist including it.

Let $\mathcal{C}$ be a simple closed curve in the plane $\mathbb{R}^2$. We can assume that $\mathcal{C}$ is given in parametric form by means of two equations

$$x = f(t) \quad , \quad y = g(t)$$

with continuous functions f and g of period 1. We want to compare the length of $\mathcal{C}$ (when it is rectifiable !) to the area enclosed by it. In particular, we would like to prove that for a fixed length, the circle maximizes the area (cf.14.6). Or dually, for a fixed area, the circle minimizes the length.

Here is a formulation of these facts. To be able to define the length of $\mathcal{C}$, let us assume that f and g are continuously differentiable except at a finite number of points of the period interval, with bounded derivatives (points where the derivatives fail to exist being "corners" of $\mathcal{C}$ for example). Without loss of generality, we can assume that the parameter t is proportional to the length (*special parameter*). This means that the parametrization has constant *speed*

$$\|v\|^2 = f'(t)^2 + g'(t)^2 = v^2 = \text{const. (where defined).}$$

The length of the curve $\mathcal{C}$ is then

$$L = \int_0^1 (f'(t)^2 + g'(t)^2)^{1/2} dt = \int_0^1 v dt = v$$

and the constant value of $f'(t)^2 + g'(t)^2 = v^2$ is L^2. In particular, we have found (with this special parametrization)

$$L^2 = \int_0^1 v^2 dt = \int_0^1 (f'(t)^2 + g'(t)^2) dt = \|f'\|^2 + \|g'\|^2 .$$

On the other hand, the area enclosed by $\mathcal{C}$ is given by

$$S = -\oint y\,dx = -\int_0^1 g(t)f'(t)dt = -(f'|g) = |(f'|g)|$$

(assuming that the parameter describes the curve in the *anti-clockwise orientation*).

Let us consider the expression

$$L^2 - 2\lambda S = \|f'\|^2 + \|g'\|^2 + 2\lambda(f'|g) =$$

$$= \|f' + \lambda g\|^2 + \|g'\|^2 - \lambda^2\|g\|^2 \qquad (\lambda \in \mathbb{R}).$$

I claim that this quantity is positive for $\lambda \leq 2\pi$, i.e.

$$L^2 \geq 4\pi S$$

and equality holds *only when $\mathcal{C}$ is a circle*. Let us write

$$g = c_o + g_o \quad \text{with } g_o \perp 1$$

(i.e. c_o = mean value of g). Since f is continuous and periodic, we infer

$$f' \perp 1 \quad \text{and} \quad \|f' + \lambda g\|^2 = \|f' + \lambda g_o\|^2 + c_o^2$$

(Pythagoras theorem). Since $g' = g_o'$ we have

$$L^2 - 2\lambda S = \|f' + \lambda g_o\|^2 + \|g_o'\|^2 - \lambda^2\|g_o\|^2$$

and it only remains to prove the following result.

Proposition. When g is a continuous function which has a piecewise continuous bounded derivative (or more simply $g' \in L^2$)

$$g \perp 1 \implies 4\pi^2\|g\|^2 \leq \|g'\|^2. \qquad \square$$

Proof. Fourier series are only needed for the proof of this statement. If $g \perp 1$, $c_o(g) = 0$ and more generally

$$c_k(g) = \int_0^1 g(t)e^{-2\pi ikt}dt \qquad (k \in \mathbb{Z}).$$

The assumptions allow us to integrate by parts and we obtain easily

$$c_k(g') = 2\pi ik c_k(g)\ .$$

Without any consideration on the convergence of the Fourier series of g', Parseval's identity gives

$$\|g'\|^2 = \sum |c_k(g')|^2 = \sum_{k\neq 0} 4\pi^2k^2|c_k(g)|^2\ ,$$

$$\|g\|^2 = \sum_{k\neq 0} |c_k(g)|^2.$$

Hence

$$\|g'\|^2 - 4\pi^2\|g\|^2 = \sum_{|k|>1} 4\pi^2(k^2 - 1)|c_k(g)|^2 \geq 0. \qquad \blacksquare$$

Equality in the preceding general inequality will hold only when all $c_k(g) = 0$ for $|k| > 1$, and hence equality will occur only for the particular functions

$$g(t) = c_{-1}e^{-2\pi it} + c_1e^{2\pi it} = a\cos 2\pi t + b\sin 2\pi t.$$

The equality $L^2 = 4\pi S$ can only hold when moreover $f' + 2\pi g = 0$. In this case, f and g provide the parametrization of a circle (having center on the Ox-axis if g has been replaced by g_o as indicated).

It is even possible to proceed in a more symmetrical way. The double of the difference $D = L^2 - 2\lambda S$ can be written

$$2D = \|f' - \lambda g\|^2 + \|g' + \lambda f\|^2$$
$$+ (\|f'\|^2 - \lambda^2\|f\|^2) + (\|g'\|^2 - \lambda^2\|g\|^2) .$$

As before, we can assume that f and g are orthogonal to the constant 1 (replacing them resp. by f_o and g_o if necessary). The proposition can then be applied to f and g and shows that $2D \geq 0$ if $\lambda \leq 2\pi$. For $\lambda = 2\pi$ and $D = 0$, the four constituents of 2D must vanish and in particular

$$\begin{cases} f' - 2\pi g = 0 \\ g' + 2\pi f = 0 . \end{cases}$$

This is a coupled linear system for f and g. Its solutions are parameterized circles... but we can observe that the velocity vector

$$\begin{pmatrix} f \\ g \end{pmatrix}' = \begin{pmatrix} f' \\ g' \end{pmatrix} = 2\pi \begin{pmatrix} g \\ -f \end{pmatrix}$$

is constantly orthogonal to the radius vector

$$\mathbf{r} = {}^t(x,y) = {}^t(f,g).$$

It implies

$$(d/dt)\|\mathbf{r}\|^2 = \mathbf{r}' \cdot \mathbf{r} + \mathbf{r} \cdot \mathbf{r}' = 0$$

and $\|\mathbf{r}\|$ is constant on the trajectories... The *centered* solutions of $L^2 = 4\pi S$ are the concentric circles

$$x^2 + y^2 = r^2.$$

APPENDIX A: CONVERGENCE DEFINITIONS AND RESULTS

A.1 NUMERICAL SEQUENCES

Let $(a_n)_{n\in\mathbb{N}}$ be a sequence of real or complex numbers. We say that this sequence (a_n) tends to the limit a, and we write $a_n \longrightarrow a$ when the sequence $a_n - a$ tends to zero, namely when for each positive $\varepsilon > 0$, there exists an integer N_ε (rank depending on ε) for which $|a_n - a| \le \varepsilon$ for all $n \ge N_\varepsilon$.

a) The series $\sum_{n\in\mathbb{N}} a_n$ converges when the sequence (s_n) of partial sums $(s_n = \sum_{i\le n} a_i)$ converges : if so, we also denote by $s = \sum_{n\in\mathbb{N}} a_n$ its limit.

b) The series $\sum_{n\in\mathbb{N}} a_n$ converges absolutely when the series $\sum_{n\in\mathbb{N}} |a_n|$ converges (this implies that $\sum_{n\in\mathbb{N}} a_n$ converges).

The sum of an absolutely convergent series is independent of the order of summation (this is not the case for the convergent series $\sum_{n\ge1}(-1)^n/n$ —alternating harmonic series— for which term reordering can lead to convergence to any real number, or even divergence).

c) The product $\prod_{n\in\mathbb{N}} a_n$ converges when the sequence (p_n) of partial products $p_n = \prod_{i\le n} a_i$ has a limit p. We still denote by $\prod_{n\in\mathbb{N}} a_n = p$ this limit. By definition $|p_n - p| \longrightarrow 0$ $(n \longrightarrow \infty)$.

In the case of infinite products, it is reasonable to assume that $a_n \longrightarrow 1$ and to put $a_n = 1 + a'_n$ with $a'_n \longrightarrow 0$. We have

$$\sum_{n\in\mathbb{N}} a'_n \text{ converges absolutely} \implies \prod_{n\in\mathbb{N}} a_n \text{ converges}.$$

It is particularly important to give convergence criteria which do not refer specifically to the value of the limit, usually unknown.

The simplest one states that any *bounded increasing* sequence has a limit. Consequently, a series with positive terms converges as soon as its partial sums are bounded, and e.g.

$$\sum_{n\in\mathbb{N}} a_n \text{ converges absolutely} \iff \sum_{n\in\mathbb{N}} |a_n| < \infty .$$

More generally, Cauchy's criterium (expressing the fact that both $\mathbb{R}$ and $\mathbb{C}$ are *complete* fields) asserts that the series $\sum_{n\in\mathbb{N}} a_n$ converges when the sequence of partial sums (s_n) is a Cauchy sequence, i.e. when

$$|s_n - s_m| = \sum_{n<i\le m} a_i \longrightarrow 0 \quad \text{for } n \text{ and } m \longrightarrow \infty.$$

Let us also recall **Abel's result** (16.2) :

when $a_n = b_n c_n$, $b_n \searrow 0$ and $|c_1 + \dots + c_m| \le C$ (all m),
then $\sum_{n\in\mathbb{N}} a_n = \sum_{n\in\mathbb{N}} b_n c_n$ converges (not nec.absolutely!).

A.2 SEQUENCES OF FUNCTIONS

Let us consider a sequence of functions $u_n : X \longrightarrow \mathbb{R}$ (or $\mathbb{C}$) where X is a set (X will often be an interval, or a compact subset of a Euclidean space, or even a metric space). We say that this sequence (u_n) converges simply (or pointwise) on X when, for each $x \in X$, the sequence $(u_n(x))$ converges (in the numerical space $\mathbb{R}$ or $\mathbb{C}$). If this is the case, for each fixed $x \in X$, call u(x) the limit of the convergent sequence $(u_n(x))$. The association $x \longmapsto u(x)$ defines a mapping $u : X \longrightarrow \mathbb{R}$ (or $\mathbb{C}$) which is the simple limit of the sequence (u_n).

a) The series of functions $\sum_{n\in\mathbb{N}} u_n$ is said to converge simply (on X) when for each fixed $x \in X$, the numerical series $\sum_{n\in\mathbb{N}} u_n(x)$ converges.

b) The series of functions $\sum_{n\in\mathbb{N}} u_n$ is said to converge uniformly (on X) when there is a function $u : X \longrightarrow \mathbb{R}$ (or $\mathbb{C}$) such that for each positive $\varepsilon > 0$, there is an integer N_ε with

$$|u(x) - \sum_{i\le n} u_i(x)| \le \varepsilon \text{ for all } x \in X, \text{ as soon as } n \ge N_\varepsilon.$$

(In other words, all partial sums are *ε-uniformly near* u for $n \ge N_\varepsilon$.)

If we introduce the *uniform norm* on the set of numerical functions on X by

$$\|f\| = \operatorname*{Sup}_{x\in X} |f(x)| \le \infty$$

the uniform convergence of $\sum_{n\in\mathbb{N}} u_n$ to u means that

$$\forall\, \varepsilon > 0,\ \exists\, N_\varepsilon \quad \|u - \sum_{i\le n} u_i\| \le \varepsilon \quad \text{for all } n \ge N_\varepsilon.$$

c) The series of functions $\sum_{n\in\mathbb{N}} u_n$ is said to converge normally to u if the series (with positive terms) $\sum_{n\in\mathbb{N}} \|u_n\|$ converges (this implies that all $\|u_n\| < \infty$: each u_n is a *bounded* function). The following implication holds (by completeness of $\mathbb{R}$ and $\mathbb{C}$)

$$\sum_{n\in\mathbb{N}} u_n \text{ converges normally} \implies \sum_{n\in\mathbb{N}} u_n \text{ converges uniformly}$$

$$(\implies \sum_{n\in\mathbb{N}} u_n \text{ converges simply}).$$

But all three notions are distinct (examples prove that they are not equivalent).

d) An infinite product $\prod_{n\in\mathbb{N}} u_n$ <u>converges uniformly</u> if there exists a function $p : X \longrightarrow \mathbb{R}$ (or $\mathbb{C}$) such that the partial products p_n ($= \prod_{i\le n} u_i$) converge uniformly to p, i.e. if

$$\forall\ \varepsilon > 0,\ \exists\ N_\varepsilon \quad \|p - p_n\| \le \varepsilon \quad \text{for all } n \ge N_\varepsilon.$$

Here is a basic result

$$\sum_{n\in\mathbb{N}} |u_n| \ \underline{\text{converges uniformly}} \implies \prod_{n\in\mathbb{N}} (1 + u_n) \ \underline{\text{converges uniformly}}.$$

The assumption that the series of moduli converges uniformly does not imply that the series converges normally (it is weaker, but it implies that the series converges uniformly on X). Uniform convergence on X is <u>not</u> sufficient to imply convergence of the corresponding infinite product. Sometimes, a related notion of *normal convergence* for an infinite product is introduced.

e) The infinite product $\prod_{n\in\mathbb{N}} (1 + u_n)$ is said to <u>converge normally</u> when the series $\sum_{n\in\mathbb{N}} u_n$ converges normally (namely when $\sum_{n\in\mathbb{N}} \|u_n\| < \infty$ is finite).

Hence

$$\prod_{n\in\mathbb{N}} f_n \ \underline{\text{converges normally}} \implies \prod_{n\in\mathbb{N}} f_n \ \underline{\text{converges uniformly}}.$$

Let us not repeat the Cauchy and D'Alembert *criteria* for convergence of series (cf.16.1) and their applications to power series (and determination of their convergence radii). But let us remind ourselves of Abel's criterium. Consider a series $\sum_{n\in\mathbb{N}} u_n v_n$ of functions with the following properties

i) For each $x \in X$, $u_n(x) \searrow 0$ when $n \longrightarrow \infty$,

ii) There is a constant C with

$$|v_o(x) + \ldots + v_p(x)| \le C \text{ for all } p \in \mathbb{N} \text{ and } x \in X,$$

iii) The numerical sequence $\|u_n\| \longrightarrow 0$ $(n \longrightarrow \infty)$.

Then

i) and ii) imply that $\sum u_n v_n$ <u>converges simply</u>,

i), ii) and iii) imply that $\sum u_n v_n$ <u>converges uniformly</u>.

It is important to realize that the properties i), ii) and iii) are independent. Consequently, these three properties have to be checked completely in all applications.

On a metric space X (more generally any *topological space* X), any uniform limit of continuous functions is continuous. In particular, if the series $\sum_n u_n$ consists of continuous functions and converges uniformly, the sum is a <u>continuous function</u> (this is one of the main reasons for the interest in uniform convergence).

When D is an open subset of the complex plane $\mathbb{C}$, and when a series $\sum_n u_n$ consisting of *holomorphic* functions on D converges uniformly on all compact subsets of D,

$\sum_n u_n$ is also holomorphic on D and $(\sum_n u_n)' = \sum_n u'_n$ (termwise differentiation one or many times allowed!).

Let us also recall that the norms of convergence in a mean of order p (for numerical functions defined on an interval of $\mathbb{R}$) are defined by

$$N_p(f) = \|f\|_p = \left(\int |f(x)|^p dx \right)^{1/p} \quad (\text{if } 1 \le p < \infty),$$

$$N_\infty(f) = \|f\|_\infty = \text{Sup ess } |f(x)| \; .$$

f) The sequence (f_n) of functions converges in mean of order p to the function f, in other words $f_n \longrightarrow f$ *in* L^p, if

$$N_p(f_n - f) = \|f_n - f\|_p \longrightarrow 0 \quad (\text{for } n \longrightarrow \infty).$$

When p = 1, we simply say $f_n \longrightarrow f$ *in mean*, and when p = 2, we say $f_n \longrightarrow f$ *in quadratic mean.*

A.3 CONVERGENCE OF IMPROPER INTEGRALS

Instead of series, integrals may be considered (integrals in the sense of Cauchy, Riemann or Lebesgue...). Take for example the case of a continuous (or locally integrable) function f on $[0,\infty[$.

a) We say that $\int_0^\infty f(x)dx$ converges, or exists as an improper integral when $\int_0^{t_n} f(x)dx$ has a limit for each sequence $t_n \longrightarrow \infty$.

(When this is the case, the limit is automatically independent of the particular chosen sequence $t_n \longrightarrow \infty$.) If $\int_0^\infty f(x)dx$ converges, the common limit of the sequences $\int_0^{t_n} f(x)dx$ is also denoted by $\int_0^\infty f(x)dx$ and this number can be computed using the special sequence $t_n = n$: $\int_0^\infty f(x)dx = \lim_{n\to\infty} \int_0^n f(x)dx$.

b) We say that $\int_0^\infty f(x)dx$ converges absolutely when $\int_0^\infty |f(x)|dx$ converges.

This happens precisely when the increasing numerical sequence having general term $a_n = \int_0^n |f(x)|dx$ is *bounded* hence converges. An integral is often proved to converge absolutely by comparison to another existing integral. For example

$\int_1^\infty f(x)dx$ converges absolutely if $f \sim 1/x^s$ $(x \to \infty)$ for some $s > 1$.

Existence theorems for (non absolutely) convergent integrals are more delicate to establish. Let us recall **Abel's result** (cf.16.3).

The assumptions

i) $f \in \mathcal{C}^1$ and $f(x) \searrow 0$ when $x \longrightarrow \infty$,

ii) g is continuous,

iii) $|\int_0^a g(x)dx| \leq C$ for all $a \geq 0$

imply that the integral $\int_0^\infty f(x)g(x)dx$ converges (exists).

This theorem shows that the integral $\int_0^\infty \frac{\sin x}{x}dx$ exists, although the function $x^{-1}\sin x$ is not integrable in the Lebesgue theory (because it is not absolutely integrable).

Let us finally observe that an improper integral which depends on a parameter $\lambda \in \Lambda$ (where Λ is any set or topological space) is said to <u>converge uniformly</u> when the functions $F_n(\lambda) = \int_0^{t_n} f(x,\lambda)dx$ converge uniformly for every sequence $t_n \longrightarrow \infty$. Similarly, the integral $\int_0^\infty f(x,\lambda)dx$ is said to <u>converge absolutely uniformly</u> when the functions $F_n^*(\lambda) = \int_0^n |f(x,\lambda)|dx$ converge uniformly.

A.4 VARIATIONS ON THE SAME THEME

It occurs that indices in a sum may vary in the set $\mathbb{Z}$ instead of $\mathbb{N}$ (improper integrals may also be considered over $\mathbb{R}$ instead of $\mathbb{R}_+ = [0,\infty[$). The corresponding definitions of convergence are then

$\sum_{-\infty}^{\infty} a_n$ <u>converges</u> (resp.converges absolutely) when

both $\sum_0^\infty a_n$ and $\sum_{-\infty}^{-1} a_n$ <u>converge</u> (resp.converge absolutely),

$\prod_{-\infty}^{\infty} a_n$ <u>converges</u> when both $\prod_0^\infty a_n$ and $\prod_{-\infty}^{-1} a_n$ <u>converge</u>,

$\int_{-\infty}^{\infty} f(x)dx$ <u>converges</u> when $\int_{-\infty}^{0} f(x)dx$ and $\int_0^\infty f(x)dx$ <u>converge</u>.

Similar definitions for normal, uniform convergence are given for series (products or parametric integrals).

It may happen that $\sum_0^{\infty} a_n$ and $\sum_{-\infty}^{-1} a_n$ diverge but $\lim_{n\to\infty} \sum_{-n}^{n} a_n$ exists. In this case, we say that $\sum_{-\infty}^{\infty} a_n$ semi-converges. For example,

$$\sum_{-\infty}^{\infty} \frac{1}{x-n} \quad \text{semi-converges to} \quad f(x) = \pi\cot\pi x \quad \text{(cf.21.1).}$$

The limit of symmetric sums is called principal value of the sum and denoted by

$$\text{vp} \sum_{-\infty}^{\infty} a_n = \lim_{n\to\infty} \sum_{-n}^{n} a_n .$$

Similar definitions hold for infinite products and integrals

$$\text{vp} \prod_{\mathbb{Z}} a_n = \lim_{n\to\infty} \prod_{-n}^{n} a_i , \qquad \text{vp} \int_{-\infty}^{\infty} f(x)dx = \lim_{n\to\infty} \int_{-n}^{n} f(x)dx$$

(when they exist!).

In the same vein, existence of an integral in the neighborhood of a point where the function is discontinuous (not locally integrable) can be given as follows. Assume that f is continuous on]0,1]. We say that $\int_0^1 f(x)dx$ exists when $\int_{\varepsilon_n}^1 f(x)dx$ tend to a limit for every sequence $\varepsilon_n \searrow 0$ (the common limit to these is by definition the value of the integral between 0 and 1). For example

$$\int_0^1 x^s dx \quad \text{exists for } s > -1$$

and more generally, an integral $\int_0^1 f(x)dx$ exists (converges absolutely) as soon as f is continuous on]0,1] with $|f| \underset{\sim}{\ } x^s$ $(x \longrightarrow 0)$ for some $s > -1$. When the function f ceases to be continuous at an interior point of the integration interval, say $0 \in [-1,1]$ to fix ideas, we say

$\int_{-1}^1 f(x)dx$ exists when both

$$\int_{-1}^0 f(x)dx \quad \text{and} \quad \int_0^1 f(x)dx \quad \text{exist (in the preceding sense).}$$

When only symmetric limits can be formed, we say that the integral semi-converges and its principal value is

$$\text{vp} \int_{-1}^1 f(x)dx = \lim_{\varepsilon \searrow 0} \left(\int_{\varepsilon}^1 f(x)dx + \int_{-1}^{-\varepsilon} f(x)dx \right).$$

For example $\text{vp} \int_{-1}^1 x^{-1}dx = 0$. The notation $\text{vp} \int_{-\infty}^{\infty} f(x)dx$ similarly represents the value of $\lim_{n\to\infty} \int_{-n}^{n} f(x)dx$ when it exists.

It may happen that all difficulties occur simultaneously. In this case, a notation like vp $\int_{-\infty}^{\infty} f(x)dx$ should mean that symmetric limits have to be formed *at all points where* f ceases to be locally integrable (*and* at ∞ !). In each particular case, the context should make clear which points are troublesome.

A.5 A FEW IMPORTANT CONVERGENCE THEOREMS

The first results are concerned with the possibility of *permuting limits and integrals*. The **Fatou theorem** corresponds to the existence of a limit for a bounded increasing sequence of real numbers :

Let $(f_n)_{n\in\mathbb{N}}$ be an increasing sequence of (real valued) integrable functions with bounded integrals, say $\int f_n dx \le M$ ($M < \infty$) for all $n \in \mathbb{N}$. Then these functions have a limit a.e. $f(x) = \lim f_n(x)$ ($x \notin N$ negligible) which is *integrable*. Moreover, $f_n \longrightarrow f$ in mean, hence $\int f_n dx \longrightarrow \int f dx$ and in other words

$$\lim \int f_n \, dx = \int \lim f_n \, dx$$

A similar statement concerning permutation of integrals with limits is furnished by the **basic theorem** (cf.15.2)

If a sequence of continuous functions $f_n : I \longrightarrow \mathbb{C}$ converges uniformly (to a continuous function f) on the compact interval I (or piece $I \subset \mathbb{R}^n$), then $\lim \int_I f_n \, dx = \int_I \lim f_n \, dx$.

More generally, **Lebesgue's dominated convergence theorem** is the following.

Let $1 \le p < \infty$ and $(f_n)_{n\in\mathbb{N}} \subset L^p(\mathbb{R}^n)$ be a sequence of pth-power integrable functions. Assume that $f_n(x) \longrightarrow f(x)$ a.e. (say for $x \notin N$ negligible) *and* there exists a fixed $g \in L^p(\mathbb{R}^n)$ with $|f_n| \le g$ (all $n \in \mathbb{N}$). Then $f \in L^p(\mathbb{R}^n)$ and $f_n \longrightarrow f$ in the mean of order p.

One important case of the preceding result is $p = 1$, in which case the conclusion gives $\lim \int f_n \, dx = \int \lim f_n \, dx$. Let us also observe that a mean convergence of order p, $f_n \longrightarrow f$ in L^p (i.e. $\|f_n - f\|_p \longrightarrow 0$) does not imply convergence of $f_n(x)$ to $f(x)$ a.e. : however, it is always possible to find a subsequence $(f_{n_k})_{k\in\mathbb{N}} \subset (f_n)_{n\in\mathbb{N}}$ for which convergence a.e. holds.

The corollaries of Lebesgue's theorem are numerous and important. Let us quote a few. Here is a **first one**

When a series $\sum_{n\geq 0} u_n$ consists in integrable terms $u_n \in L^1$ and converges a.e. with partial sums $|\sum_{i\leq n} u_i| \leq g$ for a fixed integrable function $g \in L^1$, then

▷ $\sum_{n\geq 0} u_n$ converges in mean to an integrable function,

▷ $\int \sum_{n\geq 0} u_n(x)dx = \sum_{n\geq 0} \int u_n(x)dx.$

Here is a **second one** which is concerned with the continuity of a parametric integral.

Let Λ be a metric space (parametric space) and $f : \mathbb{R} \times \Lambda \longrightarrow \mathbb{C}$ with

▷ $x \longmapsto f(x,\lambda)$ integrable for all $\lambda \in \Lambda$,

▷ $\lambda \longmapsto f(x,\lambda)$ continuous (on Λ) a.e. in x,

▷ $\forall\ \lambda$, $\exists$ nbd V_λ of λ and a fixed integrable $g \in L^1$ with $|f(x,\mu)| \leq g(x)$ a.e.in x and *all* $\mu \in V_\lambda$.

Then the function F defined by $F(\lambda) = \int f(x,\lambda)dx$ is *continuous* (on Λ).

The **third one** generalizes **Leibniz' rule** (16.5) for the differentiation of a parametric integral.

Let I be an open interval and $f : \mathbb{R} \times I \longrightarrow \mathbb{C}$ be a numerical function satisfying

▷ $x \longmapsto f(x,t)$ integrable for all $t \in I$,

▷ for nearly all $x \in \mathbb{R}$ (say $x \notin N$ negligible in $\mathbb{R}$),

$t \longmapsto D_2 f(x,t) = \frac{\partial f}{\partial t}(x,t)$ exists and is continuous,

▷ $\forall\ t$, $\exists$ nbd I_t of t and a fixed integrable $g \in L^1$ with $|D_2 f(x,s)| = |\frac{\partial f}{\partial t}(x,s)| \leq g(x)$ all $s \in I_t$ and a.e. in x

Then the numerical function F defined by $F(\lambda) = \int f(x,\lambda)dx$ is $\mathcal{C}^1$ on I with

$$F'(t) = \int D_2 f(x,t)dx = \int \frac{\partial f}{\partial t}(x,t)dx.$$

Let us recall that in all these statements, no compactness is made on the integration space and all integrals are taken in the Lebesgue sense over $\mathbb{R}$: in particular, they represent *absolutely convergent integrals*. Similar results for improper integrals (A.3) can only be used when suitably justified (case by case) !

A.6 COMPLEMENTS

There are a few relations between *simple convergence* (for numerical functions) and *uniform convergence*. Let us summarize the most important ones.

Consider a sequence (f_n) consisting of continuous numerical functions (defined on some metric space X) and consider the following properties

a) $f_n(x) \longrightarrow f(x)$ for all $x \in X$ (*simple convergence*),

b) $(f_n)_{n\in\mathbb{N}}$ is an *equicontinuous set* of functions,

c) $f_n \longrightarrow f$ *uniformly* on X,

d) f is continuous,

e) $(f_n)_{n\in\mathbb{N}}$ is *monotonous increasing* (or *decreasing*).

Then, clearly c) $\Longrightarrow$ a) and the **basic theorem** (15.2) shows that c) $\Longrightarrow$ d). More precisely

c) $\Longrightarrow$ both a) and b),

a) with b) $\Longrightarrow$ d) .

Conversely,

a) with b) $\Longrightarrow$ $f_n \longrightarrow f$ uniformly on every compact subset of X hence c) if X is compact,

a) with d) and e) $\Longrightarrow$ $f_n \longrightarrow f$ uniformly on every compact subset of X hence c) if X is compact

(when X is compact, this is precisely **Dini's lemma**, cf.15.3).

APPENDIX B
SAS PROGRAMS FOR A FEW FIGURES

For convenience of the reader, we now list the programs used for a couple of figures. They were made for the widely known and used package SAS (Statistical Analysis System). Even without previous knowledge of this programming language, the reader will be able to adapt data and/or equations to get similar figures in other cases. Let me thank J. Moret who made the original programs and introduced me to the SAS package.

B.1 GRAPH OF THE SURFACE OF $\sqrt{|xy|}$: Fig.1.12

```
OPTIONS LT;
GOPTIONS DEVICE=LN03 NOCHARACTERS;
GOPTIONS VSIZE=5 HSIZE=8;
TITLE ' ';

DATA SURF1;
  DO X=-10 TO 10 BY 0.5;
    DO Y=-10 TO 10 BY 0.5;
      Z=SQRT(ABS(X*Y));
      OUTPUT;
    END;
  END;
RUN;

PROC G3D DATA=SURF1;
  PLOT Y*X=Z/GRID
             ZMIN=0
             ZMAX=10
             XTICKNUM=5
             YTICKNUM=5
             ZTICKNUM=3
             TILT=60
             ROTATE=45;

RUN;

ENDSAS;
```

B.2 FOURIER SERIES OF THE PENTAGON : Fig.21.3

```
OPTIONS LT;
GOPTIONS DEVICE=LN03;
GOPTIONS VSIZE=7.5 HSIZE=9;
TITLE ' ';

DATA PENTA;
  ARRAY X{9} X0-X8;
  ARRAY Y{9} Y0-Y8;
  DO I=0 TO 1000 BY 1;
    T=I*2*3.141593/1000;
    X0=COS(T);
    Y0=SIN(T);
    DO K=1 TO 8 BY 1;
      A=1-5*K;
      B=1+5*K;
      X{K+1}=X{K}+COS(A*T) / A**2 + COS(B*T) / B**2
    END;
    X2=X2+0.6;
    X3=X3+1.2;
    X4=X4+1.8;
    X8=X8+2.4;
    OUTPUT;
  END;
  KEEP X1 Y1 X2 Y2 X3 Y3 X4 Y4 X8 Y8;
RUN;

SYMBOL1 W2 I=JOIN L=2 V=NONE;
SYMBOL2 W2 I=JOIN L=3 V=NONE;
SYMBOL3 W2 I=JOIN L=4 V=NONE;
SYMBOL4 W2 I=JOIN L=5 V=NONE;
SYMBOL5 W2 I=JOIN L=1 V=NONE;

PROC GPLOT DATA=PENTA;
  PLOT Y1*X1=1
       Y2*X2=2
       Y3*X3=3
       Y4*X4=4
       Y8*X8=5 /OVERLAY NOLEGEND NOAXES;

RUN;

ENDSAS;
```

B.3 FOURIER SERIES OF THE SNOWFLAKE : Fig.21.4

```
OPTIONS LT;
GOPTIONS DEVICE=LN03;
TITLE ' ';

DATA SNOWFLAKE;
  X=0;
  Y=0;
  S=1;
  OUTPUT;
  S=2;
  PI=3.1415926536;
  A=1+1/SQRT(3);
  B=1-1/SQRT(3);
  DO I=0 TO 4800 BY 1;
    T=I/4800;
    X=0;
    Y=0;
    PSI=6*PI*T;
    DO N=1 TO 7 BY 1;
      PHI=2*PI*T 4**N;
      ALPHA=PSI-PHI;
      BETA=PSI-2*PHI;
      XNUM=A*COS(ALPHA) + B*COS(BETA);
      YNUM=A*SIN(ALPHA) - B*SIN(BETA);
      X=X + XNUM/3**N;
      Y=Y + YNUM/3**N;
    END;
    IF X>1 THEN X=1;
    IF X<-1 THEN X=-1;
    OUTPUT;
  END;
  KEEP X Y S;
RUN;

SYMBOL1 W=2 I=NONE H=1.5 V=PLUS;
SYMBOL2 W=2 I=JOIN V=NONE;

PROC GPLOT DATA=SNOWFLAKE;
  AXIS1 ORDER=-1 TO 1 BY 1;
        LENGTH=15 CM;
  PLOT Y*X=S/HAXIS=AXIS1
            VAXIS=AXIS1
            NOLEGEND
            NOAXES;

RUN;

ENDSAS;
```

B.4 GIBBS PHENOMENON : Fig.23.3

```
OPTIONS LT;
GOPTIONS NOCHARACTERS DEVICE=LN03;

DATA GIBBS;
  PI=3.141592654;
  K=1; X=0;  Y=0;    OUTPUT;
       X=0;  Y=PI/2; OUTPUT;
       X=PI; Y=0;    OUTPUT;
       X=0;  Y=0;    OUTPUT;
  DO K=4 TO 9 BY 1;
    K2=2**K;
    DO I=0 TO K2 BY 1;
      X=I*PI / K2;
      Y=0;
      DO N=1 TO K2 BY 1;
        Y=Y+SIN(N*X)/N;
      END;
      X=X+(K-4)*0.36;
      Y=Y+(K-4)*0.21;
      OUTPUT;
    END;
  END;
  KEEP X Y K;
RUN;

TITLE ' ';
SYMBOL1 V=NONE I=JOIN W=2;
SYMBOL4 V=NONE I=JOIN W=2;
SYMBOL5 V=NONE I=JOIN W=2;
SYMBOL6 V=NONE I=JOIN W=2;
SYMBOL7 V=NONE I=JOIN W=2;
SYMBOL8 V=NONE I=JOIN W=2;
SYMBOL9 V=NONE I=JOIN W=2;

PROC GPLOT DATA=GIBBS;
  AXIS1 ORDER=0 TO 5.4
        LENGTH=18 CM;
  AXIS2 ORDER=0 TO 3.3
        LENGTH=11 CM;
  PLOT Y*X=K/NOLEGEND NOAXES
             HAXIS=AXIS1
             VAXIS=AXIS2;

RUN;

ENDSAS;
```

EXERCISES FOR PART I

Chapter 1

1. a) Let f and g be two differentiable functions $\mathbb{R} \longrightarrow \mathbb{R}$ with $f(0) = g(0) = 0$. Using limited expansions of f and g at the origin, prove that $\lim_{x\to 0} f(x)/g(x) = f'(0)/g'(0)$ if $g'(0) \neq 0$ (*L'Hospital's rule*). More generally give conditions for

$$\lim_{x\to 0} f(x)/g(x) = \lim_{x\to 0} f'(x)/g'(x).$$

b) Apply the preceding method to determine the limit of

$$\frac{a^x - b^x}{c^x - d^x} \quad \text{when} \quad x \longrightarrow 0$$

(Hint : $a^x = 1 + (\log a)x + o(x)$).

c) Notice that it may be easier to work with limited expansions than to apply blindly L'Hospital's rule. The following examples will show what we mean :

- Compute $\lim_{x\to 0} \frac{\sin(1/x)}{1/x}$ using L'hospital rule
  (on the other hand, $\lim_{\xi\to\infty} \frac{\sin\xi}{\xi} = 0$ since $|\sin\xi| \leq 1$),
- Evaluate $\lim_{x\to 0} \frac{x^3\cos x - \sin^3 x}{x^5}$.

2. Compute the limits for $x \longrightarrow 0$ of the following expressions

$$\frac{\mathrm{Ch}x - 1}{\sin x}, \quad (1/x)^{\sin x}, \quad \frac{\sin(\mathrm{Sh}x)}{x^2},$$

$$\frac{\sin(x + \mathrm{Sh}x)}{\log(1+x)}, \quad \frac{x^n(1-x)^n - x^n}{(\mathrm{Sh}x)^{n+1}} .$$

(Hint : use limited expansions of higher order.)

3. Let $(a_n)_{n\geq 0}$ denote a numerical sequence. We write

- $a_n = o(n^\alpha)$ for $a_n/n^\alpha \longrightarrow 0$ (when $n \longrightarrow \infty$),
- $a_n = O(n^\beta)$ if $|a_n| \leq cn^\beta$ for some constant $c > 0$.

Give a description of all implications between the following statements

A) $a_n = O(1/n)$, B) $a_n = O(1/n^2)$, C) $\sum a_n$ converges absolutely,

D) $a_n = o(1/n)$, E) $n^2 a_n \longrightarrow \alpha$ (also written $a_n \sim \alpha/n^2$).

Hint : E) implies X) for all X (!). To show that C) does <u>not</u> even imply A), consider the sequence (a_n) defined for $n \geq 1$ by

$$a_n = \begin{cases} 1/\sqrt{n} & \text{when } n = m^4 \text{ for some } m \in \mathbb{N} \\ 0 & \text{otherwise} \end{cases} .$$

4. Let $F : \mathbb{R}^2 \longrightarrow \mathbb{R}^2$ be the map $\mathbf{r} \longmapsto \mathbf{r}\cos(\mathbf{v}\cdot\mathbf{r})$ where $\mathbf{v}$ is a given (constant) vector. Give a limited expansion of F near a vector $\mathbf{a}$ and deduce the value $F'(\mathbf{a})$. In particular, what is $F'(0)$?

5. What is the derivative of the map

$$F : \mathbb{R}^3 \longrightarrow \mathbb{R}^3 , \mathbf{x} \longmapsto \mathbf{x} \wedge (\mathbf{x} \wedge \mathbf{v})$$

where $\mathbf{v}$ is a given (constant) vector.

Answer : $F'(\mathbf{x})$ is the map $\mathbf{h} \longmapsto \mathbf{h} \wedge (\mathbf{x} \wedge \mathbf{v}) + \mathbf{x} \wedge (\mathbf{h} \wedge \mathbf{v})$.

6. Let F denote the map $g \longmapsto g^{-1}$ in the set of 2 by 2 real invertible matrices. In other words $F\begin{pmatrix} a & b \\ c & d \end{pmatrix} = \begin{pmatrix} a & b \\ c & d \end{pmatrix}^{-1} = \Delta^{-1}\begin{pmatrix} d & -b \\ -c & a \end{pmatrix}$ where $\Delta = ad - bc$. Compute the derivative of F at the point $g_o = \mathrm{Id} = \begin{pmatrix} 1 & 0 \\ 0 & 1 \end{pmatrix}$. Determine similarly the derivative of $g \longmapsto g^2$.

7. Give the mathematical nature of the following expressions

$$\int_a^b f(x)dx,$$

$$\mathbf{v}\cdot\mathbf{curl\ u},$$

$$adx + bdy + cdz,$$

$$\iint_\Sigma \varphi(x,y,z)d\vec{\sigma}.$$

For example

$\mathbf{r} = {}^t(x,y,z)$ is a *vector*,

$\mathbf{u} = \mathbf{u}(x,y,z)$ is a *vector field,...*

Chapter 2

1. The following equations

$$x = t\cos t,\ y = \log(1+t),\ z = e^t - 1$$

represent a $\mathcal{C}^1$ parametrized curve (say for $t > -1$) in $\mathbb{R}^3$. The equation

$$e^{x+y} + \sin(x+z) - 1 = 0$$

represents a surface in $\mathbb{R}^3$. Using limited expansions near the origin, find the angle of intersection of the curve and the surface at their intersection point (0,0,0).

2. Assume that $t \longmapsto M(t)$ is a $\mathcal{C}^1$ parametrized curve in the group of n by n invertible matrices (with real coefficients). Compute the derivative of $t \longmapsto M(t)^{-1}$.

3. Let R(t) be the matrix of the rotation of angle t around a fixed vector $\mathbf{v} \in \mathbb{R}^3$ having unit length. Show that the derivative of $t \longmapsto R(t)$ at $t = 0$ is the matrix $M_\mathbf{v}$ of the map $\mathbf{h} \longmapsto \mathbf{v} \wedge \mathbf{h}$ (cf.8.1). Hint : Use a change of basis bringing the vector $\mathbf{v}$ on the third basic vector and make very few computations...

4. Consider the $\mathcal{C}^1$ parametrized curve in $\mathbb{R}^2$ given by

$$x = \log\sqrt{(1 + t^2)}\ ,\ y = \text{Arctan}\ t\ .$$

Compute the length of the arc $0 \leq t \leq 1$.

5. Consider the $\mathcal{C}^1$ parametrized curve —also called a **helix** as in 2.1—

$$t \longmapsto \varphi(t) : x = \cos t\ ,\ y = \sin t\ ,\ z = at\ .$$

a) Compute the length of an arc $0 \leq t \leq a$. Why is the result so simple ? (Observe that the helix is situated on a cylinder which can be "unwrapped" on a plane; this map preserves distances and sends the helix on a line.)

b) Show that the vector difference $\varphi(2\pi) - \varphi(0)$ is not a multiple of the velocity vectors $\varphi'(t)$ (for any value of t !). More generally, show that if $s > t$, $\varphi(s) - \varphi(t)$ cannot be written in the form $(s-t)\varphi'(\tau)$ for any intermediate value $t \leq \tau \leq s$. In other words, prove all statements given in the example of 2.1.

6. Find the length of the arc $-1 \leq t \leq 1$ on the curve in $\mathbb{R}^{2n+1}$:

$$t \longmapsto (t, \cos t, \sin t, \cos 2t, \sin 2t, \ldots, \cos nt, \sin nt).$$

7. Consider the following "spiral" $\mathcal{C}$ on the unit sphere (in $\mathbb{R}^3$) : the velocity **v** of $\mathcal{C}$ makes a constant angle $\pi/4$ with the meridians (and also with the parallels, i.e. the horizontal circles on the sphere). Prove that the length of the arc of $\mathcal{C}$ linking an equatorial point and the north pole is finite.

8. Consider the curve $\mathcal{C}$ on the unit sphere in $\mathbb{R}^3$ given in spherical coordinates by the equation $\vartheta = \varphi$ (c.f.9.5 for the definition of the spherical coordinates). Determine the normal of the curve at the north pole. Hint : the curve $\mathcal{C}$ projects on a circle in the base horizontal plane and thus represents the intersection of a vertical cylinder with the unit sphere; find a regular parametrization near the north pole. Also observe that the projection of $\mathcal{C}$ on the vertical plane yOz is a parabola.

9. The parametrized curve $\varphi : \mathbb{R} \longrightarrow \mathbb{R}^2$ given by

$$\varphi(t) = \begin{cases} (t^2, t^2) \text{ if } t \geq 0 \\ (-t^2, t^2) \text{ if } t < 0 \end{cases}$$

is $\mathcal{C}^1$ but not $\mathcal{C}^2$. Prove it (cf. 2.1).

Chapter 3

1. Let $F : M_n(\mathbb{R}) \longrightarrow M_n(\mathbb{R})$, $X \longmapsto e^X$ denote the exponential map. Determine the tangent linear maps of F at 0 and $I = 1_n$ and deduce the values of F'(0) and of F'(I). Prove that the image of the exponential map is contained in the set of invertible matrices. More precisely, prove : $\det e^X = e^{\mathrm{Tr}(X)}$. (Hint : If X is diagonal, the formula is nearly obvious ; when X is in upper triangular form, it is also easily checked ; in general, there exists an invertible matrix S such that $X = SYS^{-1}$ with Y upper triangular and use the identity $e^X = e^{SYS^{-1}} = S\,e^Y S^{-1}$ implying $\det e^X = \det e^Y$.) Let n = 2 and take $X = \begin{pmatrix} 0 & 1 \\ 0 & 0 \end{pmatrix}$, $Y = \begin{pmatrix} 0 & 0 \\ 1 & 0 \end{pmatrix}$; show that with these matrices, $e^X \circ e^Y \neq e^{X+Y}$ (also compare $e^X \circ e^Y$ to $e^Y \circ e^X$).

2. Solve the following coupled linear system

$$x' = 4x - y \ , \ y' = x + 2y \quad \text{(where } x' = dx/dt, \ldots).$$

(Answer : $\begin{pmatrix} x \\ y \end{pmatrix} = a\begin{pmatrix} 1 \\ 1 \end{pmatrix} e^{3t} + b\begin{pmatrix} t \\ t-1 \end{pmatrix} e^{3t}$.)

3. Solve the system $x' = x - y$, $y' = x + 3y$.

(Answer : $\begin{pmatrix} x \\ y \end{pmatrix} = \begin{pmatrix} a-bt \\ b-a+bt \end{pmatrix} e^{2t}$.)

4. Solve the system $x' + 2y = 3t$, $y' - 2x = 4$.

(Answer : $\begin{pmatrix} x \\ y \end{pmatrix} = \begin{pmatrix} a \\ -b \end{pmatrix} \cos 2t + \begin{pmatrix} b \\ a \end{pmatrix} \sin 2t + \begin{pmatrix} -5/4 \\ 3t/2 \end{pmatrix}$.)

5. Solve $tx' + 2x = t$, $ty' - (t+2)x - ty = -t$ with the initial condition $x(1) = y(1) = 1$.

(Answer : $x = (t + 2/t^2)/3$, $y = e^{t-1} - (t + 2/t^2)/3$.)

6. Solve the vector linear equation $X' = \begin{pmatrix} 0 & 1 \\ -4 & 4 \end{pmatrix} X + \begin{pmatrix} 0 \\ t+1 \end{pmatrix}$.

(Answer : $X = a\begin{pmatrix} 1 \\ 2 \end{pmatrix} e^{2t} + b\begin{pmatrix} t \\ 1+2t \end{pmatrix} e^{2t} + \frac{1}{4}\begin{pmatrix} t+2 \\ 1 \end{pmatrix}$.)

7. Determine the evolution operator corresponding to the system

$$X' = A(t)X \text{ in } \mathbb{R}^2 \text{ where } A(t) = \frac{1}{1+t}\begin{pmatrix} 1 & 0 \\ 1 & 1+t \end{pmatrix}.$$

8. Give the general solution of $X' = A(t)X + \mathbf{b}(t)$ in $\mathbb{R}^2$ where

$$A(t) = \begin{pmatrix} 0 & 2t \\ 2t & 0 \end{pmatrix} \quad \text{and} \quad \mathbf{b}(t) = \begin{pmatrix} t \\ 0 \end{pmatrix}.$$

9. Give a non trivial (i.e. not identically zero) solution of the equation $\mathbf{x}' = (\mathbf{a}\cdot\mathbf{x})\mathbf{b}$ where $\mathbf{a}$, $\mathbf{b}$ are non zero vectors (in $\mathbb{R}^3$).

10. Solve the system

$$x' = 6z\ ,$$

$$y' = -x + 11z\ ,$$

$$z' = -y + 6z + e^{-t}.$$

(Answer : $\begin{pmatrix} x \\ y \\ z \end{pmatrix} = a\begin{pmatrix} 6 \\ 5 \\ 1 \end{pmatrix}e^{t} + b\begin{pmatrix} 3 \\ 4 \\ 1 \end{pmatrix}e^{2t} + c\begin{pmatrix} 2 \\ 3 \\ 1 \end{pmatrix}e^{3t} + \frac{1}{24}\begin{pmatrix} 6 \\ 17 \\ -1 \end{pmatrix}e^{-t}$.)

11. Solve the system

$$x' + y' = e^{t}(x + y)\ ,\ x' - y' = e^{-t}(x - y)$$

using the general method. Check your answer using the change of functions $u = x + y$, $v = x - y$.

12. Let M be an invertible constant matrix. Check that the primitive of e^{tM} that vanishes for $t = 0$ is $M^{-1}(e^{tM} - 1)$. For example, the primitive of $A(t) = \begin{pmatrix} \mathrm{Ch}t & \mathrm{Sh}t \\ \mathrm{Sh}t & \mathrm{Ch}t \end{pmatrix}$ that vanishes for $t = 0$ is

$$B(t) = \begin{pmatrix} \mathrm{Sh}t & \mathrm{Ch}t - 1 \\ \mathrm{Ch}t - 1 & \mathrm{Sh}t \end{pmatrix} .$$

In this case, check that B(t) commutes with A(t) (observe more generally that two symmetric matrices A and B commute when their product AB is symmetric). Check that $A(t+s) = A(t)\circ A(s)$ $(t,s \in \mathbb{R})$ and $(d/dt)_{t=0}\ A(t) = M = \begin{pmatrix} 0 & 1 \\ 1 & 0 \end{pmatrix}$ and consequently $A(t) = e^{tM}$. (Compare with the preceding exercise 11).

13. Consider the following matrix valued map

$$M(s,t) = \begin{pmatrix} 1+(s^2+t^2)/2 & -(s^2+t^2)/2 & s & t \\ (s^2+t^2) & 1 - (s^2+t^2)/2 & s & t \\ s & -t & 1 & 0 \\ s & -t & 0 & 1 \end{pmatrix}.$$

Compute the following (partial) derivatives (at the origin)

$$A = \partial M/\partial s(0,0) \quad , \quad B = \partial M/\partial t(0,0)\ .$$

Check that these matrices commute and that $M(s,t) = \exp(sA + tB)$. (Compare with the preceding exercise 12.)

14. Give the real solutions of the system

$$x' = 2x + 2z - y\ ,$$

$$y' = x + 2z\ ,$$

$$z' = y - 2x - z\ .$$

15. Let C denote the evolution operator corresponding to the vector system $X' = A(t)X$. Assume that a matrix M commutes with all matrices $A(t)$ $(t \in \mathbb{R})$. Show that M also commutes with all $C(t)$. (Hint : Compute $(d/dt)(MC - CM)$.)

16. Let $t \longmapsto A(t) \in M_n(\mathbb{R})$ be a continuous curve. Prove that all matrix solutions $t \longmapsto H(t)$ of the linear differential system

$$dH/dt = A \circ H - H \circ A$$

have a constant trace. More precisely, prove that if M is a fixed (constant) matrix, the solution of $dH/dt = A \circ H - H \circ A$ with initial value $H(0) = M$ is $H(t) = C(t) \circ M \circ C(t)^{-1}$ where C is the evolution operator corresponding to the linear equation $X' = A(t)X$. Conclude that all eigenvalues of H are constant (and resp. equal to the eigenvalues of M). In this way, not only is the trace of H constant, but also the determinant of H, and all symmetric functions of these eigenvalues ($pc_H = pc_M$).

17. Compute e^{tA} where A is the matrix

$$A = \begin{pmatrix} 0 & 0 & 6 \\ -1 & 0 & 11 \\ 0 & -1 & 6 \end{pmatrix}$$

For this purpose, solve the system $X' = AX$ and find the particular solutions going through the vectors $\mathbf{e}_i$ $(i = 1,2,3)$ for $t = 0$.

18. Consider the following two integrals

$$u(x) = \int_0^{\infty} e^{-t} t^{-1/2} \sin(tx)dt ,$$

$$v(x) = \int_0^{\infty} e^{-t} t^{-1/2} \cos(tx)dt$$

where x is a real parameter. Admitting that u and v are $\mathcal{C}^1$ and that u' and v' can be computed by differentiating under the integral sign (the techniques of Chap.16 will be relevant to this situation!), show that u and v satisfy the coupled linear differential system of the first order

$$u' = av - xau \quad , \quad v' = -au - xav$$

where $a = a(x) = \dfrac{1}{2(1+x^2)}$. Find the evolution operator of this system. Conclude that

$$u(x) = \sqrt{\pi}\,(1 + x^2)^{-1/4} \sin\left(\frac{1}{2} \operatorname{Arctan} x\right) ,$$

$$v(x) = \sqrt{\pi}\,(1 + x^2)^{-1/4} \cos\left(\frac{1}{2} \operatorname{Arctan} x\right) .$$

(Hint : Use the value of the Eulerian integral

$$\int_0^{\infty} e^{-t}/\sqrt{t}\, dt = 2 \int_0^{\infty} e^{-t^2} dt = \sqrt{\pi} .)$$

Chapter 4

1. Give the general solution of the system

$$x'' = 2(x + y) \ , \ y' = x + y + 2x'.$$

What is the particular solution for which $x(0) = 1$, $y(0) = -1$ and $x'(0) = 0$?

2. Reduce the order of the following differential equation

$$y''' - 5y'' + 8y' - 4y = e^t$$

and solve the corresponding first order system in $\mathbb{R}^3$. Deduce the general solution of the original third order equation.

(Answer : $y = (a+t)e^t + (b+ct)e^{2t}$.)

3. Solve the first order system corresponding to the equation

$$y''' + y'' - y' - y = 0.$$

Find the evolution operator $C(t)$ of the corresponding first order equation in $\mathbb{R}^3$. Also deduce the particular solution of the initial equation satisfying $y(0) = y''(0) = 1$, $y'(0) = 0$.

(Answer : $y = \text{Ch}t$.)

4. Give the evolution operator of the first order vector system corresponding to

$$y'' + y' - 6y = 0.$$

Chapter 5

1. a) Which points of the curve $x^2 + y^4 = 2$ are furthest from the origin ?

b) A particule moves on the curve $x^2 + y^4 = 2$ according to $x = f(t)$, $y = g(t)$; assuming $f(1) = g(1) = f'(1) = 1$, compute $g'(1)$.

2. Let A, B and C be three given points in the plane $\mathbb{R}^2$ and not colinear. Find the point P which minimizes the sum of distances to A, B and C.

(Hint : compute the gradient of the sum of distances, using for example the fact that the gradient of $r = \|OP\|$ is $\vec{r}/r$ for $r \neq 0$; conclude that the point P is the point for which the three angles APB, BPC, CPA are all equal to $2\pi/3$.)

3. Let A and B be two distinct points in the plane $\mathbb{R}^2$. Find a geometric construction of the tangent at a point P of the ellipses, hyperbolas and the lemniscate with foci at A and B. Recall that these curves are resp. defined by the conditions

$$\|PA\| + \|PB\| = 2a, \ \|PA\| - \|PB\| = 2a, \ \|PA\| \cdot \|PB\| = a^2.$$

(Hint : Use the gradient, as determined in the preceding exercise, to determine the normal to the curves.)

4. Compute the gradient of the function

$$\mathbb{R}^6 \cong \mathbb{R}^3 \times \mathbb{R}^3 \longrightarrow \mathbb{R}\ ,\ (\mathbf{x},\mathbf{y}) \longmapsto \mathbf{x}\cdot\mathbf{y} \text{ (scalar product).}$$

Use first a limited expansion, and alternatively, try to work in coordinates (i.e. in components). Similarly, if $S = {}^tS$ is a symmetric matrix, compute the gradient of $\mathbf{x} \longmapsto \mathbf{x}\cdot S\mathbf{x}$.

5. Let f denote the scalar field in $\mathbb{R}^3 - \{0\}$ given by

$$f(x,y,z) = xyz/\sqrt{(y^2z^2 + z^2x^2 + x^2y^2)}.$$

Discuss the possibility of extending f continuously at the origin. Is this continuous extension differentiable ? Observe also that f vanishes identically on all three coordinate planes (and in particular, the three partial derivatives of f vanish at the origin).

6. Take three points A, B and C on the unit circle $x^2 + y^2 = 1$ in the plane $\mathbb{R}^2$. Find a condition on the angles α, β and γ so that the area of the triangle is maximal. (First, observe that the surface S is proportional to $\sin 2\alpha + \sin 2\beta + \sin 2\gamma$ where the angles are linked by the relation $\alpha + \beta + \gamma = \pi$. Either use the Lagrange multiplier method to handle symmetrically the three angles, or eliminate one angle and find a maximum for

$$\sin 2\alpha + \sin 2\beta - \sin 2(\alpha+\beta).\ )$$

7. Let $\mathcal{C}$ denote the parabola $y = x^2$ in $\mathbb{R}^2$. Find the extrema of the distance function between $A = (0,a)$ and $P \in \mathcal{C}$ where $a > 0$ is a given positive number.

8. Find the extrema of the scalar function $f(\mathbf{x},\mathbf{y}) = \mathbf{x}\cdot\mathbf{y}$ (scalar product function) when the variables are submitted to the conditions $\|\mathbf{x}\| = \|\mathbf{y}\| = 1$.

(Hint : write the conditions in the form $\mathbf{x}\cdot\mathbf{x} - 1 = \mathbf{y}\cdot\mathbf{y} - 1 = 0$.)

Chapter 6

1. Let $F : \mathbb{R}^n \longrightarrow \mathbb{R}^n$ be the map $\mathbf{x} = (x_i) \longmapsto \mathbf{y} = F(\mathbf{x}) = (\exp(x_i))$. What is the derivative of F ?

2. Compute the derivative of $f(x,y) = {}^t(x^2+1, 2y^2+x)$ at a general point $(x,y) \in \mathbb{R}^2$. Find a condition for the nonvanishing of its functional determinant.

Chapter 7

1. For $\mathbf{x} \in \mathbb{R}^n$, the norm of $\mathbf{x}$ is $r = \|\mathbf{x}\| = \sqrt{(x_1^2 + \ldots + x_n^2)}$. Compute $\mathbf{grad}(r^k)$ and $\Delta(r^k)$ (Laplace operator) for $k \in \mathbb{N}$. When $\mathbf{a}$ is a fixed vector of unit length, compute $\mathbf{grad}(\mathbf{a}\cdot\mathbf{r})$ and $\Delta(\mathbf{a}\cdot\mathbf{r})$. More

generally, if the scalar function f in $\mathbb{R}^n$ only depends on r, show that **grad** f(r) = f'(r)**r**/r (outside the origin). Conclude that $\Delta f = f'' + \frac{n-1}{r} f$. Determine all harmonic functions in $\mathbb{R}^n$ which depend only on r (treat the cases n = 2 resp. n > 2 separately).

2. Come back to exercises 4 and 5 of Chap.5 and discuss the extremum type using the second differential. In the first case, prove that the area function of the triangle has a maximum for equilateral triangles. In the second case, prove that the distance function has a local minimum for P = (0,0) *if* a is small, and has a local maximum *if* a is large (the limiting case being that of a equal to the curvature radius of the parabola at P).

3. A polygon has four sides with prescribed lengths. Find a condition on its angles so that its area is maximal. (Recall that the four vertices will lie on a circle exactly when two opposite angles have sum π.)

Chapter 8

1. Consider the map $T : \mathbb{R}^n \times \mathbb{R}^m \longrightarrow \mathcal{L}(\mathbb{R}^n,\mathbb{R}^m)$, $(\mathbf{a},\mathbf{b}) \longmapsto \mathbf{a} \otimes \mathbf{b}$. Is it differentiable ? What are the tangent linear maps ? On $\mathbb{R}^n \times \mathbb{R}^m$ we consider the norm coming from the identification to $\mathbb{R}^{n+m}$, i.e. $\|(\mathbf{a},\mathbf{b})\|^2 = \|\mathbf{a}\|^2 + \|\mathbf{b}\|^2$, and on $\mathcal{L}(\mathbb{R}^n,\mathbb{R}^m)$ we can consider the norm $\|u\| = \sup_{\|\mathbf{x}\|\le 1} \|u(\mathbf{x})\|$.

2. Let $\mathbf{a} \in \mathbb{R}^3$ be a constant vector of unit length. Compute

$$\mathbf{a}[\mathbf{grad}(\mathbf{a}\cdot\mathbf{v}) + \mathbf{curl}(\mathbf{a} \wedge \mathbf{v})].$$

3. Let **v** be a vector field with components

$$v_1 = e^x(\cos y - \cos z),\ v_2 = e^y\cos z - e^x\sin y,\ v_3 = f(x,y,z).$$

For which functions f does **v** derive from a vector potential ? Are there such functions f with $f(x,y,0) \equiv 0$? If so, give a field **u** with **v** = **curl u**.

4. Let **u** be the vector field given in components by expressions $u_1 = -xy$, $u_2 = yf(y)$, $u_3 = z(1 - f(y))$. Show that there is only one function f with f(1) = 1 and such that **u** derives from a vector potential **u** = **curl v**.

5. Let f be a $\mathcal{C}^1$ function depending only on $r = (x^2 + y^2 + z^2)^{1/2}$. Show that **curl**(**r**f) = 0 and give the form of the derivative of the vector field **r**f (i.e. give the form of Grad(**r**f)).

Chapter 9

1. Prove the Cauchy-Schwarz inequality $|\mathbf{x}\cdot\mathbf{y}| \le \|\mathbf{x}\|\|\mathbf{y}\|$ for two vectors $\mathbf{x}$ and $\mathbf{y} \in \mathbb{R}^n$. (Hint : It is enough to consider the case where $\mathbf{x}$ and $\mathbf{y}$ are linearly independent, and hence generate a two dimensional subspace $E \subset \mathbb{R}^n$; the determinant

$$\begin{vmatrix} \mathbf{x}\cdot\mathbf{x} & \mathbf{x}\cdot\mathbf{y} \\ \mathbf{y}\cdot\mathbf{x} & \mathbf{y}\cdot\mathbf{y} \end{vmatrix}$$

is then positive as seen in the lemma of 9.1.). Observe that in $\mathbb{R}^3$ we have more precisely

$$\|\mathbf{x}\|^2\|\mathbf{y}\|^2 - |\mathbf{x}\cdot\mathbf{y}|^2 = \|\mathbf{x} \wedge \mathbf{y}\|^2 \ge 0 \ .$$

2. Study the following coordinate system in $\mathbb{R}^2$ (i.e. determine the coordinate curves, the matrix (g_{ij}), $\sqrt{g}$, ds^2,...)

$$x = c \operatorname{Ch}\xi \cos\eta \ ,$$
$$y = c \operatorname{Sh}\xi \sin\eta \ .$$

In particular, is it an orthogonal curvilinear coordinate system ?

3. Same questions for the coordinate systems in $\mathbb{R}^3$

A) $$x = \xi\eta \cos\varphi \ ,$$
$$y = \xi\eta \sin\varphi \ ,$$
$$z = (\xi^2 - \eta^2)/2 \ .$$

B) $$x = c(\xi^2 - 1)^{1/2}(1 - \eta^2)^{1/2}\cos\varphi \ ,$$
$$y = c(\xi^2 - 1)^{1/2}(1 - \eta^2)^{1/2}\sin\varphi \ ,$$
$$z = c\, \xi\eta \ .$$

C) $$x = c(\operatorname{Ch}\alpha + \cos\beta - \operatorname{Ch}\gamma)/2 \ ,$$
$$y = 2c \operatorname{Ch}(\alpha/2) \cos(\beta/2) \operatorname{Sh}(\gamma/2) \ ,$$
$$z = 2c \operatorname{Sh}(\alpha/2) \sin(\beta/2) \operatorname{Ch}(\gamma/2) \ .$$

4. Let $\Pi \subset \mathbb{R}^2$ denote the compact piece limited by the four arcs on

$$y = x^2 \quad \text{and} \quad y = 2x^2 \ ,$$
$$xy = 1 \quad \text{and} \quad xy = 2 \ .$$

Introduce curvilinear coordinates u and v by

$$u = xy \quad \text{and} \quad v = y/x^2$$

to compute the area of Π.

(Answer : the area of Π is $(\log 2)/3$.)

5. Let u and v be the curvilinear coordinates in the plane $\mathbb{R}^2$ defined by

$$y = \sqrt{(uv)} \quad \text{and} \quad x = \sqrt{(u/v)} \ .$$

What are the local frames $\mathbf{e}_u$, $\mathbf{e}_v$? Let f be the scalar field defined in the curvilinear coordinates u, v by the expression

$$f(u,v) = u^2 - \sin(uv) .$$

Compute the gradient of f (determine the components of grad f in the local frame $\mathbf{e}_u$, $\mathbf{e}_v$, say $\mathbf{gradf} = X\mathbf{e}_u + Y\mathbf{e}_v$).

6. Show how one would compute an integral

$$\iiint_V \left[z + \sqrt{(x^2+ y^2+ z^2)}\right]^{-1/2} dxdydz$$

using the curvilinear coordinates given in the exercise 3 A).
(Partial answer : $\sqrt{g} = \xi\eta(\xi^2 + \eta^2)$, $z + \sqrt{\ldots} = \xi^2$.)

7. Let us denote by Σ the unit sphere in $\mathbb{R}^3$ and by C the vertical cylinder tangent to the equator of Σ. Let also f denote the horizontal projection from Σ to C (hence f is only defined outside the north and south poles of Σ). Show that f is area preserving.

Chapter 10

1. Let f denote a central function in $\mathbb{R}^3$, say $f(\mathbf{r}) = g(r)$ (where $r = \|\mathbf{r}\|$ as usual). Show that the vector field $\mathbf{u} = \mathbf{r}f$ derives from a scalar potential (compute $\mathbf{curl}(\mathbf{r}f)$, recalling that $\mathbf{gradf} = f'\mathbf{r}/r$ as in ex.5 of Chap.8). When does $\mathbf{u} = \mathbf{r}f$ derive from a vector potential ? Conversely, show that if f is a scalar field in $\mathbb{R}^3$ and $\mathbf{u} = \mathbf{r}f$ derives from a scalar potential, then f is a central function (observe that the level surfaces of f are the spheres with center at the origin since the gradient of f must be parallel to $\mathbf{r}$).

EXERCISES FOR PART II

Chapter 11

1. Let f and g be two scalar functions in $\mathbb{R}^3$. Compute $df \wedge dg$ and its exterior differential $d(df \wedge dg)$.

2. Let $\omega = u\mathbf{dr}$ and $\eta = v\mathbf{dr}$ denote 1-forms in $\mathbb{R}^3$. Compute the following exterior products

$$\omega \wedge \eta \ , \ \omega \wedge d\omega \ , \ \omega \wedge d\eta \ .$$

If $\zeta = \mathbf{w}d\sigma$ is a 2-form, compute the exterior product $\omega \wedge \zeta$.

3. An integrating factor for a differential form $\omega \in \Omega^p(\mathbb{R}^n)$ is a scalar function f (i.e. a 0-form) such that $f\omega = d\eta$ is exact (cf. 10.3). Give a necessary condition for ω to admit an integrating factor.

4. Let $\omega = Pdx + Qdy \in \Omega^1(\mathbb{R}^2)$ where P and Q are affine linear functions. Give a condition on the coefficients of P and Q in order that ω be closed. Show that if this condition is satisfied, ω is exact, and determine a primitive in this case.

5. The 1-form $\cot x \, dx - y^{-1}dy$ is exact (why ?). Find a primitive outside the Ox-axis.

6. Is the following 2-form in $\mathbb{R}^3$ closed ? exact ?

$$\omega = x^2 dy \wedge dz - y^2 dz \wedge dx + 2(xy + zy - xz)dx \wedge dy \ .$$

(Answer : $-y^2zdx + (x^2y - x^2z)dy$ is a primitive.)

7. For two $\mathcal{C}^1$ scalar fields u and v in $\mathbb{R}^3$, consider the differential form

$$\omega = ydx - xdy + dz - vdu.$$

Give conditions on u and v for ω to be closed. Show that in this case, u and v are independent of z.

Chapter 12

1. Compute the flow of the vector field $\mathbf{u} = \mathbf{r}$ through the three faces of the cube $0 \le x,y,z \le 1$ which do not contain the origin. Same question through the rectangle with vertices

$$(0,0,1), \ (1,0,1), \ (1,1,0) \text{ and } (0,1,0).$$

2. Let $\omega = xdy + ydz + zdt + tdx \in \Omega^1(\mathbb{R}^3)$. Is ω exact ? Integrate $d\omega$ on the portion of surface

$$t = 0, \ x + y + z = 1 \text{ and } x \ge 0, \ y \ge 0, \ z \ge 0.$$

3. Let $\omega = xdy - ydx \in \Omega^1(\mathbb{R}^2)$. Compute $d\omega$ and deduce Riemann's

formula $S = 1/2 \oint_{\partial\Pi} (xdy - ydx)$ for the area of a compact piece Π in $\mathbb{R}^2$. Can you generalize this formula for volumes of 3-dimensional pieces in $\mathbb{R}^3$? (Hint : Consider any 2-form $\omega = \mathbf{u}d\sigma$ where div$\mathbf{u}$ = 1, for example $(1/3)(xdy \wedge dz + ydz \wedge dx + zdx \wedge dy = \mathbf{r}d\sigma/3.)$ What is the value of $\int_\Pi \mathbf{r}d\sigma/3$ extended to a planar piece $\Pi \subset \mathbb{R}^3$ (in geometrical terms) ?

4. Compute the flow of the field $\mathbf{u} = {}^t(yz,zx,xy)$ through the oblique simplex having the three basic vectors $\mathbf{e}_i$ for vertices (with the orientation given by the choice of parametrization x,y). Does this vector field derive from a scalar potential ?

5. Compute the integral of $\omega = (yx+x)dy + y^2dx$ on the oriented boundary of the piece Π defined by $|x| + |y| \le 1$ (canonically oriented by $\mathbb{R}^2$). (Observe that ω is not exact !)

6. What is the flow of a field $\mathbf{u} \wedge \mathbf{grad}f$ crossing the boudary $\Sigma = \partial V$ of a 3-dimensional piece $V \subset \mathbb{R}^3$. In particular, show that this flow vanishes when the vector field $\mathbf{u}$ derives from a scalar potential.

7. Compute the flow of **curl u** through the northern hemisphere of the unit sphere (oriented by the choice of normal $\mathbf{e}_3$ at the north pole, or equivalently, oriented by the parametrization x,y). Here $\mathbf{u}$ denotes the vector field with respective components

$$u_1 = 2x - y\ ,\ u_2 = -yz^2\ ,\ u_3 = -y^2z.$$

Use first a direct computation of **curl u** and integrate **curl u**$\cdot d\sigma$ on the proposed 2-dimensional piece. Secondly, use Stokes' formula and evaluate the corresponding curvilinear integral on the equator (suitably oriented).

8. Let $\mathbf{u}$ be the vector field (defined in the complement of the Oz-axis in $\mathbb{R}^3$) $\mathbf{u} = (x/r^2, y/r^2, z)$ where $r^2 = x^2 + y^2$ (hence r denotes the distance to the Oz-axis). a) Compute **curl u** ; does $\mathbf{u}$ derive from a scalar potential ? b) Compute the circulation of $\mathbf{u}$, i.e. $\oint \mathbf{u}d\mathbf{r}$ along circles orthogonal to the Oy-axis (centered on this axis) and along circles orthogonal to the Oz-axis (centered on this axis) ; is it possible to guess the results without computation ? (Answer : "Yes" for the circles orthogonal to the Oy-axis, but "No" for the circles orthogonal to the Oz-axis.)

9. Does the vector field $\mathbf{u} = {}^t(2x-y, -yz^2, -y^2z)$ derive from a scalar potential ? Compute $\iint_{x^2+y^2+z^2=1} d\sigma \wedge \mathbf{u}$.

Chapter 13

1. Let the function $f(z) = \text{Log } z = \log|z| + i\text{Arg}z$ be defined in the right half plane $x = \text{Re}(z) > 0$. Check the Cauchy-Riemann relations for the real and imaginary parts of f. (Use $r^2 = x^2 + y^2$, hence $\partial r/\partial x = x/r, \ldots$ and $\text{Arg}z = \varphi = \text{Arctan}(y/x)$.)

2. Check the formulas

$$(\partial/\partial z)\left(z^n \bar{z}^m\right) = nz^{n-1}\bar{z}^m, \quad (\partial/\partial \bar{z})\left(z^n \bar{z}^m\right) = mz^n \bar{z}^{m-1}.$$

(Recall that by definition $\partial/\partial z = (\partial/\partial x - i\partial/\partial y)/2$ can be computed on any $\mathbb{R}$-differentiable function $f : \mathbb{R}^2 \longrightarrow \mathbb{R}$ (or $\mathbb{C}$).)

3. Compute $\oint_{\mathcal{C}} (1/z)d\bar{z}$ where $\mathcal{C}$ denotes the circle $|z| = 2$ (oriented positively, i.e. anticlockwise).

4. Compute $\oint_{|z|=1} \dfrac{dz}{z - i/2}$ (anticlockwise).

5. Compute $\oint_{\mathcal{C}} \bar{z}dz$ where $\mathcal{C} = \partial\Pi$ is the oriented boundary of the compact piece Π defined by $x \le 1$, $y \le 1$ and $x^2 + y^2 \ge 1$.

Chapter 14

1. Let $\varphi \ge 0$ be a continuous function on $\mathbb{R}$, vanishing outside the interval $[-1,1]$, and define $\varphi_n(x) = n\varphi(nx)$ (so that φ_n vanishes if $|x| \ge 1/n$). Prove that for any continuous function f on $\mathbb{R}$,

$$\lim_{n\to\infty} \int_{\mathbb{R}} f(x)\varphi_n(x)dx = cf(0)$$

where $c = \int\varphi dx$ (the most interesting case is $c = 1$: φ normalized). (Hint : Observe that $cf(0) = \int f(0)\varphi_n(x)dx$ and estimate the difference $\int_{|x|\le 1/n} [f(x)-f(0)]\varphi_n(x)dx$ using the change of variable $nx = t$.)

2. Generalize the situation of the preceding exercise to $\mathbb{R}^k$ by defining $\varphi_n(x) = n^k\varphi(nx)$.

3. Express as a simple integral the double integral

$$\int_0^t dx \int_0^x f(y)dy \ .$$

(Make a picture of the domain of integration in the x,y plane, and reverse the order of integration ; the legitimacy of interchanging the order of integration is Fubini's theorem, cf.15.6.) Generalize the result to n variables x_i instead of x and y. As an application, deduce the general solution of the differential equation $y^{(n)} = f(x)$.

4. Let $E = \{f \in \mathcal{C}^2_{\mathbb{R}}([1,2]) : f(1) = 0,\ f(2) = 1\}$ and $\Phi : E \longrightarrow \mathbb{R}$

$$\Phi(f) = \int_1^2 x^2(3f + f' + (x^2 - 2x + 2)f'^2)dx.$$

Find $f \in E$ such that $\Phi(f)$ is minimal (Hint : Express the vanishing $(d/dt)_{t=0}\ \Phi(f+th) = 0$ whenever $h \in \mathcal{C}^2$ vanishes at the limits, so that $f + th \in E$, and integrate by parts as in the theory).

5. Among all curves $\mathcal{C}$ linking the points $A = (0,0)$ and $B = (1,1)$ with a prescribed length L ($\geq \sqrt{2}$) and which are graphs of functions $y = f(x)$, find one for which

$$\int_0^1 y\ ds = \int_0^1 y \cdot \sqrt{(1 + y'^2)}\ dx$$

is minimal (such a curve has the *lowest possible gravity center*!).

6. Solve the Euler equation corresponding to the extremal problem for

$$I = \int_a^b \left(xy' + (y'^2 - 1)^2\right) dx .$$

Hint : The differential equation

$$x + 4y'(y'^2 - 1) = k$$

can be solved in parametric form by putting $y' = -t$, namely

$$x - 4t(t^2 - 1) = k ,$$

$dy = -tdx = -t(12t^2 - 4)$, and

$$y = -3t^4 + 2t^2 + k'.$$

7. Find the shortest curves (geodesics) linking two points in the plane $\mathbb{R}^2$.

8. Find the paths of light in the upper half plane having a refraction index $n = n(x,y) = 1/y$: in other words, find the curves for which

$$\int_A^B \frac{ds}{y} = \int_A^B y^{-1}\sqrt{(1 + y'^2)}\ dx$$

is minimal.

(Answer : these curves are circles orthogonal to the Ox-axis.)

EXERCISES FOR PART III

Chapter 15

1. Let E be the subspace of $\mathcal{C}_b(\mathbb{R})$ (with the uniform norm) consisting of uniformly continuous functions. Is E a subalgebra of E ? Is E a complete space (i.e. is E closed in $\mathcal{C}_b(\mathbb{R})$) ?

2. For $n \geq 1$, let $f_n : I = [-1,1] \longrightarrow \mathbb{R}$ be the function defined by $f_n(x) = (x^2 + \frac{1}{n})^{1/2}$. Show that f_n is indefinitely differentiable and $f_n \longrightarrow |x|$ uniformly on I when $n \to \infty$ (but $x \longmapsto |x|$ is not $\mathcal{C}^1$).

3. Let f be a continuous function $[0,1] \longrightarrow \mathbb{R}$. If $\int_0^1 x^n f(x)\, dx = 0$ for all $n \in \mathbb{N}$, show that $f = 0$ (Hint : Use the Stone-Weierstrass theorem 15.5 and lemma 1 of 14.1.)

4. Let f be a continuous function $[0,1] \longrightarrow \mathbb{R}$. Show that

$$\int_0^1 x^n f(x)\, dx \longrightarrow 0 \quad \text{when} \quad n \longrightarrow \infty .$$

(Hint : Separate $\int_0^1 = \int_0^\varepsilon + \int_\varepsilon^1$. Variant : f is bounded !)

5. Let (f_n) be a sequence of numerical functions on $\mathbb{R}$ such that

a) $f_n(x) \longrightarrow 0$ when $x \longrightarrow \infty$,

b) $f_n \longrightarrow f$ uniformly when $n \longrightarrow \infty$.

Is it true that $f(x) \longrightarrow 0$ when $x \longrightarrow \infty$?

6. Let (f_n) be a sequence of numerical functions on $\mathbb{R}^2$ with $f_n \longrightarrow 0$ uniformly when $n \longrightarrow \infty$. Is it true that

$$\lim_{n\to\infty} \iint_{D_n} f_n(x,y)dxdy = 0$$

in the following two cases

a) $D_n = \{ n \leq \sqrt{(x^2 + y^2)} \leq n+1 \}$,

b) $D_n = D = \{ |y| \leq 1/(1+x^2) \}$?

Conclude that the part b) of the basic theorem of 15.2 is still true if Π is a non compact p-dimensional piece having finite p-dimensional area.

7. Let $\varphi \geq 0$ be a continuous function vanishing for $|x| \geq 1$ and define $\varphi_\varepsilon(x) = \varepsilon\varphi(\varepsilon x)$. What is the value of $\lim_{\varepsilon\to 0} \int f\varphi_\varepsilon dx$ in the following two cases : $f \equiv 1$ or f vanishes outside a bounded set.

8. Let (f_n) be a sequence of continuous functions on $[0,1]$ such that $f_n \longrightarrow 0$ uniformly on $[\delta,1]$ for every $\delta > 0$. Is it true that $\int_0^1 f_n(x)dx \longrightarrow 0$? (Proof or counterexample.)

Chapter 16

1. Consider the binomial series for the square root

$$(1 - t)^{1/2} = 1 - t/2 - t^2/(2\cdot 4) - 1\cdot 3t^3/(2\cdot 4\cdot 6) - \dots$$

and take in particular $t = 1 - x^2$. Discuss the convergence of

$$|x| = 1 - (1-x^2)/2 - (1-x^2)^2/2\cdot 4 - \dots$$

2. Let f, g and h be $\mathcal{C}^1$ functions. What is the derivative of

$$\Phi(t) = \int_{f(t)}^{g(t)} h(t,x)\, dx \ ?$$

(Hint : Define $F(u,v,w) = \int_u^v h(w,x)\, dx$ and use the fundamental theorem of calculus 5.4 together with Leibniz' rule 16.5 to compute the derivative of $\Phi(t) = F(f(t),g(t),t)$.) In particular, what is the derivative of a function of the form

$$\Phi(t) = \int_0^{\sin t} f(e^t,x)\, dx \ ?$$

3. Let $I \subset \mathbb{R}^n$ (or more generally any metric space, or even any topological space) and Φ a set of functions $I \longrightarrow \mathbb{R}$ (or $\mathbb{C}$). We say that Φ is **equicontinuous at a point** $a \in I$ when

$$\forall\ \varepsilon > 0\ \exists\ \delta > 0 \text{ such that } |x-a| < \delta \Longrightarrow \begin{cases} |f(x)-f(a)| < \varepsilon \\ \text{for all } f \in \Phi \end{cases}.$$

a) Show that if the sequence (f_n) consists of continuous functions and converges uniformly on I, then the set $\Phi = \{f_n\}$ is equicontinuous at every point.

b) Show that if a sequence (f_n) is equicontinuous at a point $a \in I$ and converges simply to a function f, then f is continuous at a.

4. Let F and f_n be continuous functions $\mathbb{R} \longrightarrow \mathbb{R}$ (or $\mathbb{C}$). Assume that $\int|F|dx < \infty$, $|f_n| \le F$ and $f_n \longrightarrow f$ uniformly on all bounded intervals I. Prove that $\int f_n dx \longrightarrow \int f dx$.

(Hint : For $\varepsilon > 0$, choose a large enough M to ensure that the integral of F over $|x| \ge M$ is $\le \varepsilon$; the integral of $|f_n - f|$ over the same set will certainly be $\le 2\varepsilon$, and conclude with a 3ε proof.)

5. Construct a sequence (u_n) of continuous positive functions on the interval $I = [-1,1]$ with $\|u_{n+1}/u_n\| = 2$ for all $n \ge 0$ and such that the series $\sum u_n$ converges normally, i.e. $\sum \|u_n\| < \infty$.

(Hint : Show first that there exists a sequence (I_n) of open disjoint non empty intervals in I; choose then $0 < u_n \le 1/2^n$ in the following manner : $u_n = 1/2^n$ outside I_n, but u_n has a minimum at a point $x_n \in I_n$ with $u_n(x_n) = 1/2^{n+2}$.)

6. For each of the following three sequences (f_n)

$$n^{-1}e^{-x/n}\ ,\ (1 + nx)^{-1}\ ,\ e^{-nx}$$

determine the limit for $n \longrightarrow \infty$ (and $x \geq 0$), and the type of convergence on $I = [0,\infty[$ and $I_c = [c,\infty]$. What is $\lim_{n\to\infty} \int_0^\infty f_n(x)dx$?

7. Study the following sequence of functions (on $t \geq 0$)

$$f_n(t) = nt^2/2 \text{ for } 0 \leq t \leq 1/n\ ,\ f_n(t) = t - 1/2n \text{ for } t > 1/n.$$

Does f_n converge uniformly to a $\mathcal{C}^1$ function f ? Does the sequence of derivatives $f'_n(0)$ converge to $f'(0)$?

8. Study the sequence of functions (on $x \geq 0$)

$$f_n(x) = nx^2 \text{ for } 0 \leq x \leq 1/2n\ ,\ f_n(x) = x - 1/4n \text{ for } x > 1/2n\ .$$

Are the functions f_n , f'_n continuous ; do the sequences (f_n) and (f'_n) converge (uniformly) ?

9. Show that the integral $\int_0^\infty e^t \sin\left(e^{e^t}\right)dt$ exists. What is its value ? (Hint : introduce $y = e^t$ and then $x = e^y$.) Observe that an integral $\int_0^\infty f(t)dt$ may exist even if the function f is unbounded. Give an example where f is unbounded and $\int_0^\infty |f(t)|dt$ exists (we say that $\int_0^\infty f(t)dt$ is *absolutely convergent* in this case).

10. Let $F(x) = \int_0^\infty e^{-t\sin x}dt$. Can you compute the limits of F when either $x \longrightarrow 0$ or $x \longrightarrow \infty$?

11. Let $(x_n) \subset \mathbb{R}$ be a convergent sequence, say $x_n \longrightarrow x$ and assume that $f_n \longrightarrow f$ uniformly on $\mathbb{R}$. Prove that $f_n(x_n) \longrightarrow f(x)$. For example, $\left(1 + (x + 1/n)/n\right)^n \longrightarrow e^x$.

12. Consider the sequence of functions $f_n(x) = \text{Arctan}(x/n)$. Study the convergence of (f_n) and (f'_n). Also observe $f_n(n) = \pi/4$ for all $n \geq 1$.

13. Study the convergence of the series $\sum_{n\geq 0} u_n$ where $u_n(x) = x/(1 + x^2)^n$.

14. Let g be a continuous function with $\int_0^\infty |g|dx < \infty$. For each uniformly convergent sequence (f_n) of continuous functions (on $[0,\infty[$) show that $\int_0^\infty f_n g\ dx \longrightarrow \int_0^\infty fg\ dx$.

15. Let $f \in \mathcal{C}_b(\mathbb{R})$. What is the limit when $n \longrightarrow \infty$ of $\int_0^\infty e^{-nx}f(x)dx$.

16. Consider the parametric integral $F(x) = \int_0^\infty e^{-t}\sin xt\,dt$ as limit of the $F_n(x) = \int_0^n e^{-t}\sin xt\,dt$.

a) Examine the convergence $F_n \longrightarrow F$.

b) Compute $G(x) = \int_0^\infty te^{-t}\cos xt\,dt$.

c) Does one have $G(x) = F'(x)$?

d) Find the value of $\int_0^\infty \frac{1}{t}\, e^{-t}(1 - \cos t)dt$.

(Hint for the last point : Compute $\int_0^1 F(x)dx$.)

17. Compute the integral $I(x) = \int_0^\infty \frac{tdt}{e^{xt} + 1}$ for $x > 0$ (expand the quantity under the integral sign in powers of e^{-xt}). By derivation of the result, deduce the value of $\int_0^\infty \frac{t^2e^tdt}{(e^t + 1)^2}$. Justify all steps involved. (Hint : Recall that $\sum_{k\geq 1} 1/k^2 = \pi^2/6$.)

18. Compute the definite integrals

$$I_n = \int_{-\infty}^\infty \frac{dx}{(1 + x^2)^n}$$

by successive derivation of

$$I(t) = \int_{-\infty}^\infty \frac{dx}{t + x^2} = \pi/\sqrt{t} .$$

Justify all steps involved.

19. Let $f : [0,\infty[\longrightarrow \mathbb{R}$ be $\mathcal{C}^1$ on $x > 0$ and assume $f(x) \longrightarrow 0$ for both $x \longrightarrow 0$ and $x \longrightarrow \infty$ (e.g. $f(x) = x^ne^{-x}$ and $n > 0$). Compare the integrals

$$\int_0^\tau dt \int_0^\infty dx\ f'(tx) \quad \text{and} \quad \int_0^\infty dx \int_0^\tau dt\ f'(tx) .$$

Why is Fubini's theorem of 16.5 not applicable ?

20. In the integral $\int_0^A x^{-1}\sin x\,dx$, replace x^{-1} by $\int_0^\infty e^{-tx}\,dt$. Show that Fubini's theorem is applicable and find another proof of

$$\int_0^\infty x^{-1}\sin x\,dx = \pi/2.$$

21. Prove the following generalization of Abel's convergence test. Let C and σ be two positive constants with

1) $|\varepsilon_n| \searrow 0$ 2) $|\varepsilon_n - \varepsilon_{n+1}| \leq C(|\varepsilon_n| - |\varepsilon_{n+1}|)$ and

3) $|\sum_p^q u_n| \leq \sigma$.

Then $\sum_{n\geq 0} \varepsilon_n u_n$ converges and $|\sum_{n\geq p} \varepsilon_n u_n| \leq \sigma|\varepsilon_p|$.

Chapter 23

1. Prove the formula

$$x(\pi - x) = (8/\pi)\sum_{k\geq 0} (2k+1)^{-3}\sin(2k+1)x \quad \text{for } 0 \leq x \leq \pi .$$

Examine the case $x = \pi/2$: what do you get ? Use Parseval's equality for the preceding function and derive

$$\sum_{k\geq 0} (2k+1)^{-6} = \pi^6/960 .$$

(Compare with the value of $\zeta(6) = \pi^6/945$ found in 21.3 .)

2. Obviously $\sum_{k\in\mathbb{Z}} k|c_k|^2 = 0$ when $|c_k| = |c_{-k}|$. What happens geometrically to the curve when $c_{-k} = c_k$ (or $\bar{c}_k$) ?

3. Let f be a function having fast decreasing derivatives of all orders

$$\text{for all integers } i,j \geq 0, \ |x^i f^{(j)}(x)| \longrightarrow 0 \quad \text{when } |x| \longrightarrow 0.$$

Prove that the Fourier transform $\hat{f}$ of f also decreases fast and consider the 1-periodic function $x \longmapsto \sum_{n\in\mathbb{Z}} f(x-n)$. Show that

$$\sum_{n\in\mathbb{Z}} f(x-n) = \sum_{k\in\mathbb{Z}} \hat{f}(k)e^{2\pi ikx},$$

$$\sum_{n\in\mathbb{Z}} f(n) = \sum_{k\in\mathbb{Z}} \hat{f}(k) \quad \text{(Poisson' summation formula).}$$

Prove that the particular function $f(x) = e^{-\pi\varepsilon x^2}$ satisfies the assumptions (for $\varepsilon > 0$) and prove in this case

$$\hat{f}(\xi) = \varepsilon^{-1/2}e^{-\pi x^2/\varepsilon}.$$

Using the first formula proved above, prove that the kernel corresponding to the filter with characteristic $e^{-\varepsilon k^2}$ (cf.22.4-5) is given by

$$\Theta_\varepsilon(t) = \sum_{k\in\mathbb{Z}} e^{-\varepsilon k^2}e^{2\pi ikt} = \varepsilon^{-1/2} \sum_{k\in\mathbb{Z}} e^{-(t-n)^2/\varepsilon} .$$

Hint: To compute the Fourier transform

$$\hat{f}(\xi) = \int_{-\infty}^{\infty} f(x)e^{-2\pi i\xi x}dx,$$

compute $(\hat{f})'$ by derivation with respect to ξ under the integral sign. Integrating by parts, show that $\hat{f}$ is a solution of the differential equation $y' = -2\pi\xi y$ whence $\hat{f}(\xi) = ce^{-\pi\xi^2}$. The integration constant c is of course given by

$$c = \hat{f}(0) = \int_{-\infty}^{\infty} e^{-\pi x^2} dx = 1.$$

Chapter 24

1. Let $f : \mathbb{R} \longrightarrow \mathbb{C}$ be continuous, 2π-periodic and one-to-one into. Then, $f(\mathbb{R})$ is a simple closed curve in $\mathbb{C}$. We have proved in 23.5 that if the Fourier series of f is $\sum_{k\in\mathbb{Z}} c_k \mathbf{e}_k$, then the area enclosed by this curve is $\sum_k k|c_k|^2 = (f|f')/i$. On the other hand, if we write $f = g + ih$ with two real periodic functions g and h, we have seen in 24.5 that the same area is given by $(h|g')$ or by $-(h'|g)$. Can you deduce directly e.g. that $(f|f') = i(h|g')$?

2. Compute the area of the regular polygons using their Fourier expansions (check that your result agrees with the formula found in the elementary geometrical way).

3. Let T denote the equilateral triangle with vertices 1 and $e^{\pm 2\pi i/3}$ and T' its symmetric with respect to the imaginary axis. The union $T \cup T'$ is a star-shaped piece Π with six vertices (at the points $e^{2\pi ik/6}$, $0 \leq k < 6$). Find the Fourier expansion of the uniformly parametrized boundary $\partial\Pi$. Check your result by estimating the area of Π as in the preceding exercise.

Find the Fourier expansion of the uniformly parametrized star having six vertices $e^{2\pi ik/6}$ $(k = 0, \ldots, 5)$

BIBLIOGRAPHY

Prerequisites in calculus

M. Spivak : Calculus, Publish or Perish (2nd ed. 1980).

S. Lang : Undergraduate Analysis, Springer-Verlag (GTM 1983).

Prerequisites in linear algebra

H. Anton : Elementary linear algebra, John Wiley (1987).

For proofs of Stokes' theorem, implicit functions theorem,...

W. Rudin : Principles of Mathematical Analysis, McGraw-Hill (3rd ed. 1976).

M. Spivak : Calculus on Manifolds, Benjamin Inc. (1965).

S. Lang : Analysis I, Addison-Wesley (Paperback 1971).

Exterior differential forms, abstract approach

H. Cartan : Cours de Calcul Différentiel, Hermann (Coll.Méthodes 1967).

Special topics

E. Nelson : Topics in Dynamics I : Flows, Princeton University Press (1969).

A.J. Weir : Lebesgue integration and measure, Cambridge University Press (1973).

Higher Analysis

J.Dieudonné: Foundations of Modern Analysis, Acad. Press (1960).

W. Rudin : Real and Complex Analysis, McGraw-Hill (2nd ed. 1974).

Tables of special functions

M. Abramowitz, I. Stegun : Handbook of Mathematical Functions, Dover Publ. (1970).

INDEX

D

E

F

G

H

S

T

U

V W

Z

MAIN THEOREMS

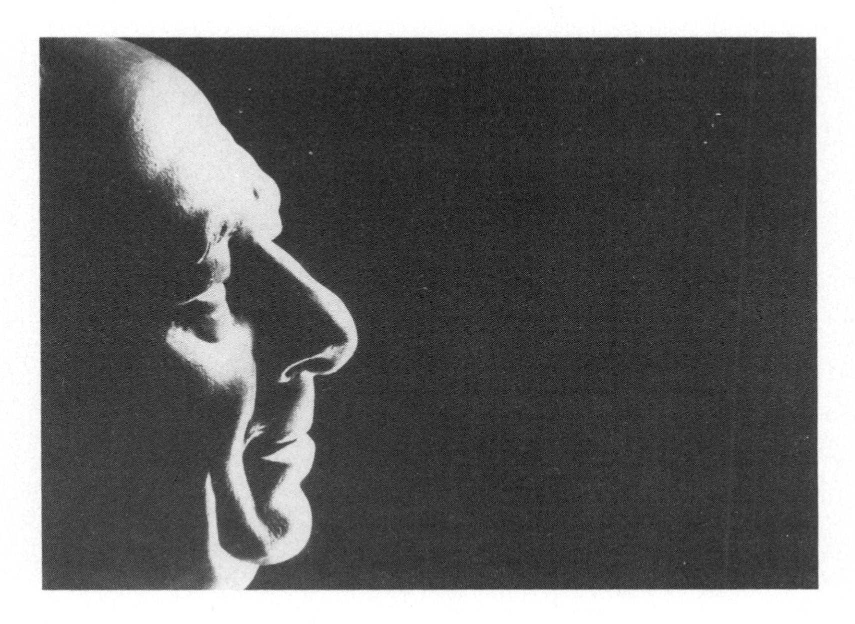

The Rachette bust of Euler *(Photo A. Robert)*